GHOSTS OF THE ADOBE WALLS

GHOSTS OF THE ADOBE WALLS

Human Interest and Historical Highlights
from 400 Ghost Haunts of Old Arizona

by

NELL MURBARGER

Treasure Chest Publications, Inc.

P. O. Box 5250
Tucson, Az. 85703

Library of Congress Catalog No. 64-18546

PRINTED IN THE UNITED STATES OF AMERICA BY WESTERNLORE PRESS

"A faithful friend is a strong defense,
and he that hath found such a one
hath found a treasure . . ."

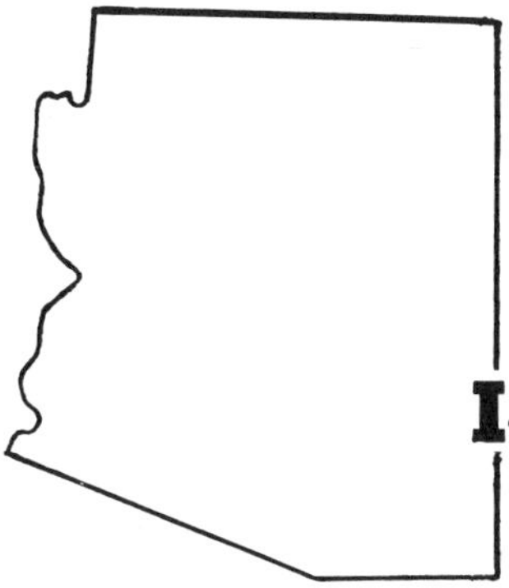

Foreword

It would be interesting to know how much of Arizona's long and varied history has been made within sight of adobe walls.

Back in the 9th Century, when Charlemagne was battling the Saracens, stout earthen walls already were rising in Arizona's great prehistoric pueblo of Casa Grande. The Jesuit missions of the 17th Century were fabricated of this same natural construction material. So were most of Arizona's frontier Army posts, mining boom towns, shipping ports bordering the Colorado River, Overland Stage stations, and nearly all other pioneer buildings from the baronial haciendas of Spanish dons to the rude shanties of dryland homesteaders.

Adobe, actually, is nothing but plain clay mud—one of the most ignoble elements known to man—with straw or grass usually added for a binder. Molded into bricks weighing around 30 pounds each, or poured into forms and baked hard in the desert sun, it is transformed into one of the world's best and cheapest building materials. Granted a small amount of attention and maintenance, it is virtually as enduring as stone. The years roll over it and are gone, and still it stands—sturdy, honest, and foursquare.

But even adobe cannot long endure abandonment and abuse, twin evils that have laid low so many of Arizona's historic buildings.

Missions and haciendas were burned by marauding Indians. The need for stage stations was eliminated by coming of the railroads. Power and irrigation dams on the Colorado River brought an end to its shipping industry and the ports it served. Mining boom towns, after producing their millions in gold, silver, copper, or other metals, at last failed and faded, and shriveled to ghost towns.

In nearly every instance, the result has been the same.

With the settlement abandoned, the buildings comprising it were left to the destructive forces of man and nature. Amid the adversities and austerities presented by a stern, ungenerous frontier land, where frugality is not only a virtue but is inescapable, little thought is given to what may, in some future era, be considered "romantic" and historical. When a man's family or his cattle stand in need of shelter, or a bridge needs repair, or his mine tunnel lacks timbering, about the only value he can see in a deserted building is what it can supply to serve his immediate wants.

In consequence, the dust had barely settled behind departing owners before the roofs and beams of their buildings were pirated for other use. The abrasive teeth of wind and weather began gnawing at the unroofed walls. Wandering prospectors and saddle-tramps tore out door- and window-frames to feed their lonely cookfires, and treasure seekers began their insidious probing.

Time after time, for more than 200 years, this program of willful destruction and natural deterioration has been repeated until most of Arizona's historic adobes have crumbled away. Inch by inch, literally one grain of sand at a time, the old mud walls have vanished, and are vanishing, like gallant ships going to their doom with flags flying.

And now, the end is near.

In many of the smaller and longer-abandoned settlements the forces of destruction have done their work so thoroughly that few men living today can pin-point the site of these former habitations. The only remaining evidence of a once-extensive Spanish mission or hacienda may be two or three long-vacant adobes, a few fast-disappearing walls, and an old graveyard filled with thistles and nameless dead.

Vanished, near vanished, and vanishing missions, forts, stage stations, ranches and, most of all, mining camps, are the places of which I have written in this book.

Collecting the material for these chapters has been a true labor of love. Every hour spent doing it has been fun. Especially enjoyable have been the priceless days and months devoted to traveling the unpaved roads of Arizona's back-country, exploring desert and mountains, forming friendships with her wonderful old timers, and visiting the abandoned settlements where so much of the state's history has been made. On many of these trips, over the past 35 years, I have had as companion my precious mother, Bessie N. Lounsberry. Mom, like myself, has a great fondness for cooking over campfires, sleeping under the stars, and delving into the secrets of persons and places long vanished.

Though such places commonly are designated "ghost towns," few of us prowlers-into-the-past believe that restless spirits of the departed actually come back to haunt the scenes of their earthly triumphs and tribulations. What we *do* believe is that the atmosphere of these old places—the blessed stillness and peace, the splendid absence of everyday distractions—gives our jaded minds a better chance to function properly, to think, interpret, and thus to understand and to recreate in fancy the storied happenings of Long Ago.

In preparing these true accounts of Arizona ghost haunts, I have drawn information from personal interviews, pioneer diaries and letters, unpublished manuscripts, early magazines, and reference works of various descriptions. Most of all I have drawn from the columns of early-day newspapers of Arizona and adjacent states. In all this world, I believe, there is no greater source of authentic Western American history than in the local newspapers published when that history was being made, day by day and blow by blow.

For long extended use of priceless historical files, library facilities, help in research, and other personal assistance, I am indebted to the Arizona Pioneers' Historical Society and its staff, at Tucson; to the State Department of Library and Archives, Phoenix; the Sharlot Hall Memorial Museum, Prescott; and *Mohave County Miner*, at Kingman.

To the University of Arizona Press goes my sincerest gratitude for kind permission to draw heavily on three of its publications: *Arizona Place Names*, Will C. Barnes, 1935; Will C. Barnes' *Arizona Place Names*, revised and enlarged by Byrd H. Granger, 1960; and *A Guide to the History of the U.S. Army Installations in Arizona, 1849-1886*, by my good friend, Ray Brandes. First published in 1959 in the University's excellent historical quarterly, *Arizona and the West*, Mr. Brandes' all-inclusive *Guide* later was republished in booklet form by Dale Stuart King.

I also wish to express appreciation to all the wonderful Westerners who have loaned early maps and historical photographs, as well as to those who have helped me obtain needed information from old settlers or their descendants, or have given their time to guide me to nearly-vanished townsites which, otherwise, I might not have found. To all the writers and publishers whose material I have borrowed, and to everyone, everywhere, who has helped to bring to reality this tribute to Arizona's Yesterdays, my heartfelt thanks.

As the Western editor, Rollin Daggett, wrote many years before I was born:

I drink alone and in silence
To the builders of the West:
Long life to the hearts still beating,
And peace to the hearts that rest.

NELL MURBARGER.

Costa Mesa, California.
June 1963.

CONTENTS

ILLUSTRATIONS

ILLUSTRATIONS (*Continued*)

GHOSTS OF THE ADOBE WALLS

"Of the great extent of the gold region of Arizona there can be no doubt. The late discovery of placers, or surface diggings, on the Gila, has long been anticipated. Emory, in 1849, expressed his belief in its existence . . . Senator Gwin informs me that he heard of gold on the Gila from emigrants at San Diego in 1849."

—Sylvester Mowry, *address before The American Geographical and Statistical Society, New York, February 3, 1859.*

Gila City Opened the Game

Though Arizona's mining industry would not hit its full stride until after the Civil War years, this was not because her desert hills and mountain canyons had been overlooked as possible sources of gold and silver. Some of the earliest Anglo-American travelers through the region had recognized its mining potential, and early as the spring of 1850 the *Alta Californian,* of San Francisco, tossed out a teaser concerning gold in that little known land to the immediate east of the Colorado River:

"Mr. N. B. Coons, of Texas, an old freighter between St. Louis and Chihuahua, Mexico . . . communicates a very favorable opinion relative to . . . gold upon the Gila River," set forth this brief news story. "Mr. Coons saw at El Paso several gentlemen who had been in the interior of the State of Sonora and report the Mexicans to be busily engaged in gathering gold, such as is taken from the dry diggings of California. He thinks that if our government will only extend its protection over the Gila country, a large population will soon be found along its banks and territories."

Less than four years later that desired protection came in the form of the Gadsden Purchase, under terms of which the United States added to its real estate holdings all the territory between the Gila River and the present U.S.-Mexico boundary—some 30 million acres acquired at the marked-down, once-in-a-lifetime, bargain counter price of roughly 35 cents an acre!

In time to come, those bargain-basement lands would produce the busy modern cities of Tucson and Yuma, the richly productive mines of Tombstone and Bisbee, cattle ranches, citrus groves, alfalfa and lettuce fields, and cotton plantations almost without end. But, for those days of more than a century ago, it was enough that the door to the Gila country and its potential riches at last was open. The table was spread—and it wasn't necessary to send engraved invitations!

Within a short time the banks of the Gila were swarming with gold-seekers and in 1858 there came into being Arizona's first mining boom camp of the Anglo-American era.

Situated about 18 miles upriver from the present site of Yuma, it was known as Gila City. Founder of the camp was Jacob Snively, a native of Pennsylvania, who fought in the Texas War for Independence and later served as personal secretary to General Sam Houston, first president of the Republic of Texas.[1]

Describing the first glorious effulgence of the Gila City boom, when more than 1000 miners were combing the gold from nearby ravines and any man with sufficient means to purchase a shovel could envision himself a potential millionaire, J. Ross Browne wrote in 1863: "Enterprising men hurried to the spot with barrels of whiskey and billiard tables; Jews came with ready-made clothing and fancy wares; traders crowded in with wagonloads of pork and beans; and gamblers came with cards and monte tables. There was everything in Gila City within a few months but a church and a jail, which were accounted barbarisms by the mass of the population . . ."

Before the boom was a year old, however, the Los Angeles *Star* was openly skeptic concerning the future of the Gila mines.

"We have seen a number of persons during the week who have returned and speak in the most disparaging manner of the whole affair," declared the *Star* in its issue of January 8, 1859. "Those fond of packing dirt and washing it for two or three dollars a day can have plenty of employment that about pays expenses. Large invoices of goods have been shipped to the river, by the Gulf, but the owners were glad to sell them to the resident storekeepers, Hinton and Hooper, rather than have to stay and dispose of them by retail."

More optimistic was the view expressed in a letter written from Fort Yuma and published in the *Alta Californian*, June 17, 1860: "Yesterday I saw 152 ounces of gold from the Gila mines deposited with the agent of Wells Fargo & Company," stated this peripatetic and anonymous reporter. "Fine, very beautiful it was; and there is more in the same country and about 400 men digging for it . . ."

After visiting the actual scene of operations, however, the reporter's enthusiasm diminished. Writing in the *Alta* of June 24 he described Gila City as "a sporadic camp of miners, Sonorenses and dogs—some 400 of them—who are here today and somewhere else tomorrow. I find quite a difference of opinion as to the richness of the mines," he hedged. "The gold, which I see in bags at the express offices, excites cupidity and confidence; but the forlorn looking men I meet on the road and about the stage offices, do not justify very high expectations of wealth."

Gila City, obviously, was going the way of most placer gold camps and it was probably a kindness when Mother Nature took steps to hasten its demise.

During a period of unusually high water on the river, in 1862, nearly every man-made structure in the town was destroyed, and when J. Ross

[1] Slain by Indians in Arizona's White Picacho mountains, March 18, 1871, Snively is buried in the old ghost town graveyard at Gillett.

Browne visited its site, a year later, he reported that the former boom-camp then consisted of "three chimneys and a coyote."

With the short, merry boom on the Gila swept into oblivion, the restless-footed ones were free to turn their attention to the new excitement at Laguna de la Paz, some 65 miles to the north, in the same county.

Although Mexicans residing in the region presumably knew of the occurrence of gold at La Paz, its "discovery" at that point generally is credited to the frontiersman, Pauline Weaver, whose effeminate name apparently had no ill-effect on his hardihood and all-around ability.

"Captain Pauline Weaver and others, in the month of January 1862, were trapping on the Colorado River and, at times would stray off into the mountains for the purpose of prospecting for gold," wrote the Honorable A. McVey, a member of Arizona's Territorial Legislature, who supplied the following resumé to the U.S. Government's 1868 *Report on Resources of the States and Territories West of the Rocky Mountains.* "They had discovered what was then named and is still called El Arollo (*sic*) de la Tenaja, which is about two miles north of El Campo Ferra (*sic*) and about seven miles east from La Paz. In this gulch they had discovered gold in small quantities . . . which Capt. Weaver kept in a goose quill . . .

"Don José M. Redondo, having heard of the discovery, at once set out to visit the newly found El Dorado . . . He arrived a few days afterward at the camp of Capt. Weaver, who pointed out to him . . . the . . . gulch from which he had taken the gold . . . Within less than a mile from Weaver's camp, south, Redondo took a pan of dirt to prospect and when he had dry-washed it, to the astonishment of himself and the party with him, he found that he had one *chispa* which weighed two ounces and one dollar, besides other small pieces."

Not having brought with them equipment for mining or camping, Redondo's party returned to Laguna (a settlement about 20 miles above Fort Yuma), for supplies. Soon as they showed their gold and spread the remarkable tidings, nearly half a hundred persons immediately set off for the new district. Discoveries, thereafter, were made almost daily and it soon was known that every gulch and ravine over a ten-mile radius was rich with gold. American Camp, Campo en Medio, La Plumosa, Los Chollos, and many smaller camps all had their rich diggings; but the prize discovery was made in Ferrera Gulch where Don Juan Ferrera took out a nugget weighing 47.2 ounces! Its sale value, at that time, was close to $1000.

With the Los Angeles *Star* publicizing the new bonanza, and dust and nuggets worth thousands of dollars pouring into the Pueblo of the Angels, everyone from Sonora to San Francisco went wild with excitement. Every gold-hungry son who had scored a goose-egg in the Big Bustle of 1849—as well as those who had chalked up misses in the Washoe and Holcomb Valley excitements, then still going their delirious way—scrambled for a foothold on the new merry-go-round of La Paz.

First-comers from the West were ferried across the Colorado River on crude tule rafts manned by Indians, but by midsummer regular ferry service had been instituted between Providence Point and the little settlement of Oliva, or Olive City, seven miles down-river from La Paz. Owners of the ferry were William D. Bradshaw, freighter, guide and gold hunter, and William A. Warringer. By authorization of Arizona's First Territorial Legislature the Bradshaw Ferry, as it was known, charged four dollars for each team and wagon taken across the river; three dollars for a carriage and one horse; one dollar each for saddle-horses; 50 cents each for pedestrians, cattle and led horses, and 25 cents per head for sheep.

During that same summer of 1862 the town of La Paz, point of supply for the entire area, consisted of less than 100 primitive brush ramadas built mainly of mesquite and arrowweed. Two of these were outfitted as stores; 12 were saloons. Before the onset of winter many of the brush shanties had been replaced with more substantial adobe structures, and the town had grown to number around 1500 inhabitants.

"La Paz had the heterogeneous population characteristic of Western mushroom towns," states the Arizona Writers' Project Guide. "Few men brought their families with them. Mojave squaws, comely and vigorous women, did the housework and, as one writer put it, 'met the domestic needs of the camp.' "

The foregoing observation is echoed by a California news reporter who, upon visiting La Paz in the autumn of 1863 found its streets thronged with travelers and trade brisk. "The population," he wrote, "is the worst mixture of Indians, Mexicans, Pikes, and white men from all parts of the earth, I ever saw."[2]

Among the more prominent early residents of La Paz was Michael Goldwater, head of the three-generation Goldwater dynasty of Arizona mercantilers, and Isaac Goldberg, also a pioneer merchant and freighter. Mr. Goldberg, in 1894, wrote for Arizona Pioneers' Historical Society a manuscript covering his recollections of La Paz and the early-day life of a desert boom-camp.

"When I arrived there, flour cost from $30 to $40 per hundred," recalled the former storekeeper. "People had little to eat besides mesquite beans and river fish. I sold empty drygoods boxes to get back my freight expenses of 15 cents per pound. Those boxes were used for coffins, furniture and other purposes. Every evening the miners and gamblers would spread their blankets upon the streets and play card games for the nuggets which were daily found in abundance . . . Water for drinking and culinary purposes was sold at one dollar the ten-gallon kegfull.

"Shortly after my arrival a thief, who had been stealing from stores and other places, was caught. There was, of course, no law officer to confine and try the culprit, so the miners and citizens held a meeting and sentenced him to receive 25 lashes. These were promptly and lustily

[2] *Alta California*, Oct. 14, 1863

given. After the affair was over they handed him five dollars in cash, telling him that if he dared to again visit the settlement he would receive a double dose of the same 'back medicine.' You may be sure that the rascal did not return, and that the community was no more troubled by thieves. We could leave all our property unguarded and yet not miss a single cent's worth of anything . . ."

Petty thievery, however, does not seem to have been the No. 1 civic problem in La Paz. Of greater importance to peace of mind and the mortality rate was the fact that the town was overrun with gangs of Mexican and American toughs, including those headed by the notorious Boss Danewood, later hanged in Los Angeles. Street and saloon brawls were of common occurrence. An ounce of golddust is said to have had a greater value there than a human life, and throughout the Civil War years—which just about spanned the life of La Paz—the camp was a hotbed of secession. The story is told of two volunteers slain there by a rebel sympathizer in the early '60s. After the war was over, a former comrade of one of the murdered men returned to La Paz, found the slayer of his friend at work in the mines, and shot him down without warning.

Almost immediately after its founding La Paz had been made the governmental seat of Yuma County—a nice feather in the cap of any early day boomcamp. Two years later, the new mining town on the river narrowly missed an honor even more significant when certain members of Arizona's first Legislature attempted to have re-located there the Territorial capital. The ambitious plan failed, assertedly, by only two votes. At that time, and continuing until 1865, La Paz was one of the largest towns in Arizona with a claimed population of 5000, many of whom were "in the chips" for the first time in their lives.

How much gold was produced by the placers of La Paz and its environs is something no man will ever know. *Los Angeles Mining Review* estimated the output of Ferrera and Catarina Gulches, alone at $5,000,000.

"Captains Polhamus and Mellon, who have navigated the Colorado for 40 years, tell of the large consignments of gold they frequently carried down the river. More than once, in lieu of a safe, the mattresses of their bunks concealed bags of dust valued at $100,000," stated the *Review*.[3]

In those pre-inflation days of a century ago, a million dollars—even $100,000—was still regarded with respect. Any mining camp that produced two or three millions was outstanding, and a placer gold camp where a miner could take out $50 to $100 a day was almost a phenomenon. La Paz was such a camp.

"Hundreds of dollars per day to the man was common and, now and then, a thousand or more per day," it was stated of the La Paz placers in the J. Ross Browne government report of 1868. In addition to the prize Ferrera nugget, which Browne credits with a weight of "47 ounces

[3]Quoted in the *Weekly Yuma Sun*. Mar. 29, 1901.

and six dollars," mention is made of two other nuggets weighing 27 and 26 ounces. "Many others found pieces of from one or two ounces up to 20 . . . The gold was large and generally clear of foreign substances. The largest piece (above mentioned) did not contain an apparent atom of quartz or any other base matter. The gold from the different camps varied a trifle in its worth at the mint of San Francisco and brought from $17.50 to $19.50 per ounce . . .

"I have failed to find anyone whose opinion it is that less than $1,000,000 were taken from these diggings within the first year . . ." stated the report.

But a placer gold camp is, at best, an ephemeral thing. By close of the Civil War the cream of the La Paz placers had been skimmed. Miners were finding it necessary to work farther and farther from sources of water. The gold-bearing sand and gravel was loaded in bags and packed to the river for washing; or water was carried to the diggings in rawhide containers slung over the backs of burros. Either method was laborious and time-consuming. Dry washers were introduced and worked with some success, but the big boom was past.

The final blow fell in 1868 when an Act of the Territorial Legislature switched the site of Yuma County's seat from La Paz to Arizona City. Two years later the county records were shipped down the river on the paddlewheeler, *Nina Tilden*.

With this crushing defeat, the City of Peace became a ghost town.

At the turn of the century, I have been told, the site of La Paz was still marked by many ruined adobe buildings, and the iron doors and shutters that had dignified the front of banks and mercantile establishments still lay partially embedded in the heaps of crumbled rock and brick and scaling rust.

But the iron doors and shutters had disappeared before my first visit to the site of La Paz in 1940. Or, if they were there, I failed to find them. About the only indications of the old town that came to my notice were a few low adobe walls almost worn away by wind and weather, a couple of stone chimneys, and a little miscellaneous rubble. Even less remains today. As a matter of fact, it takes a better-than-average ghost town sleuth to even identify the former townsite . . .

In 1867, with the placer fields diminishing in richness and La Paz on the wane, a new town had begun taking form on the Arizona side of the Colorado River, about six miles south of the decadent boom city. Founded by the pioneer mercantiler, Michael Goldwater, and named for his friend, Herman Ehrenberg, the new town was better situated in relation to the steamer landing and the Prescott-San Bernardino stage and freight road, and quickly began pulling trade and population from the earlier camp. Ehrenberg, unfortunately, never saw the town named in his honor. He had been murdered about a year previous to its founding. The unsavory deed had taken place at Dos Palmas station, on the Bradshaw stage road, a little way south of the present town of Mecca,

California. Ehrenberg, assertedly carrying $3500 in gold, had stopped to spend the night at the Dos Palmas station. He had spread his bedroll under a brush *ramada* and retired. Sometime during the night he was shot through the heart and robbed of his gold. The stationkeeper—a man named Smith—was suspected; but proof was lacking, and no one was held to answer for the crime.

The new town named for Arizona's ill-fated pioneer was not as spectacular in its growth as La Paz had been, because Ehrenberg was not a boomtown—no thousand dollar nuggets were being found in its environs. As a steamboat landing, an important supply point and terminal of several freight lines and stage lines plying to Wickenburg, Prescott and other interior centers, its growth, however, gave promise of being more substantial than that based on the chimera of gold dust.

By 1870, according to records of the United States Census Bureau, the town of Ehrenberg included some 87 dwellings inhabited by 96 families. Slightly less than half the inhabitants were foreign born (principally Mexicans); there were nearly four men to each woman, and only five children under ten years of age. It was about this time that poor Martha Summerhayes, genteely griping, as usual, wrote of Ehrenberg; "Of all the dreary, miserable looking settlements that one could possibly imagine, it was the worst."[3]

Another correspondent, writing from Ehrenberg under date of August 17, 1871, took a rather different view.

"This is a nice, cool, healthy place," he declared. "It is not afflicted with flies and but few mosquitoes. It is not sultry, a breeze is always felt from midday to next morning. It is . . . the most pleasant town on the Colorado. Several good and large buildings are now up and occupied, and more will soon follow. Hooper, Whiting & Company recently opened here an immense stock of goods, and their large building on Front Street will soon be completed. This firm has all the transportation contracts to the northern military posts, and their business is managed by J. F. Barney, who knows how. It is surprising to walk through their establishment and note the magnificent stock of staple groceries, liquors, cigars, tobaccos, wagon material, mining apparatus, etc., and they do a large wholesale business . . ."[4]

Less than a month later, however, the *Arizona Citizen* published another bit of news from the river town—this not so laudatory:

"Ehrenberg, A.T., Sept. 10, 1871:—We have had no overflow of the Colorado for four years, and everything is very much burnt up. Hay is very scarce and high priced. A large quantity of mesquite beans will be used this year . . . They are now selling at $50 per ton . . ."

The spring of 1872 brought to Ehrenberg Miss Mary Elizabeth Post, who had come to open the first school in the frontier outpost.

[3]*Vanished Arizona,* Salem Press, Salem, Mass., 1911.

[4]*Arizona Citizen,* Aug. 26, 1871.

One of the last of Ehrenberg's original walls, 1947.

24

Boothill, Ehrenberg, 1947.

"Ehrenberg was a lively little town of about 500 inhabitants," recalled Miss Post.[5] "My schoolroom was a building formerly used as a saloon. It consisted of a roof resting upon a series of arches. These arches were closed at night by wooden doors and barred from the inside—all but one, which was locked from the outside . . . The floor was of earth and every night was carefully sprinkled and then swept in the morning . . .

"I had 15 pupils, not not of whom knew any English, and I knew nothing of Spanish. One of the trustees was of an old California family and was master of both English and Spanish. He helped me organize my school the first morning and told me that if I ever needed any help to send for him as his store was quite near . . .

"I taught through the months of May, June, July and August. We had no electric lights in those days and a lighted lamp added much to the heat and attracted too many insects. So I used to sit outside evenings and gather the children around me . . ."

The riverfront shipping town continued to grow, both in population and importance. New stores were established and existing merchants enlarged their buildings, expanded their stocks. In its issue of January 16, 1878, *Arizona Enterprise* reported that the Goldwater store in Ehrenberg kept constantly on hand "a large and fresh stock of groceries, provisions, tobaccos, cigars, Yankee notions, paints, oils, crockery, steel, wallpaper, tinware, linens, matting, fleece twine, wool sacks, sheep shears, carpets, furniture, harness and brushes."

Though the fortunes of her merchant princes and the town's affluence might wax and wane, the prosperity of Ehrenberg's boothill cemetery was assured. Graveyards never lacked patrons in a land where men packed the law on their hips and made up the rules as they went along.

Pioneers laid to rest in Ehrenberg's cemetery during the first decade after the town's founding included many buried with genuine reverence and sorrow. Others were buried with satisfaction. Saint and sinner, slayer and slain, most of those interred here had died by violence. Here were bandits fallen before the guns of lawmen, peace officers slain by outlaws, desperados killed by their own confederates. Here were laid teamsters, stage drivers, ranchers and prospectors dry-gulched by hostile Indians, men trampled to death beneath the hooves of wild mustangs, men drowned beneath the gray waters of the Colorado, men shot to death over card games and fickle dancehall girls. Ehrenberg, in the 1870s, provided many avenues by which one might acquire six feet of space in boothill.

With marble and granite suitable for engraving almost as scarce as platinum, and professional engravers scarcer still, it was only in exceptional cases that stone markers were erected over the graves. Most of the earlier burials throughout the Western mining country were

[5]*Pioneer Days in Arizona,* Frank C. Lockwood.

marked only by wooden headboards or crosses on which was crudely carved, burned or painted the appropriate dates and the name by which the victim was known at time of death. The name that appeared on his grave marker, needless to say, was not always the one bestowed on him at birth by a loving mother, back in The States. In the case of transient strangers, the identity of a victim might not even be known at time of burial; or, for that matter, ever known.

How true the tale I cannot say—it may be only a bit of Arizona folklore—but I have been told, several times, the story of a well-mounted stranger who rode into Ehrenberg one sizzling day. Every article of the man's equipment, as well as his general bearing and behavior, suggested refinement and affluence. His garments were fashionably cut and of excellent material; and though he carried no luggage and must have ridden many miles across the desert, his linen was as spotless as if he had donned it only minutes earlier. His saddle and bridle were of expensive make. The horse he rode, a magnificent animal, exhibited good blood lines and equally good grooming. On the animal's rump was burned a neat brand—JC. It was a brand wholly unfamiliar to cattlemen of that region.

Dismounting in front of an Ehrenberg saloon, the stranger entered the building, briefly eyed the men draping the bar, ordered whiskey and paid for it—from a roll of greenbacks half as large as a sofa pillow. As the man hoisted the liquor to his lips a volley of gunshots ripped through the saloon and when the excitement and smoke had cleared away, the mysterious stranger lay dead on the barroom floor.

No one seemed to know who had done the shooting. Some swore the shots had come from within the room; others declared they had come through an open window from the street. Chances are there were plenty of men in the saloon who knew perfectly well where the shots had originated. But in those days—even as now—it was more advantageous to be the close-mouthed friend of a live gunman than to bear witness for a dead stranger.

Search of the man's personal effects revealed only the substantial bankroll and the usual articles most men carry in their pockets—a penknife, a pencil, a few coins. There was nothing whatever to offer any clue to his identity.

After a grave had been chipped out of the hard, unfriendly bosom of boothill, and the rock and earth mounded over the victim's remains, there a rose the question: What should be carved on the wooden headboard? Someone suggested that the board be lettered "The Stranger." Then, someone else remembered the man's fine horse, still drowsing at the hitchrail in front of the saloon—the horse branded JC. For frontier Ehrenberg that was identity enough. Besides, it was short and easy to carve.

With the wooden marker quickly inscribed "J.C. 1867" everyone returned to the saloon for another drink.

Everyone except the stranger.

According to the story the body was never claimed, nor was anything ever learned concerning the man's identity or the reason for his mysterious assassination.

First time I heard this strange little tale was a couple of years before I made my first visit to Ehrenberg, about 1930. The story appealed to me and I hoped that I might locate the "J.C." marker and so substantiate the yarn—at least, in part. But though I searched the weedy graveyard from stem to stern I found no marker so inscribed.

The fact that I failed to locate the marker is not proof, of course, that such a marker never existed. Even before the discontinuance of river shipping and subsequent abandonment of the town, around the turn of the century, Time had taken heavy toll of the older headboards. Some had been knocked down and trampled by cattle; from others, every trace of inscription had been sandblasted by the desert wind.

Of the 200-or-more graves contained in the old cemetery scarcely half a dozen, I believe, bore legible epitaphs at the time of my first visit to Ehrenberg. To partially compensate for this neglect by man and the toll of the years, the Arizona State Highway Department, in 1935, had erected in the Ehrenberg cemetery a fitting memorial to all those anonymous dead. Built in obelisk form from choice minerals, ores, crystals and agatized wood, the monument carried on its westerly face a bronze plaque setting forth its purpose "To perpetuate the memory of the pioneers and adventurers resting in these unmarked graves."

In a concrete apron encircling the obelisk, about three feet above the ground, was embedded a choice collection of pioneer relics—all articles of general usage in the days when these nameless mounds were being filled with their human dead. Picks and shovels, gold pans and iron candlesticks, were embedded in honor of all the unknown miners and prospectors whose earthly trails reached their conclusion in this lonely spot. Stirrup irons, cinch rings, Spanish bits and branding irons paid homage to cowboy and rancher dead; wheel wrenches and harness buckles bespoke the frontier importance of stage driver and freighter. An old muzzle-loading rifle and a rusted bowie knife hinted of mountain scout and trapper. Even the ghosts of Ehrenberg's pioneer housewives—should they return to their former haunts—would find here the tools of their earthly endeavors: a Dutch oven, a couple of iron skillets, a stove-lid lifter, a primitive coffee mill . . .

As for Ehrenberg, herself, she has fared scarcely better than La Paz or Gila City. The name, to be sure, still appears on highway maps; but the burgeoning tourist way-station known as Ehrenberg, refulgent with Neon signs and painted stucco, has nothing in common with Old Ehrenberg except its name and general location. Almost nothing remains to mark the site of the old town; though as late as 1950 there were rather extensive adobe ruins of several commercial buildings on the hillslope below the Arizona Agricultural Inspection Station at the Colorado River crossing of U.S. 60-70. One of the structures still standing at that time had been a large dancehall and saloon, I was told. Another had housed a general mercantile. One of the doorframes was still in place, and several of the walls were standing to roof height. But not any longer. The site of Old Ehrenberg will soon be completely swallowed by encroaching motels, trailer parks, air-conditioned cafes and Bar-B-Ques.

Since the realignment of U.S. 60-70, even the monument in the old cemetery is no longer readily visible from the road. Which is just as well. Its significance, apparently, was lost on quite a large segment of the traveling public. Tourists, sighting the old graveyard in the days when the highway passed immediately alongside it, often stopped to give the place a "once over." Some of these visitors were honestly impressed by the significance of these nameless mounds. Others wrote or craved their names on the remaining headboards and the monument. Some even pried the relics out of the concrete; others scanned the place in the hope of finding "something amusing."

They might as well have saved their time. There was nothing "amusing" about Ehrenberg's boothill. No funny epitaphs, no ludicrous tombstones, no phony boots sticking out of phony graves.

Every weathered cross, every sandblasted headboard, every unmarked mound in that old burying ground, was—and still is—as starkly simple and straightforward, as honest and unpretentious, as the vanished days of Ehrenberg's youth.

Monument to Unknown Dead, Ehrenberg, 1947

"We are in the pleasantest region we have seen since we left the Choctaw country. Here are clear rivulets, with fertile valleys and forest trees. The wide belt of country that borders the Black Forest, and probably extends along the Rio Verde to the Salinas and Gila, bears every indication of being able to support a large agricultural and pastoral population. The Valley of the Rio Verde is magnificently wooded with firs and oaks, affording excellent timber . . . Besides the advantages already enumerated, the mountains in this vicinity bear indications of mineral wealth."

—*Report of Lt. A. W. Whipple, 1853-54.*

Bonanza in the Bradshaws

SCARCELY MORE than a century ago the lands now embraced within Arizona's Yavapai County were any man's plum, ripe for the harvesting. Saw timber, straight and tall and all but limitless, darkened the mountains. The canyons teemed with wild game. Seas of native grass blanketed the valleys, the hills were fat with silver and copper, and almost every stream bed sparkled with virgin gold.

Yet, to the white man of 1860, this was virtually *terra incognita.* A few trappers had sought beaver on the Salt and Verde Rivers. A few rugged mountaineers, in the course of their migrations between Santa Fe and California, had dared cross this wild expanse—but not one of these few had left more impress on the land than the fading print of a deerhide moccasin, the blackened embers of a dead campfire.

In all that vast territory between the present sites of Flagstaff on the north and Phoenix on the south—a realm as large as several Eastern states combined—there was not one settlement save a few makeshift Apache villages. There were no roads other than the soft, brown trails trodden smooth by the pads and hooves of wild beasts and the bare feet of the Redman, and opening of the 1860 decade found in these richly mineralized lands not one known mine, one working miner.

Even the lanky, 65-year-old mountaineer, Capt. Joseph Reddiford Walker, who had discovered gold in these mountain fastnesses in 1861 and, two years later, was leading back to the scene of that discovery some three-dozen prospectors and gold seekers, must have been stirred emotionally by the potential significance of this expedition he headed. As he and the members of his party urged their mounts across wide, unnamed valleys, through trackless canyons and ever deeper into the high ranges, the big Tennessean must have known that some of these impassioned adventurers riding with him were destined to return home defeated, empty handed, disillusioned. Others among them would come to grips with the land, bend it to their will, and remain to carve new homes from its bosom. Where one found honor and riches, another's sole gain from the wilderness ahead would be a lonely death and an unmarked cairn.

George Lount, riding the big roan yonder, would establish one of the first sawmills in the Bradshaws. Dan Conner would help lay out the new town of Prescott and would later serve two terms in the Territorial legislature. Sam Miller would become a prominent rancher. Jackson McCraken, the tall, drawling, 42-year-old South Carolinan, would go on to discover the rich Del Pasco mine and the even-more-famous McCracken mine, and when he died at his sumptious California home, past 80 years of age, he would still be wealthy and highly respected. But young Jack Swilling, riding at his side—the handsome, dashing, straightforward chap on the big bay—would not fare so well. After years of successful mining and ranching operations, after opening the Salt River Valley to irrigated farming and building the first near-palatial home in the new city of Phoenix which he had helped to found, Swilling would be falsely arrested for stage robbery and would die in disgrace in the Territorial prison at Yuma, a broken man at the age of 47.

Who could have guessed what drama, what tragedy, what reams of future history, were riding into the heart of Arizona's highlands along with the 30-odd goldseekers of Capt. Joseph R. Walker!

After discovering the Hassayampa river the Walker party followed it to its source on the north side of Mt. Union, a short distance southeast of the spot where Arizona's new Territorial capital of Prescott would be founded a year or two later. At nearly every point where the party halted to sample the riverbed gravels, gold was found, but not in large quantity. Continuing over the divide, the goldseekers started down a small, bright mountain stream, which they later named Lynx creek.

When the first pan of gravel taken from this creek yielded nearly five dollars in gold, Walker's men were satisfied to halt their explorations and settle down to work.

Organizing the Walker Mining District the party apportioned the ground by drawing lots, each man receiving 300 running feet of the stream bed. Some of the claims, naturally, proved better than others, but even the poorest held golden treasure beyond any reasonable dreams of man.

Not again in Arizona would there be found another placer gold-bearing stream so phenomenally rich as Lynx Creek.

Though this section of the wilderness then had no means of rapid communication save the smoke signals of the hostile Apaches, news of the great strike was not long reaching the outside world, and other gold seekers soon were pouring in from La Paz and Tucson, even as far distant as California.

By midsummer, 1863, a visitor at the Walker Diggings reported in the *Alta Californian* that approximately 60 men were at work in one camp, making up to fifty dollars a day with rockers, while about 150 were busily engaged on Lynx Creek.

"The laws of the district forbid the presence of Mexicans and unnaturalized foreigners in the district," it was stated. "The mines are doing well. All making good wages and some big strikes. Everyone has his buckskin purse of dust, shaking it around, and the place generally reminds one of California in her harvest days. The diggings, so far as prospected, extend 14 miles down Lynx Creek, in a northerly direction, and 25 miles in an easterly direction on tributaries of a stream called Big Bug. Provisions are scarce. The mines are supplied from Altar[1] and the Pima villages. Flour sells at $20 per hundred . . . The miners subsist chiefly on turkey, elk, deer, and other game . . ."[2]

Soon as production of the rich surface placers showed indication of languishing, these first-comers of Sixty-Three probed deeper into the ranges, seeking out and claiming the gold-bearing quartz veins whose treasure had fed the streams. Almost everywhere they turned they seem to have found precious metals!

"All along the Hassayampa, upon the Agua Frio [sic] . . . Lynx Creek, Big Bug, Turkey Creek, and indeed upon nearly all the streams of central Arizona, lodes of gold, silver, and copper were found," states Browne.

In virtually every canyon where encouraging prospects were located, a mining camp was born. Many such camps grew into sizable towns, some even gave promise of becoming cities.

But all this was a long while ago and most of those incipient cities are ghost towns today. Sites of the older and shorter-lived camps are all but lost to man's knowledge. Other camps, by enduring longer, became more firmly established, more substantially built. These, consequently, are a trifle less obliterated. Still others, such as Walker and Poland, have been reborn as summer home colonies where fugitives from the torrid lowlands spend a few months each summer luxuriating in the blessed coolness of expensively-rustic cottages equipped with clever names, backyard barbecues, first and second mortgages, and No Trespass notices.

My introduction to two of these last mentioned towns took place in the fall of 1959.

As Arizona *aficionados* of long standing, my mother and I, for many years, have spent the major portion of each October camping in the National Forests of Arizona and exploring the state's highways and byways. Especially the byways—the little sun-splashed, inviting, back country roads so delightfully lacking in traffic jams, high speed, carbon monoxide, and the other insidious by-products of pavement. On each of

[1]A small town in Sonora, about 300 miles south of Lynx creek.

[2]*Alta Californian*, Sept. 23, 1863.

our junkets for quite a number of years, we allotted at least one of our four precious autumn weeks to the fascinating sport of tracking down ghost towns in the Bradshaw Mountains .In view of all the roads traveled (and the wrong turns taken), I don't believe it is stretching the truth to say we have covered at least 1000 miles in this one relatively small range and its environs; yet, the Bradshaws still hold ghost towns we have not visited.

Walker and Poland—our ghost town targets in 1959—were two famous old mining camps situated in the northern end of the range, not more than a couple of miles apart.

Traveling the Black Canyon highway east out of Prescott, we soon turned right on a bladed but unpaved road signposted to Walker. As the afternoon had been drawing to a close even when we left Prescott, we halted for the night as soon as we found a pleasant camping place in the pines and oaks of the National Forest. All the while we were preparing supper we had as non-paying spectators two shaggy gray burros with long, expressive ears and wise faces. They looked almost old enough to have come in with the Walker party in '63. If they had been able to speak English I'm quite sure they could have told us a lot about the churned acres of gravel tailings visible along Lynx Creek, just down the hill from our campsite. They might even have been able to tell us which of these tens of thousands of boulders had hidden the $5000 in gold nuggets assertedly found at the base of a big rock by one early-day prospector. But as the burros couldn't talk English and Mom and I couldn't talk burro, we contented ourselves with feeding them cookies and carrots, and scratching their silken muzzles.

We had traveled only a short way the next morning, through an almost unbroken aisle of tall pines, when we came to an old stone house with a wooden addition. It had been a nice house in its day, but probably had been deserted many years. Cobwebs and long, ghostly banners of muslin hung from the ceiling; scraps of faded wallpaper still clung to the walls. All the windows lacked glass, and no doors covered the doorways. Shading the house were some very large black walnut trees, and at one corner of the building twined a grapevine, gnarled with age. Lynx Creek—its banks grown green with watercress and mint, and flanked by a tumbled chaos of tailing piles—tinkled pleasantly only a few rods from the house.

An ancient pickup rattled to a halt beside our parked car. The driver an old man with a long, thin gray beard and a cheerful voice, called out, "Find what you're lookin' fur?"

When we inquired about the old house our visitor said the stone portion of it had been built by members of Joe Walker's party, in 1863, but that the wooden part had been added later—about 1875, he thought. He also told us about three men who were attacked by Indians while rocking out gold along the stream.

"They finally managed to escape to this old stone house. One of the men soon died from his wounds but the other two were well protected behind those thick walls and were able to stand off the attack. Other miners on the creek finally heard the commotion and came to their rescue."

"How long ago was this?" asked Mom.

The old man shook his head. "I don't rightly know," he said. "It was a good while before I came to these parts. I'm just tellin' you what others have told me . . ."

A little farther up the road we stopped to prowl around another former building site where only foundations, rock-edged paths, and rubble testified to man's one-time presence. We passed an old power plant, falling to ruin; and several more bends in the mountain road brought into view a wooden building with a false front—an architectural gimmick popular in the pioneer West for stores, saloons and shops. By giving additional height to a structure, the false front made a single-story building look almost as impressive as one with two floors. The square front also provided an admirable place for the owner to paint or hang a sign identifying his business.

Directly across the road from the store building was a deserted cabin and a little orchard of long-neglected apple trees. Half hidden in the soft yellow mat of leaves that carpeted the grove were a few remaining fruits. They were small and somewhat withered but we gathered a little sackful of them and a couple of evenings later Mom made a pan of apple dumplings which we baked in our reflector oven in front of the campfire.

Continuing upgrade from the deserted store and orchard we passed other old ruins and assorted rubble identifiable as the residue of former mining and milling operations. Eventually our road reached a three-way fork. As we sat discussing which of the three ways might be the one we wanted, we heard voices issuing from a small house a little way up the slope above the road forks. It was a pretty place enveloped in shrubbery and flowers, with the wild growth of the mountain sweeping down almost to the back door.

The man and woman who greeted me were friendly and genuine-seeming—like their pleasant living room with its Navajo rugs and books and desert paintings and rock specimens. When I asked where the main part of Walker had been located, the lady laughed.

"You're standing in the middle of it, right now!"

Looking up and down the canyon and over the rough heights beyond, it was hard to imagine how a town of any pretensions could have clung to these steep slopes. Yet, records show that even into the present century Walker was a busy place with some 2700 inhabitants, a high school, jail, hospital, hotel, several pool halls, eight saloons, sundry stores and a post office—established in 1879 and discontinued in 1940.

"According to what we've been told, more than $2,000,000 in gold has been taken from Lynx Creek—and it still isn't all gone," said the woman. "Just the other day one of our neighbors found a nice little gold nugget right in her own back yard!"

Old smelter chimney, Humboldt, 1950.

Early hydraulic placer mining operations, near Prescott.
—Courtesy Arizona Pioneers' Historical Society.

During our chat the man introduced himself as Jimmie James, second cousin of the notorious Jesse—his grandfather having been Jesse's uncle. He said he had been over much of the territory made famous by the James boys and that he owned many pictures of Cousin Jesse.

Although our maps showed a road leading over the mountain from Walker to Poland, about a mile and a half, Jimmie James warned that it was completely impassable to anything on wheels.

"We tried it the other day with a four-wheel drive and had to give it up," he said. "You can walk over the mountain, of course; or you can drive around by Humboldt and Big Bug. That way it's about 30 miles . . ."

Mom and I had gone only a little way back down the road ascended that morning when a big buck deer, with a fine set of antlers, trotted across the road ahead of us. The two fat does comprising his harem broke out of a clump of manzanita brush, regarded us curiously for a moment, then bounded away.

After returning to State Highway 69 we drove southeasterly down the valley, past Dewey and Cherry Creek, to Humboldt. It was the first time we had been there in a couple of years and we were pleased to see that the big Val Verde smelter chimney was still standing, tall and impressive. The company offices, hospital, boarding house, and several other large empty buildings which we had explored at the time of our first visit to Humboldt, about 15 years earlier, were no longer in evidence.

During that initial visit to the old smelting town we had asked one of the storekeepers if there were any ghost towns in the vicinity?

"Why, yes!" he had replied. "There's a dandy ghost town down the road about nine miles. Lots of old empty buildings. It's called Mayer . . ."

Nine miles from Humboldt, in the direction indicated, had brought us to a nice little town, perhaps not in the very best economic circumstances, but certainly far from dead. When Mom suggested that our ghost town was probably off the road a little way, I drove into a gasoline station to inquire.

"We're looking for a ghost town in this vicinity," I said innocently. "It's called 'Mayer.' "

"Who in the billy hell says Mayer's a ghost town?" the station attendant demanded.

When I said the information had come from one of the merchants at Humboldt, the air grew bluer still.

"Listen!" shouted the station keeper. "You go back and tell that blankety-blank so-and-so he'd better be careful who he's callin' a ghost town! This is Mayer, *right here*—and we can still lick th' pants off that washed-up smelter camp of Humboldt!"

Thinking it would be interesting to stoke the flames of this Yavapai county feud, Mom and I had carried the message back to the prankish storekeeper in the rival town. By the way he was laughing when we took our departure I'm certain the episode has brightened his life immensely, and every time since then that Mom's and my ramblings have taken us through the old smelter town on the Agua Fria, our recollection of the incident has given us a good chuckle.

On this occasion we did not stop at Humboldt but continued south four miles and turned off the highway on a reasonably good sideroad that led toward the blue mountains hemming the valley on the west. Somewhere in those mountains was the old mining camp of Poland where we expected to camp that night.

Following the course of Big Bug Creek, a small stream named by placer miners who had drifted down here from Walker's diggings, about 1863, we remembered that on this creek had been the mining camp of Big Bug. Although the place never had a large population, it had a post office as early as 1879—and as late as 1910; and its commercial district possessed most of the business concerns common to frontier villages.

"Big Bug is a lively camp," *Arizona Journal-Miner* had reported on January 5, 1898. "The town is new and growing, new structures going up right along . . . The country is full of prospectors, and developed mines make a fine showing. New Year's Eve was properly observed with a big dance in the schoolhouse . . . Kingsley and Oliver celebrated the occasion by opening their new saloon, which is one of the finest places in town."

Other Big Bug saloons, according to the same story in the *Journal-Miner,* were operated by Mead and McMahon, and Johnson and Trenberth; general stores were owned by R. F. Burney, Garrett and Avery, and Mrs. Trenberth "assisted by her daughter, Miss Fannie." Dan Reams and a Mr. Vasser, were the town barbers.

Huge tailing piles of gravel and cobbles, extending back for several rods on either side of the stream bed, testified to the prodigious amount of labor expended here by man in his long years of searching for the yellow dust.

As our road continued to climb, the desert brush and scrub oak were succeeded by piñon pines. These, in turn, yielded to tall, long-leaf pines and fir. The canyonsides began to pinch together, the air grew cooler, and Mom started watching for a nice campsite along the creek.

It was just about at this stage of the game that we saw the first advertising signs for a land development called Breezy Pine Acres. As the signs increased, our blood pressures followed suit.

Every direction we turned we were confronted by signs extolling the benefits of living in the canyon; also a plethora of signs advising us that this desirable site was "Reserved for Future Store,"—or Future Church, or Future Schoolhouse—all in the tried and true formula of the tract promoter. But, most of all, we resented the signs warning us to Keep Out, No Trespassing, No Shooting, No Hunting, No Camping, No Parking. There didn't seem to be one open, unposted spot—and I do mean NOT ONE—where we were welcome to pull off the road even long enough to consult our roadmap. In our upward climb through the summer homes tract we passed some fine old stone ruins, presumably the foundation of the old Poland mill. But, even here, we were kept moving by the bared fangs of the ubiquitous Keep Out signs.

With the radiator of the old Mercury and our tempers boiling with equal fierceness, we at last gained the upper limit of the Breezy Pines development—and here the road took a sudden turn for the worse. Frequently so narrow that meeting cars could not have passed, it was further cursed by sharp turns, sharp rocks, loose rocks, and dust. Creeping around blind turns where my right hand never left the horn button, we continued to labor up the mountainside in low gear, the throttle pressed hard against the floorboards, until the poor old Mercury refused to climb even one foot higher.

Since we couldn't go ahead, and there was no possible way to turn a vehcle around on that narrow road, the only choice left us was to let the car roll back down that crooked, winding grade. With our hearts in our throats as we negotiated those sharp, blind turns *backward,* we at last reached a place where, with much jockeying to and fro, I managed to get the car turned around and we started to the flatlands and the highway.

We had arrived about 50 years too late to see Poland at its best. We should have come along sooner, before Keep Out signs became standard pieces of equipment. It was a real nice place, then, so I have been told...

The town of Poland owes its name to Davis R. Poland who arrived in Arizona soon after the discovery on Lynx Creek and discovered the Poland mine in 1872. A small lively camp bearing his name began taking form near the mine and eventually grew to around 800 population, with several boarding houses, a general store and post office, a school, and seven saloons.[4]

Most of my information on Poland, the mining camp, has come from Clarence E. Wager, of Paradise, California, who was employed at Poland back in 1909 and 1910. Fifty years later, when Mr. Wager learned I was writing a book on Arizona ghost towns, he graciously contributed several long, thoroughly interesting letters concerning his recollections of this old camp in the Bradshaws.

"A railroad running from Prescott to Mayer and Crown King, ran a spur up Big Bug creek to a landing platform below the entrance of the big Poland Tunnel," wrote Mr. Wager. "This tunnel ran all the way through the mountain to Walker, about a mile distant. Freight destined for Walker, after being unloaded at the bottom of the hill below the Poland end of the tunnel, was pulled up a rather steep incline to the tunnel entrance, from which point it was hauled through the tunnel to Walker in regular mine ore cars drawn by mules. Original purpose of the tunnel, of course, was for ore development and delivery of ore to the Poland mill. About halfway through the mountainside, tunnels were driven to the right and left. The main producing tunnel was called the

[4]Ida Smith, "Ghosts of the Old Town of Poland," *Arizona Republic,* June 14, 1959.

Accidental.[5] From these side tunnels the ore cars were shuttled to the main line and on out to the mill. In cleaning the mill batteries we sometimes found fair size gold nuggets, though the exact source was never known.

"Poland lies at an elevation of more than 6000 feet and the winters are quite cold. When working on the afternoon shift the men came out of the mine near midnight. Their clothes usually were wet and were frozen stiff before they got to their bunks. When I was there, there was no change room and no place to get warm except in bed. The mine ores were processed by a 20-stamp mill. The eight men who worked in the mill put in 12-hour shifts, changing shifts once each month. For these long hours the millmen received four dollars per day; the helpers, a dollar less.[6]

"Poland was a lively camp, but generally law-abiding. When the railroad was under construction there were more people in the town, and a much wilder element. The railroad company employed a large crew of Greeks who built very smooth, mortarless retaining walls along the right-of-way.

"About a mile below Poland there was a smaller camp named Providence, populated during the operation of the Lottie Tunnel on the south side of the canyon, and the Great Belcher mine on the north side. Another mile down the canyon, on the south side, was the Stirling mine.

"The man who tended the mules used in the Poland Tunnel—a Negro named Brooks—once worked at the Stirling as top man, dumping the ore buckets as they were hoisted. About this time, many of the smaller mines were being infiltrated by Austrian and Italian workers who were hotly resented by the old timers. One day a friend of mine applied for work at the mine and stopped to exchange a few words with Brooks, the Negro. In the course of their chat the Austrian-Italian subject was mentioned and Brooks got all steamed-up about it.

" 'That's one good thing about *this* mine,' declared Brooks, mopping the sweat from his ebony-hued brow. '*We* don't hire anybody but *white* men!"[7]

"About half a mile above Poland was a two-room shack known as Robber's Roost. At this point, near the head of Big Bug creek, it is but

[5]Herman Ehrenberg, reporting on mines of the Bradshaw Mountains area for J. Ross Browne's *Resources*, in 1866, stated: "The first lode upon which machinery was erected was the Accidental . . . a gold vein yielding some showy and beautiful specimens of free gold—specimens which attracted great attention in San Francisco. The ore . . . averages about $100 per ton . . ."

[6]About eight years prior to Mr. Wager's arrival at Poland, miners there had been working 12 hours a day, seven days a week, for $2.25 a day—out of which they paid a dollar a day for board and lodging. The only holidays recognized were Fourth of July and Christmas. In 1905 the workers struck for an additional 25 cents a day. The men roomed in a sheet-iron bunkhouse and ate at the company boarding house, which was operated by a crew of Chinese.

[7]Brooks, the Negro, stayed on at Poland, living alone in his little cabin long after the camp had died and its people were scattered afar. One severe winter, recalled Mr. Wager, the old darky was found frozen to death in a snowbank on the trail to his cabin.

a short distance over the hills, south, to the upper end of Turkey Creek, which runs through some very rough country to enter the Agua Fria near the old Black Canyon stage road. The stage was occasionally held up and robbed, and it was believed that the holdup men then returned to their hideout in this cabin on Big Bug.

"When I was in Poland, 50-odd years ago, this cabin was occupied by an old man who lived in seclusion, with no visitors and no companions but his two burros, of which he seemed very fond. One day he was found dead. I was told that he had first shot his burros, then himself.

"Naturally, there were many stories of possible hidden loot around the cabin and folks used to go up there looking for it. So far as I know, there was nothing to substantiate such talk . . ."

Candidly, I am almost glad there is nothing to substantiate this story of buried loot; but, even if I knew positively that treasure was there for the digging, not too deep and tax-exempt, I seriously doubt if I would go back through Poland to look for it.

I'd rather explore some other ghost town where the Keep Out signs are fewer.

Bradshaw City, for example . . .

On the topographic map of the Bradshaw Mountains appear the names of more ghost towns and former mining camps than are to be found on any other Arizona quadrangle map. But, even here, does not appear the name of Bradshaw City, which was past its prime long before the publication of that map in 1903.

Birth of the new camp was reported by the *Arizona Citizen* on May 20, 1871, in its story of a promising new gold discovery in the Bradshaw Mountains, south of Crown King.

"The lucky finders were a party of placer miners who had been working all winter in the gulches near the Del Pasco mine and, water commencing to fail, were prospecting for diggings on Humbug creek. They propose building arrastras and going to work at once on quartz. The finding of these rich veins has given quite an impetus to prospecting and much confidence to those parties just opening stores in Bradshaw, as their working will quickly give a supply of gold-dust which is good to buy bacon and whiskey with and is appreciated by dealers in those articles. The pioneer saloonkeeper from Bradshaw, while in town a few days since, was heard to say that everything there is lovely and reminds one of old times in California . . .

"Considerable difference of opinion has existed as to where the principal town should be located. Now the question seems to be settled for the present. A town has been surveyed in the pine timber on the top of the southern ridge of the Bradshaw mountains, where never-failing springs give a good supply of purest water. The town is, very properly, called Bradshaw . . . Two stores, a restaurant, two saloons, an assay office and a butcher shop, are already built and operating, and many other buildings are being erected . . . A wagon road is being built to the town from Minnehaha Flat, and it will be the point of supply for the surrounding country."

When Lieut. Wheeler visited the youthful town of Bradshaw City during the course of reconnaissances covered by his *Preliminary Report,* of 1871, he found hay selling there for $75 a ton; barley, 15 cents a pound; lumber, $100 per thousand feet, and blasting powder, $15 a keg. Miners were being paid $2.50 a day, and freight was 15 cents a pound.

"Prescott, Arizona Territory, is the nearest post office, about 40 miles distant, reached by a trail. This has a good track but is, in places, very steep," reported Lieut. Wheeler. "There is a wagon road from Walnut Grove to Minnehaha Flat, five miles from Bradshaw City. A steep trail leads from the flat to the city . . ."

Despite its isolated location, Bradshaw City grew so encouragingly that by midsummer of 1871 it was being predicted that within a year or two the town would have a populaton of 10,000 to 20,000, with many different industries and men of all professions.[8] These sanguine expectations, of course, were never realized; however, I have heard it said, many times, that the present-day ghost town of Bradshaw, high on the slopes of Mt. Wasson, once numbered as many as 5000 citizens.

Another town I have yet to see is Alexandra. I tried to get there once from Mayer, but ran into snow so deep my truck couldn't negotiate it. On another occasion I tried to reach the town by hiking down Peck Canyon from Crown King. That time I got my wires crossed, somehow, and instead of reaching Alexandra I wound up on the east side of Tuscumbia Mountain. But someday—luck willing—I'll get there.

Alexandra was no overnight ragtown but the main camp for the Peck mine, one of Arizona's greatest.

In 1875—a full dozen years after Old Joe Walker and his party had made their astounding discoveries on Lynx Creek, and more than three years after the discoveries that sparked the founding of Bradshaw City and Poland—the great Peck mine was discovered by Edmund George Peck, C. C. Bean, William Cole, and T. M. Alexander. Actually, it was Peck who discovered the first heavy silver float as he was stooping to drink from a spring.

The Peck was the sort of mine men dream about, but seldom find. A five-foot body of silver chloride carried a paystreak that was about half silver. Original locators of the mine are said to have received $13,000 for their first ten tons of ore, and at a depth of not more than 300 feet, according to Bancroft, the ore was running $10,000 to $17,000 per ton!

In the course of three years, beginning in 1875, the Peck is said to have produced around $1,200,000; and then litigation set in, and the mine stood still while its owners and operators grew poorer by the day and the bank accounts of lawyers grew fatter.

Meanwhile, in 1877, a townsite surveyed by T. M. Alexander and Col. Bigelow had been named for Mrs. Alexander, first white woman to visit the remote site high in the Bradshaws.

[8]*Arizona Citizen,* July 1, 1871.

"Alexandra, the town which is springing up between the Peck and Prince mines, is likely, from all reports, to be quite a place," declared the *Arizona Miner* on June 15, 1877. "It has now two large stores, Alexander & Company, and Anders & Rowe; three boarding houses, Conners & Company, Corbin & Hardwicke, and McPhee's; four places where spiritual refreshments are provided, two livery stables, one butcher shop, one blacksmith shop, and several parties are grading lots and preparing to build. The parties owning the mining claims on which the town is built have joined together under the name of Alexandra Town Association and . . . have expended nearly $2000 in laying out and grading the main streets . . ."

"The general impression is that this is destined to be . . . the best camp in the whole Territory, if not on the whole Pacific slope . . ." commented the *Miner* on July 26, 1878. "Times are getting better every day. L. C. Conners is still running the boarding house and will, on the first of the month, open a first class bar. A restaurant was opened on the 10th . . . by Mr. M. L. Calhoun. Everything about the establishment is neat and clean and his tables will compete with any in Prescott . . ."

A month later, (August 30, 1878) the *Miner* was still boosting Alexandra. "Mr. Frank Barnum is running the French billiard hall . . . Mr. John Ellis is still conducting the Gold Room resort . . . Mr. Dall's Palace Bar is conducted by our old-time friend, Mr. Hawley . . . Our livery stable is run by Mr. William Mulvejnon. Mr. Waters still superintends the Eureka Bar. Only one butcher shop in the city, at present, superintended by R. Gillespie. Mr. William Yancey is a partner in the business. Mr. Yancey furnishes the city with milk, to say nothing about the water. The Minger brothers, and the Brewery, still continue and are doing splendid. The city is furnished with four caterers. Mrs. Corbin runs the Summit House and has all she can attend to. The Alexandra Restaurant is doing a cash business and gets the cream of the customers. Next in order is the Palace Hotel, owned and run by Samuel McDonald . . . L. C. Conners continues to run the company's boarding house. He also has a fine bar attached to the house . . ."

In other words, everything was lovely and the goose hung high . . . except that litigation over the mine dragged on and on. Meanwhile, the mine remained closed; and since there's a limit to how long folks can make a living by "taking in each other's washings," as the saying goes, the town gradually withered and died on the vine. Poor old Ed Peck spent more than 30 years prospecting throughout the Bradshaws, but he never struck another bonanza like the Peck. When he died, in 1910, he is said to have been a poor man.

Which bears out the old adage that even a poor compromise beats a good lawsuit.

It also brings to mind a little anecdote related by the venerable prospector Mom and I had met at the old stone cabin, near Walker.

A man who was thinking of buying a certain mining property was tipped off that title to the claim was in dispute and, if he bought it, he would be buying a lawsuit along with it. One day the owner of the property encountered the other on the street and asked him if he was going to buy the claim?

"I'm afraid not," said the prospective purchaser. "I understand the property's in litigation."

"Litigation?" bellowed the old miner. "Who ever said that's a damned liar! It ain't no such thing in litigation—it's in lime and porphyry!"

"Up and down the mountain ranges . . . this parable of the abandoned mining camp repeats itself a thousand times . . . If one fixes his attention on the positive and constructive side of their life, emphasizing not what has been lost by their decline but what has been gained by their achievement, there is much in the record to stir the blood and call out the warmest admiration . . . The story of these pioneers in the gold fields is part of the greater story of the American frontiersman, the man who fought the desert and the wilderness and subdued them."

—ARLEY BARTHOW SHOW, *in Death Valley Magazine, 1908.*

Weaver Was Rich and Rowdy

SPANNING 80-odd miles in its greatest north-south diameter and about 100 miles from east to west, Yavapai County embraces an area slightly greater than that of the state of Massachusetts where 25 or more cities each has a larger population than this entire Arizona county.

That's one of the reasons I like Yavapai. I've liked it a long while. I liked it away back when I was still struggling to pronounce its name with four syllables—and liked it even better when I learned that the third "a" is not sounded.

With its broad pine forests, clear air and pure water, its cool, clear nights and sparkling days, life in Yavapai is singularly good, even in this current Space Age. It is not, however, quite as exciting today as it was 100 years ago when the golden harvest of Rich Hill was being dug out of the rocks with butcher knives . . .

Almost coincident with Capt. Joseph Reddiford Walker's return to central Arizona, another prospecting party, organized in California by Abraham Harlow Peeples, had headed for the same blue-tiered ranges and with the same avowed purpose: The finding of gold. Which of these two parties—Walker's or Peeples'—was first to make an important strike is a debatable point. It isn't really important which was first. All that matters is that either of those strikes was of such phenomenal richness that there no longer could be any doubt concerning the presence of gold in central Arizona.

After crossing the Colorado River at Arizona City, where they were joined by their guide, Pauline Weaver, nominal discoverer of the La Paz placers, the Peeples party followed the river northerly to La Paz. Turning east at this point they circled the north end of the Plumosa range and made their way up the desert valley between the Harcuvars and the Harquahalas. Upon nearing the foothills of the higher mountains Peeples shot several antelopes for meat and from this incident named a nearby stream Antelope Creek. That evening members of the party panned a few "colors" from another small stream which they named Weaver Creek in honor of their guide.

When it was discovered the next morning that part of their saddle stock had strayed during the night, four Mexicans, who had joined the party at Yuma, started in pursuit of the animals.

Any reader who has scratched even the surface of Western mining history should have no difficulty recognizing the familiar pattern of a bonanza's impending discovery. (As Muriel Wolle commented facetiously in the introduction to her *Bonanza Trail,* "I am convinced . . . that mines are discovered by accident. Had it not been for Indians who showed glittering stones to eager prospectors, and for burros which broke their hobbles and strayed away while their masters slept, the gold rushes would not have occurred.")

Be that as it may, when the Mexican wranglers returned that evening with the errant livestock, they also brought with them a large quantity of gold nuggets they had gathered from the summit of a nearby hill!

"The next morning," relates Fish, "the party went to the top of the hill where innumerable chunks and nuggets of gold were found in the sort of sloping basin. Peeples picked up $7000 worth of gold before breakfast . . . It is estimated that during the first month a quarter of a million dollars in gold was gathered . . ."[1]

Throughout that summer of 1863 reports of the miraculous new strike sped across mountains and valleys to the nearest stage and telegraph lines, thence by letter and wire to the newspaper offices of the world. With each retelling, the excitement mounted.

One of the more notable visitors to the Weaver Diggings that summer was William H. Oury, of Tucson, who on September 20, 1863, dispatched to the *Daily Alta Californian,* of San Francisco, "A Letter from the Goldfields of Arizona." Needless to say, the printing of that communication did nothing to allay the pyramiding enthusiasm of the mining world.

"I returned last month from a short trip to the newly-discovered gold mines in the northwest part of the Territory and saw some wonderful sights in the way of gold diggings," wrote Mr. Oury. "I saw a party of ten men, who had discovered gold on the top of the mountain which divides Antelope Peak from Indian Gulch, with no other tools than butcher knives, scratching and gathering up gold like pebbles, and in less than the two hours I was looking at them they took out ten pounds. One of the party . . . invited me to sit down with my knife and see what I could find for myself. In less than half an hour, scratching with knife and fingers, I picked up $74 in pieces of from four bits to half an ounce. During my stay of four or five days in that neighborhood, the party above mentioned took out on an average of $1500 per day. Another party of five Frenchmen, who had a claim near them, often took

[1]*History of Arizona,* Chapter 8. Unpublished manuscript in APHS files.

out from $600 to $1000 per day; and many others in the vicinity were getting from $60 to $100 per day . . ."

In this same issue of the *Alta* (October 14, 1863) appeared another item "From the Weaver Diggings," which fully supported Mr. Oury's glowing report and added additional fuel to the mounting flames of excitement:

"Mr. Landsberger, a prominent merchant of La Paz, and a brother of Isadore Landsberger, of San Francisco, informs me that he has seen the following sums brought in within the last few weeks: By a Mr. Webber, about twelve pounds avoirdupois; by a man named Winneger, $800; by a Mexican named Redondo, eight pounds; by Antonio Contrero, five and one-half pounds; by Mr. McMullen, 90 to 100 ounces; by Boss Denver, three weeks since, $2800; by James Leonard, ten days since, 40 ounces. Mr. Landsberger has sent a large stock of goods up to these mines and appears to be confident of their being rich and perhaps extensive . . ."

Another of the earliest merchants at Rich Hill, or Weaverville, as the new camp was known, was Michael Goldwater, whose branch store at this place was managed by Sol Barth and Aaron Barnett.

"Weaver and all his crew would come in once a week to weigh out their gold dust on the scales of Barnett and Barth's store and to purchase supplies, of which a very large item always was whiskey," relates McClintock. ". . . The Weaver party usually brought in about 25 pounds of gold a week. This was generally coarse, most of it dug out by butcher knives. With the proceeds, the miners would spend a day or two each week in the wildest dissipation afforded by the rough, tent village, which stood not far from the present settlement of Stanton. One nugget brought in by a Mexican named Clemente was valued at $900 . . ."[2]

In its issue of October 16, 1863, the *Alta Californian*, still mesmerized by the miracle of Weaver's Diggings, reported that two men, Bryant and Holcombe by name, working six men with rockers over a period of 30 days, had found and brought out with them "$10,000 to $12,000 in dust each, as a result of their operation."

It was also reported that beef was still to be had at the diggings at 12½ cents a pound; that potatoes and onions were selling for 25 cents a pound, and flour at $15 per hundred.

"Plenty of feed could be brought in by the Indians here if they were willing to do so, but many of them have now got horses of their own and wish to save the fodder for them," continued the report from the goldfields. "They are as smart at 'bulling the market' as New York stock brokers and literally 'beat the Jews' every time. They come down with a small lot of dry, last year's grass . . . and ask a dollar per bundle for it. One of these bundles will just about fill a horse. If the price is demurred at, they will set the whole lot on fire and burn it before your eyes. The green buyer now thinks he must purchase at any price when another lot is offered, and he does so. The next day . . . the same red rascals who

[2]*Arizona,* James H. McClintock, 1916.

burned the hay . . . come grunting into town, staggering under fresh loads of grass which they dispose of at their own price . . ."

Weaverville's pay gravel, unfortunately, proved to be only six inches to two feet in depth and the big excitement was soon past. Before the strike was two years old even the hitherto loyal and exuberant *Alta Californian* was employing the past tense when it referred to, "the great Weaver Diggings where only a few Mexicans remained to follow mining by dry washing on the mountain and in the neighboring gulches.

"Only a few days ago a lucky hombre found a *chispa* which weighed nine ounces; but the average pay will not exceed $1.50 per day," reported the *Alta.*

But, even though the nuggets found there were no longer big as baseballs, Weaverville, as a town, continued to flourish. Almost from the day of its founding it had been a rough, rowdy little burg. As it had grown in stature so had the number of its saloons, gambling dens, bawdy houses, outlaw hangouts and shooting scrapes, increased accordingly. Neither law nor lessening cleanups seemed to have any effect upon its deportment, and more than 30 years after the glittering miracle of the original placer strike had dwindled to a sowbelly-and-beans proposition the little mining camp, at the wedding point of mountains and desert, was still going its lawless way.

·The murder there of William Segna, a popular and respected Austrian who operated at Weaver a saloon and mercantile establishment, gave rise to an editorial blast of condemnation by the *Arizona Weekly Journal-Miner,* of Prescott.

"One of the darkest spots in Arizona is the . . . adobe town of Weaver," charged the *Journal-Miner* in its issue of September 30, 1898. "From the earliest period in the history of the Territory, up to the present, everything that has a horror to it, and crimes and criminals, has been sadly its own. Nothing good has come from it, and as long as the element it harbors and encourages in fiendish lawlessness and cold-blooded murders is permitted, just that long will Weaver thrive in its barbaric life . . .

"Weaver has reveled in its bloody misdeeds too long. Unless courts supply a remedy soon, the place should be razed to the ground and wiped away. It is the darkest spot in crimes and criminals west of the Rockies and, if not obliterated, justice will be equally as guilty as the men who support and sanction its bloodcurdling existence."

What frontier lawmen could not or would not do, Time has done.

Weaver, today, is the epitome of peace and quiet.

When Mom and I visited the old townsite in 1955 about all we could find there were a few stone corrals, a few foundations without walls, a few walls without roofs. Two or three old cabins were still occupied by hopeful snipers who daily combed the gravel, optimistically searching for any minute bit of gold overlooked by the three generations of miners who had preceded them. During the course of those 90-odd years the immediate landscape has been torn apart and tumbled upside down—and re-torn and re-tumbled—until just about the only undisturbed spot in Weaver is its old cemetery. There a few graves still bear legible markers and a lone saguaro stands guard.

One of Weaver's contemporaries, situated over a low rise about a mile to the southeast, was less spectacular and more enduring. Its name was Octave.

In June 1864, the claims that would develop into the principal mine at this point were located by Charles G. Mason, later to figure in re-discovery of the "lost" Silver King mine; Elijah Smith, William Holcombe, and Charles B. Genung. Of this quartet, Genung proved the more tenacious. Long after his partners had died or moved to other parts of the country, he still hung onto the property through bonanza and borrasca for 22 years—only to lose it in 1886 when a man entrusted to do the annual assessment work failed that trust and the ground was relocated by other persons.

In the late 1890s a group of eight men purchased the mine and organized the Octave Mining Company. At a cost of close to half a million dollars a 40-stamp mill of 120 tons capacity was placed in operation and soon after the turn of the century is said to have been netting its owners some $50,000 monthly.[3] After passing through several ownerships with varying degrees of profit and loss, the Octave was closed in 1942 by the War Production Board order blackballing gold mines. Up to that time its production had been around $4,000,000.

By the middle 1950s not much remained at the site of Octave. With exception of a few small stone cabins all buildings had been razed. Foundations and cement slabs marked the former locations of several large structures, including the machine shop, boarding house, and warehouse; and a vast tailings dump in the ravine below testified to hundreds of thousands of tons of ore passed through the mill. Though the ruins and rubble remaining at the site were of sufficient extent to interest any dedicated ghost town prowler, there wasn't much to show for a mining camp which, at its peak, is said to have numbered 3000 inhabitants

Third member of this ghostly trio at the foot of the Weaver mountains is the one-time town of Stanton, nee Antelope Station.

Stanton's history, up to a point, is a somewhat nebulous thing. First settlement made in this area, following the discovery of gold by the Peeples' party in 1863, apparently was at or near the present site of Stanton. This initial colony soon moved over the toe of the mountain to Weaver Creek. As early as 1866, however, Charles Stanton was keeper of the stage stop known as Antelope Station and operated a store at the same place. Name of the station was changed by degrees to that of its leading citizen, and when the town's first official post office was established in 1875 it was under the name Stanton, with Charles Stanton as postmaster.[4]

Before emigrating to Arizona, according to Dame Rumor, Stanton had been expelled from a monastery on a charge of immorality. It would

[3]*Los Angeles Times,* Oct. 17, 1901.

[4]Barnes.

Guardian of the Graves, Weaverville, 1955

Main Street, Stanton, 1955

have prevented a lot of bloodshed and saved a heap of general unpleasantness if Charles also had been expelled from Antelope Station, or Stanton, before he began playing marbles for keeps.

Charles, it seems, resented the fact that the leading store at Stanton was no longer his own place of business but one operated by the partnership of Wilson and Timmerman. Jealous of their success and desirous of eliminating all competition possible—this being before the day of the Fair Trade Act—Stanton determined to rub out his rival merchants. To gain his desired end he enlisted the aid of one William Partridge, another prominent resident of the town, who followed close after Wilson and Timmerman in the long roster of persons least liked by Stanton. Partridge, quite naturally, wasn't told that he had been chosen for the role of fall guy.

The fact that ill-feeling already existed between Partridge and Wilson simplified matters a lot. All Stanton had to do was to sit back and wait for a good opportunity to play both ends against the middle. His chance came the day Wilson's hogs got into Partridge's grub supply.

Although Wilson had said before witnesses that he would pay any damage the hogs had done, Stanton allegedly bribed a Mexican to tell Partridge that Wilson was "out to get him." Result was that when Wilson approached Partridge's residence, Partridge shot him down without warning as a presumed means of self preservation.

That took care of Wilson, and when Stanton later succeeded in railroading Partridge to prison for the murder, that took care of him. With Timmerman subsequently liquidated, Stanton had the stage road changed to run by his place of business, rather than past the former Wilson & Timmerman store, and soon cornered most of the trade his victims formerly had enjoyed. The only fly remaining in his pretty little pot of ointment was Barney Martin.

Barney and his wife owned a nice, red-brick store in the heart of town. They were good, clean, honest folk, never obnoxious and never in trouble. This characteristic, alone, was enough to invoke the displeasure of Charles Stanton, as well as emnity of the outlaw gangs at Weaver.

After repeated warnings that the ruffians were out to "get" him and his family, Barney sold his store in the summer of 1886, pocketed the proceeds, loaded his personal belongings and his family in a wagon, and set out for Phoenix. Before departing Stanton he also took the precaution to notify his friend, Capt. Calderwood, that he was on his way and should arrive at the latter's place of residence by a certain specified date.

When Barney failed to put in his apparence on schedule, Calderwood organized a search party.

Backtracking along the stage road, the searchers found where a pair of fresh wagon tracks had turned off the road into the hills.

At the end of those tracks were the burned remains of the Martin wagon and household gear, together with the dead bodies of Barney and his entire family.

Responsibility for the mass murder was laid at the door of the Valenzuela gang, with Stanton named as an accessory. But, as usual, no conviction was gained.

Later that year Charles Stanton was shot and killed by one of the Lucero boys, of Weaver, assertedly to revenge an insult to the boys' sister.

"Stanton was sitting with his chair tipped back against the counter when Lucero rode up and fired one shot through the open door," wrote Muriel Wolle in *Bonanza Trail*. "The same day Tom Pierson, coming to town over the mountain from the Crown King mine, met Lucero on the trail.

"'I've just killed Stanton and I'm heading for the border,' Lucero called as he passed.

"'You don't have to pull out,' said Pierson. 'If you stick around you'll get a reward.'"

Nine years after the death of Charles Stanton the mining camp bearing his name came into the ownership of George Upton, a geologist, who took up residence there. Forty-three years later the aging Upton was joined by a niece, Miss Maurine Sanborn, of Minnesota, who had served as a nurse in World War I and later had been a supervisor of nurses in one of the larger hospitals in Minneapolis.

Due to the constant presence and vigilance of its owners the town of Stanton escaped much of the vandalism that is usually the lot of ghost towns, and my first visit there, in 1955, found its main street still flanked by 15 buildings, including an old hotel of about a dozen rooms.

Early in 1959 the *Arizona Republic*, of Phoenix, published a story which focussed upon Stanton more attention than it had enjoyed at any time since gunsmoke was in style.

The story, by-lined by Thelma Heatwole, revealed that ten acres in the heart of the townsite had been sold by Miss Sanborn to the *Saturday Evening Post* at the well-known "undisclosed figure."

"The *Post* will restore the town, including an 11-room hotel built during Civil War days, a saloon, post office, and a house," wrote Miss Heatwole. "Already the *Post* has reinstated the hitching posts, water troughs, and many of the original trappings and embellishments of the early-day mining town."

The sales agreement assertedly provided that if any mining was done on the premises during Miss Sanborn's lifetime, one-half the proceeds therefrom would be paid to her as royalty, and that no alcoholic beverages were to be sold on the property while she lived.

Foreseeing the restoration and development of Stanton as a second Virginia City or Williamsburg—a tourist attraction that would bring fame, fortune, and shekels to the state—Arizona greeted the report with delight.

The let-down came several weeks later when it was learned that the "restoration and development" was only a bit of buffoonery, a gag, a gimmick.

The *Post*, according to an Associated Press dispatch from New York, had renamed the town Ulcer Gulch and had awarded it, lock, stock, and ghosts, as first prize for writing "the best last line" in a jingle contest. The winner, 30-year-old Anne Tolstoi Foster of New York, employe of the J. Walter Thompson Advertising agency and the mother of three children, had never been to Arizona, stated the AP dispatch, and had no idea what she would do with her prize.

Hoping to obtain some first-hand information for use in this chapter I wrote to Mrs. Foster, in care of the *Post* and, later, in care of the advertising agency. Receiving no reply from either effort, I decided to revisit Stanton to learn what I might from Miss Sanborn.

At the Stanton turnoff was a wooden signboard I didn't recall seeing at the time of my previous visit to the town. Its message, which stated that this was a PRIVATE ROAD, appeared over the names of M. E. Sanborn and G. E. Upton. Feeling that my business here was sufficiently legitimate to remove me from the status of a trespasser, I drove on down the deserted main street of the old mining camp until reaching a stout gate fastened with an iron chain. Climbing through the fence I continued up the road afoot toward the Upton-Sanborn residence, situated on a little rise about 200 feet beyond the gate. In view of the general aridity of these Weaver mountain foothills, I was pleasantly surprised to find such a pretty place at the end of the road. Set down in the midst of a wide, cool yard, with a green lawn, shade trees, vines, and a great many roses and other flowers, it was like a lovely oasis.

My rap at the front door of the dwelling was answered by a slender woman whose youth had been left somewhere in the years behind. Her hair was quite gray, her eyes held a tired look, and every line in her face revealed the footprints of strain and care. Yet, she was still an attractive woman. Her movements were quick and efficient, and I feel almost certain I would have liked her immensely if we could have met under more fortuitous circumstances.

When the woman identified herself as Maurine Sanborn and I gave her my card and explained my purpose in visiting Stanton, she said she would be unable to talk with me that morning.

"I'm terribly sorry," she continued. "But you should have made an appointment if you wished to see me. I set aside two hours each evening to talk with writers and reporters, but I simply can't spare them any time before dinner . . ."

When I hastened to apologize for my intrusion and said I had not realized she would be so busy, Miss Sanborn explained that she had the sole responsibility—24 hours a day and seven days a week—of caring for her uncle, the old mining engineer, George Upton, then 96 years of age, bedfast, demanding, and quite senile.

In lieu of a personal interview she made available to me her scrapbook collection of newspaper clippings and personal letters detailing the history of Rich Hill, Antelope Station, and Stanton, from earliest times up to and including the sale of a portion of her property to the *Post,* and the subsequent jingle contest.

Though disappointed that I was unable to visit with her, personally, I found that the clippings provided answers to many of the questions I had expected to ask—and even a few minutes in the Upton-Sanborn home was sufficient to convince me that its mistress might very well find it inconvenient to interrupt her sicksoom routine in order to chit-chat with roving reporters and other chance droppers-in.

As I was returning to the highway, I saw ahead of me a late model automobile so thoroughly wrecked it looked like a fugitive from a junkyard. Two boys, standing in the road, were signalling me to stop. Although it isn't my policy to pick up hitchhikers, I was afraid someone might have been injured in the accident. Knowing it might be hours before another vehicle passed this way, I decided I'd better stop and render such aid as I could give.

One of the boys had suffered a badly-wrenched shoulder, by now swelling and stiffening, but this seemed to be the only injury. They were from Phoenix. They liked to visit ghost towns, they said, because sometimes they could find old discarded automobiles in the trash dumps and could get parts from them.

About an hour earlier, they explained, they had been driving along this road, minding their own business, when suddenly they heard a sharp report, like a gunshot, and a bullet zinged through the open front window of the car. Missing the driver's head by an alleged "fraction of an inch," it startled him so that he lost control of the vehicle. Plunging off the road the car had hit the loose sand alongside and flipped completely over, coming back to rest on its wheels in the same position—but not in the same condition—as before the accident.

After discussing the matter we decided the uninjured lad should remain with the wrecked car and their camping gear while his companion accompanied me to Arrowhead Station, on the main highway. There he could obtain first aid for his injured shoulder, could telephone for a wrecker, and report the shooting incident to the bartender who did double duty as a deputy sheriff.

As we bounced along the rough road toward Arrowhead I learned that the boy riding beside me was only 16 years old. He was, also, just about the world's worst worried kid.

"It's Dad's car," he said. "I just borrowed it for this trip. I don't know what we're going to do without it. Dad has to drive to work every day—and we sure can't afford to buy another car . . ."

I told him I didn't imagine it would take too long to have it repaired.

"I don't know," he said, half to himself. "But I know it'll cost something awful to have it towed back to Phoenix, and another fortune to have it fixed . . ."

Summer's heat was lying heavily upon Arrowhead Station when I brought my truck to a halt in the graveled yard. There was no one around the gas pumps but from inside the nearby tavern came the blaring discordance of a juke box. Two men and a woman were draped over the bar nursing cold beers. The bartender-deputy was polishing glasses.

When the kid told the deputy that someone along the Stanton road had taken a potshot at him, causing him to lose control of his car and wreck it, everyone in the room became suddenly attentive.

"Well, well!" said the deputy, rubbing his hands together in an anticipatory manner. "A shooting on the Stanton road! This is like old times!"

Then the men and the woman and the bartender were all asking questions and conjecturing, and offering suggestions and advice, and I figured it was a good time for me to do a fadeout.

I've never been back to Stanton since—but I still like Yavapai County.

Former Hotel, Stanton, 1955

The Colorado Steam Nav. Co.

Will Dispatch

THE LARGE PALATIAL and MAGNIFICENT

STEAMER MOHAVE.

which

will leave Yuma for El Dorado Cañon and **intermediate** points on SUNDAY MAY 1st., at 9 o'clock A. M., **Parties desiring** freight or passage by this STEAMER **will please call at** the Company's Office in the Railroad Depot foot of Main Street.

J. F. KNAPP Agent C. S. N. Co.

Advertisement–Colorado Steam Navigation Company

"The Colorado is, to me, an enigma. Its waters hurry by us today as they have hurried since the beginning. They bring no intelligence. They never did. They have no history and they never had one, and never have heard other notes than notes of the ferocious savages who inhabit its barren banks."

—Letter written from Fort Yuma to Daily Alta Californian, June 17, 1860.

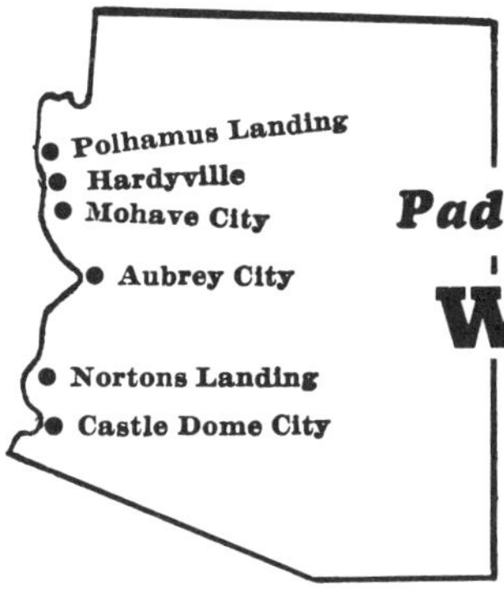

Paddlewheels On the Colorado

WITH RICH MINES coming into production, future cities sprouting in the wilderness, and diverse cargos waiting transferral to and from frontier outposts, the Colorado River—even with all its attending hazards—offered the best means of access to Arizona's vast interior. Alert shipping men of other American rivers were not slow to recognize the great potential of this desert waterway and in 1852—only four years after the navigable portion of the river and all our present Southwestern states had come under the jurisdiction and ownership of Uncle Sam—steam-driven vessels began plying the Colorado.

During the next half-century-plus the river boats hauled millions of pounds of cargo—mining and milling machinery and ore, Army troops, military supplies and civilian goods of every description, as well as passengers and mail. At each of the river ports soon converged a spiderweb of dusty roads and trails over which pounded Concord coaches, mule pack trains, horsebackers, and heavy freight wagons bound to and from the new boom camps of La Paz, Wickenburg, Weaverville, and Prescott. Even ore from mines as far distant as the Silver King, in Pinal County, and the Cerro Colorado, in Pima, was wagon-freighted to the river for water shipment to San Francisco and elsewhere.

That the treacherous Colorado ever resounded to the blast of steamboat whistles is a tribute to frontier daring and determination. Seagoing vessels, unable to negotiate the shallow river, halted at Puerto Isabel, Sonora, at the head of the Gulf of California. Here were wharfs and drydocks, and here cargos were transferred to vessels of shallower draft for the grueling upriver trip to Fort Yuma, Ehrenberg, Hardyville; even as far, on occasion, as the Mormon settlement of Callville, nearly 600 miles above the mouth of the river! Not one of those miles but presented threat to the passing ships. The lower river was beset by sandbars, hidden snags, and a channel here today and somewhere else tomorrow, while narrow, twisting canyons and submerged rocks made the upper river a mariner's nightmare.

History does not reveal which man was first to conceive the idea of putting cargo ships on this broncho river, but the first to convert such a thought into action was Capt. George A. Johnson.

After securing a contract to provision Fort Yuma via the river route, Capt. Johnson, in 1851, shipped from San Francisco to the head of the Gulf of California, by the schooner *Sierra Nevada*, materials necessary to the construction of two flatboats, each 50 feet in length with 18-foot beam. There assembled and launched near the river's mouth, one of these unwieldy little vessels filled with water and sank almost as soon as it was placed in service. The second of the two crafts, however, by performing as nobly as could be expected of a flatboat, served to convince Johnson that the lower river might be navigated by small paddle-wheel steamers patterned after those then operating on the Mississippi, Ohio, and Sacramento.

Parts for Johnson's first steamer, the *Uncle Sam*, were carried aboard the schooner *Capacity* from San Francisco to the river's mouth, where the 65-foot vessel was assembled and launched in the closing days of 1852. Despite her small size the *Uncle Sam* was a good, stout-hearted little craft, and since she could carry some 70,000 pounds of freight on only 22 inches of water, she was well suited to service on the shallow river.

Second steamer on the river, launched early in 1854, was Johnson's 108-foot sidewheeler, *General Jessup*, with a carrying capacity of 60 tons and a draft of only two-and-one-half feet. To her goes the honor of being the first steamer to penetrate the upper river, some 350 miles above Yuma. Next came the 120-foot sternwheeler, *Colorado*, put into service in 1855, and the 135-foot *Mohave*.

During his first dozen years on the river Capt. Johnson and his Colorado Steam Navigation Company operated virtually without opposition—except that offered by the river itself. Competition threatened briefly when Capt. Thomas E. Trueworthy, in 1864, steamed the 130-foot sternwheeler, *Esmeralda*, from the Golden Gate, south along the California coast, around the 800-mile finger of Baja California, up the gulf to the river and on up-river to the Mormon settlement of Callville, now drowned beneath the waters of Lake Mead. That same year the *Nina Tilden* was brought from San Francisco under her own power by Capt. Paddy Gorman for the Philadelphia Mining Company, owner of copper properties along the Colorado. Within three years, however, both of these interlopers had been purchased by Johnson's company and the CSN had no further competition on the river.

"At first, mainly supplies for the Army and, later, all kinds of household goods, mining and farming machinery, and even newsprint and the trees that were to flourish along Yuma streets and on the farms of the Salt River Valley, crowded the holds of the vessels," wrote Hazel Emery Mills.[1] "By the middle '60s, Arizona began shipping raw pro-

[1] *The Arizona Fleet*, from *The American Neptune*, a quarterly journal of Maritime history, July 1941.

ducts to San Francisco—sacks of high-grade silver, lead and copper ore, bullion, bales of wool, hides, and bundles of beaver skins . . .

"In the '70s, six boats and five barges of the CSN line chugged up and down the Lower Colorado between the Gulf and El Dorado Canyon, 540 miles above Port Isabel. The *Colorado II*, the *Cocopah II*, the *Mohave* and the *Nina Tilden* in service. The 160-foot *Gila* was added in 1872; and in 1876, the *Mohave II*, the Queen of the River, 170 feet long, with a 33-foot beam, took the place of the *Mohave I*. In general design, shape and construction, the later Colorado steamboats were similar to the boats on the Upper Mississippi, the Ohio, the Sacramento, and the Columbia, and were very like the boats on the uncertain Missouri in that they drew practically no water, light, and rarely more than two feet when loaded . . . When the leadsman called 'Scant four!' on the Colorado, the pilot did not snatch the boat back, reverse engines, pull bells, and shout down the tube to the engineer. He merely wiped his brow and reflected that the river must be rising . . ."

Nowhere is the record of the river boats better preserved than in the newspaper files of that half century, when arrival, departure, cargos, adventures and misadventures, of the Colorado fleet constituted Big News in almost every journal published in the Territory. Even the newspapers of New Mexico and California were not loath to mention outstanding accomplishments of the river fleet—such as the time in June, 1878, when the *Cocopah II* steamed downriver from Aubrey to Yuma, 240 miles, in slightly more than 11 hours, and made five landings along the way; or the time in 1873 when the powerful *Mohave*, with a barge in tow, traveled the 300 miles up-river from Yuma to Mohave City in only 60 hours, including stops.

But it wasn't in speed the little ships excelled so much as in their ability to render faithful service while operating under the most adverse conditions of navigation. Reading the record of their accomplishments, presented casually and succinctly in the news columns of their time, one is struck by the thought that the river boat and the burro had much in common—each was born to a life of hardship and daily toil.

Typical of the river shipping news of that day are these half dozen gleanings from Territorial newspapers:

"The *Newbern* arrived at the mouth of the Colorado River July 10th at 11 a.m., with 130 men, officers with their families and laundresses, of the 1st United States Cavalry, en route for Tucson. Also 9508 packages of merchandise, 740 tons, mostly for Fort Yuma and Arizona City . . . Upon the arrival of the *Newbern* there were seven steamers and barges at the mouth of the river ready to receive her cargo . . ."[2] (*Arizona Citizen*, Tucson, July 29, 1871).

[2]The *Newbern*, a 943-ton steamer, purchased early in 1871 by the CSN line, left San Francisco once monthly on the run to Puerto Isabel. Two years later the CSN company also had in service on the ocean run the 1000-ton steamer *Montana*, purchased from Pacific Mail Steamship Co.

"March 26th the steamer *Colorado* cleared for Port Isabel with . . . 146 tons galena ore, 12 tons silver ore, 22 tons copper ore, 736 hides, and a small lot of assorted freight," reported *Arizona Citizen,* April 5, 1873.

"The Colorado Steam Navigation Company's steamship, *Newbern,* is expected to arrive at the mouth of the river about Oct. 21 . . Besides passengers for Yuma, including one company of the 12th Infantry, she brings 500 tons of freight for Yuma. James M. Barney has shipped during the week, by wagon trains of Quinlan and Daguerre, to the interior, 100,000 pounds of government and citizen freight." (*Arizona Citizen,* Oct. 23, 1875.)

"The *Mohave* sailed for El Dorado Cañon and wayports on the 31st, taking up about 100 tons of freight, including part of the machinery for the Signal quartz mill . . . Before returning she will move several hundred tons of wood from Cottonwood Island to the mill in El Dorado Cañon." (*Arizona Sentinel,* June 2, 1877.)

"The *Gila* arrived from up-river Thursday in charge of Capt. Mellon. Her barge was left at Castle Dome to put on a load of lead ore . . . The *Gila* brought in over $15,000 in silver bullion from the Signal and McCracken mines." (*Arizona Sentinel,* Feb. 2, 1878.)

"The Steamer *Mohave* arrived at her berth, this city, on Monday morning last. She left El Dorado Cañon on Saturday at 3:30 p.m., and hence was only about 36 hours coming down, a distance of over 400 miles. She brought down about 37 tons of Castle Dome ore, and some 700 dry hides." (*Arizona Sentinel,* June 4, 1881.)

And so they went . . . up the river and down the river, through high water and low water, heat and storms, loading and unloading, fetching, carrying, dodging sandbars and snags and destruction, for more than 50 years.

Considering the relatively short life of the hard-pounded river boats, the high cost of transferral to and from ocean-going vessels, and the almost complete monopoly of the Colorado Steam Navigation Company, freight rates on the river do not seem to have been excessive.

In 1880 merchandise was being carried from San Francisco to Yuma at two cents a pound, with cabin passenger fare over the same run pegged at $40. Ore was out-freighted from Ehrenberg and Hardyville at $15 per ton to San Francisco, and from Yuma at $10 per ton. Passenger fare from Yuma to Castle Dome, via steamer, was $5; Yuma to Ehrenburg, $15; to Fort Mohave and Hardyville, $35, and to El Dorado Canyon, $45.[3]

Pride of the river was the twin-stacker, *Mohave II*, built in 1876 and kept in regular service on the Colorado for nearly a quarter of a century. During the many years of her active service the *Mohave* periodically cast off her role of pack-horse to become, Cinderella like, an excursion boat for a day. Scrubbed, polished, decked in bunting and ban-

[3]*Crofutt's New Overland Tourist Guide,* 1880.

ners, and with flags flying proudly, she would set off to some up-river picnic spot, her decks aswarm with Yuma's gay young blades, their ladies, and lunch hampers. Especially anticipated by Yuma's small fry was the annual May Day excursion when every child in the frontier port, dressed in his Sunday best and accompanied by teachers and parents, was taken on an all-day steamer ride and picnic, all at company expense.

On the occasion of one such outing, the *Sentinel* remarked: "The *Mohave* . . . furnishes every accommodation to passengers and is so admirably managed as to render such an excursion one of continuous pleasure the whole distance traveled."

In its issue of Sept. 21, 1878, the Yuma news medium was still singing the praises of the queenly *Mohave* and its redoubtable master: "The *Mohave* arrived from Ehrenberg about 4 p.m. last Monday . . . This trip is the 393rd made on the river by Capt. J. A. Mellon since August 17, 1864. What he has not learned about the river in that time would not make a two-line newspaper item."

After the shallow-draft steamers had been hauling freight, soldiers, supplies, and picnickers on the river for almost 40 years, Capt. Mellon began eyeing tourists as a potential source of income. Interviewed by a representative of the *San Francisco Chronicle,* early in 1893, Capt. Mellon painted a pulse-quickening picture of the thrills of Colorado River sightseeing.

With the statement that travel was "pretty brisk" on the river that year, the captain announced his plan to begin pushing his steamers forward "through the famous Black Cañon, where they have rarely ventured heretofore and, even then, only when conditions were most favorable. However," continued the published interview, "the captain has hit upon a unique plan to surmount the giant cascades. He has caused great staples to be driven in the sides of the frowning rock cliffs. In these are stout iron rings. Ropes are run through these when the steamer arrives, and by the aid of a steam capstan on the steamer, she is enabled to plough her way forward. It is a dizzy undertaking, but Capt. Mellon is convinced that it can be made a success . . .[4]

" 'The Black Cañon,' said Capt. Mellon, 'is 40 miles through one place before the sun shines on the depths . . . Where the river is 300 feet wide, the walls run up 4000 feet and are narrower at the top than at the bottom . . .' "

[4]If Capt. Mellon was responsible for the ringbolts in Black Canyon, as the interview in the *Chronicle* would lead one to suppose, he must have placed them there nearly 30 years before the time of that interview. As early as Sept. 27, 1867, G. W. Gilmore, reporting to a meeting of the merchants of San Francisco on various facts of Colorado River navigation, stated: "To pass the 200 to 300 feet of rapid broken water described, the *Esmeralda* placed a ringbolt in the rocks above . . . and ran a line 800 feet in length to it. This line was taken to the steam capstan on the single purchase and the steamboat was run up to the ringbolt in seven minutes easily, and without apparent strain." (J. Ross Browne, *Report of the Mineral Resources of the States and Territories West of the Rocky Mountains,* 1868.)

Soon after Capt. Mellon's interview appeared in the *Chronicle,* brochures advertising his projected river trips were being circulated throughout the Southwest.

Between the first and middle of June, 1894, it was stated, the "splendid river steamer, *Mohave*" would be the scene of a "grand excursion up the Colorado River to the head of navigation, through the wonderful Black Cañon and Devil's Gate Cañon.

"The foremost object of this excursion is to show the possibility of mining and agriculture of the country through which it will extend, but it will have other attractions as well," set forth the brochure. "With none of the hazardous hardships and privations of roughing it, in saddle or on foot, the trip will be through the most weird and awesome scenery on earth, nowhere else to be seen except in the abysmal chasms and gorges of the Colorado River . . ."

Round-trip fare, including meals and berth, stated the circular, would be $62.25 from San Francisco; $57.75 from San Diego.

Much less spectacular than the "abysmal chasms and gorges" and less colorful than either the gallant paddlewheelers or their masters, were the dozen-or-so drab and motley shipping ports served on the long, patience-trying voyage up-river. Even Yuma—or Arizona City as that port of call was known prior to 1873—was no prize pigeon. Actually, the original "Yuma" was Fort Yuma, on the California side of the river, the first town on the Arizona shore in this vicinity having been Colorado City. The unusual story of that settlement's survey and platting is amusingly told by Byrd Granger, in her excellent book, *Arizona Place Names*:

"Charles Debrille Poston and some friends were en route to California in 1854 seeking to raise capital for mining, but when the party reached the Colorado and the ferry belonging to Jaeger they were told that the fare would be about $25. Poston's group had no money. The men took counsel with themselves in order to meet the situation. At Poston's suggestion an engineer set to work with instruments and other members of the group ran around with signal staffs, laying out a city in squares and streets and making up a map which, incidentally, had provisions for a steam ferry landing. Jaeger's curiosity got the best of him and over he came to find out what was going on. Poston, a born promoter, convinced Jaeger that the place had a good future, with the result that the German bought several lots and the surveying party got free transportation across the river as part payment for a single lot. Poston, on arriving in San Diego, registered Colorado City as a community and in San Francisco sold its plans and acreage for $10,000 . . ."

At some period in the five years next following, Arizona City came into being and on July 30, 1859, the *Alta Californian,* of San Francisco, took occasion to remark:

"Incipient Cities Near Fort Yuma. At the junction of the river, opposite Fort Yuma, is the flourishing Arizona City, containing some half-dozen respectable adobe houses, including two stores, two saloons, and a post office. At least two houses, one of them a storehouse, are in course of erection; thus, it will be perceived, the city is improving. The Colorado Ferry and Overland Mail depot is one mile below the fort. There are two stores, several buildings of the Overland Mail company, two blacksmith shops, several dwellings, and a house where the inner man may find grub and comfort provided . . .

"Colorado City, on the Arizona side, rejoices in one adobe house, the custom house, and bids fair, I think, ultimately, to become a city of no more importance than at present. Thus, as you perceive, there are three cities and a large military post within a mile of each other, surrounded by sand . . ."

Either the *Alta's* reporter was a faulty observer or poor prophet, as Colorado City's commercial district embraced considerably more than "one adobe house" at the time of the big flood on the Gila, in February 1862. Reporting that catastrophe in its issue of March 5, the *Alta Californian* stated:

"Colorado City is entirely washed away. Mr. Hooper's dwelling house, and Capt. Johnson's, are the only buildings left standing on that side of the river . . . The principal losers are Messrs. Hooper and Hinton's store, entirely washed away. The steamboat company's machine shop and store, with all its contents, washed away. Mr. Samuel Wells' store, washed away, and but a small portion of his stock of goods saved. Mr. Sage's billiard and ten-pin alley was entirely destroyed . . . The whole country is submerged between here and Pilot Knob and impassible for anything but boats."

Colorado City never rebuilt. Arizona City did so. In 1871 it acquired the county seat from moribund La Paz and, two years later, with the importance of Fort Yuma declining, Arizona City adopted the name, Yuma, which it has since borne.

One of the first steamer landings up-river from Arizona City, or Yuma, was Castle Dome.

"On the Colorado, in a direct line west of Castle Dome mountain, a town has been laid out called Castle Dome City, the situation of which is eligible as there is a permanent landing," remarked *Arizona Miner* on April 6, 1864.

Mines of the Castle Dome area, running around 75 per cent lead and $30 to the ton in silver, were good mines. Worked since the days of Spanish or Indian occupancy, they were still going strong in 1877 when the *Arizona Sentinel* reported: "Forty men at Castle Dome are taking out more metal than ever. Besides the barge load that was taken down last week to the Steamer *Idaho,* for San Francisco, there are now over 120 tons of lead at the landing, and more than that at the mine." At 107 feet depth, the Bettis shaft was in almost solid lead, and 100 pieces of this ore, ready to break up and put into sacks, it was reported, had weighed 4000 pounds.[5]

During those productive and affluent years, the Mexican population of Yuma frequently journeyed up-river by steamer to attend special

[5]*Arizona Sentinel,* April 14, 1877.

celebrations at Castle Dome City, such events usually taking the form of general whingdings punctuated by assorted sports, horseracing, oratory, and week-long *bailes.* Typical of these jubilations at Castle Dome were the festivities staged in September, 1881, honoring the anniversary of Mexico's Independence Day.

"Large dancing halls had been erected and decorated with flags, fancy lanterns, and evergreens, and on the evening of the 15th, at 5 o'clock, the ball was opened by a grand procession in which there were about 200 horsemen and footmen," reported the *Sentinel.* "During the entire time occupied by the parade, guns and anvils were constantly fired . . . The CDM & S Company, with their usual generosity to their employes, gave all hands a week's vacation and they certainly enjoyed it. The dancing was continued every evening until the 20th, when all hands returned to their various locations . . ."

Along the horizon, a dozen miles to the east of the Yuma-Quartzsite highway, the great rock monolith that gave name to Castle Dome City still points heavenward like a directing finger, and some of the Castle Dome mines were still being worked as late as World War II. But the site of the incipient city and the old landing—originally situated about half a mile south of the present Fisher's Landing, on Martinez Lake—was long ago swallowed by the spreading waters backed up by Imperial Dam.

Next halting place on the up-river journey was Norton's Landing, at the mouth of Red Cloud Wash, some 50 miles above Yuma. Supplies for the Silver Mining District were unloaded at Norton's, and a smelter situated at this point reduced ore taken from the Red Cloud mine. Norton's seldom got much attention from the press, but on July 6, 1884, the *Mohave County Miner* noted:

"The overflow from the Colorado River is destroying a great deal of property at Norton's Landing. Upwards of 30 houses have been washed down, including all of the Red Cloud company's buildings. Water is up to the Red Cloud smelter and a determined effort is being made by Professor Julius Siebeck to save the plant. If the river continues to rise a few days more, the loss at Norton's Landing will be very heavy."

Next, in the Colorado's necklace of river ports, came Ehrenberg (treated in a separate chapter); then Aubrey, at the mouth of Bill Williams Fork, some 235 miles above Yuma.

Aubrey's Landing was named for Francois Xavier Aubrey, so-called "Skimmer of the Plains," whose lust for notoriety was greater than his regard for horseflesh, or my regard for him. In 1850 "The Skimmer" rode horseback from Santa Fe, New Mexico, to Independence, Missouri, on a wager of $1000 that he could cover the distance in eight days. By riding several horses to their death Aubrey succeeded in finishing the 800-mile course three hours short of the prescribed time. Later, at the cost of more gallant beasts, he repeated the ride in five days and 13 hours.

Aubrey's Landing, founded in 1864 (ten years after Aubrey's death in a barroom brawl in Santa Fe) soon became one of the most important shipping points on the river, freight being landed here for several big mines in the southern part of Mohave County.

"The Great Central mine . . . has recently shipped a full set of smelting furnaces with steam power and all the necessary machinery, tools, and supplies for their mine, having a large amount of rich copper ore already out and ready for smelting," noted *Los Angeles Semi-Weekly News*, February 9, 1866. "The works will be erected at Williams Fork, near its mouth . . . These are the first regular blast furnaces ever sent to Arizona."

On the year following, J. Ross Browne reported: "Greenman and Martin, of San Francisco, have erected two furnaces and other works, with a 30 h.p. steam engine . . . at Aubrey City, a town which has sprung up on the banks of the river since the opening of the mines. The company has expended nearly $100,000 in opening the mines, building a wharf, making roads, etc."[6]

Throughout the 1870s and early '80s, occasional news from Aubrey City continued to find its way into Arizona news columns. Mines opened and closed; mills and smelters followed suit. Even as late as 1884 the *Mohave County Miner* still recognized the presence of the town and in its issue of March 2 remarked: "The steamer *Gila* is on the way up the river with two carloads of lumber and one of merchandise for the Matilda Smelting Company, of Aubrey, as part of her freight."

But Aubrey eventually folded her fragile wings and, so far as I have been able to determine, almost every trace of this once-busy town has vanished.

Next in the roster of ghost ports is Liverpool Landing.

Even contemporary references to Liverpool are few and far between. Barnes notes that the town is marked on the Smith map of 1879; Hinton's map of 1886, and the General Land Office map of 1892. Site of the settlement also is indicated on the Parker Quadrangle map of 1911; but the old landing has since been lost beneath the waters of Havasu Lake, impounded behind Parker Dam.

Time was when Liverpool had a smattering of citizens, at least some of whom were concerned about their isolation from the remainder of the world.

"Mr. Hannigan, of Liverpool Landing . . . is trying to get a mail route established between Aubrey and The Needles," observed *Mohave County Miner*, June 21, 1885. "Liverpool Landing, or Godman, as it is sometimes called, lies midway between these two places and at present is without any mail communication whatever."

Next on the up-river schedule came Mohave City, a completely separate entity from Fort Mohave, situated about a mile to the south.

"The founders of the town of Mohave, Messrs. Fagen and Todd, have already built a new and large ferry boat," reported the *Alta Californian*, July 4, 1864. "The new town already boasts three stores and a

[6]Browne, *Resources.*

saloon that are stocked in a manner that would do credit to a place of higher pretensions . . ."

Again, on April 3, 1865, the *Alta Californian* remarked: "Mr. Ravenna, at La Paz, has established a general merchandise store at Mohave. He has a large and well-assorted stock."

In 1866 Mohave City was designated the first seat of newly-created Mohave County, and a post office was established there. The county seat, however, was soon transferred to Hardyville (thence, in succession, to Cerbat, Mineral Park and Kingman); and in the autumn of 1869, when the Fort Mohave military reservation was enlarged, Mohave City was engulfed and its residents ordered to decamp.

Although rather extensive ruins—foundations, broken rubble, sidewalks, street curbs, et cetera—are still visible at this site, it is all but impossible to know whether they are the remains of Mohave City, or of Fort Mohave. Since Fort Mohave, as site of an Indian school and agency headquarters, far outlived Mohave City, it is probable that all remaining ruins are of it, rather than of the one-time shipping town.

At Hardyville, several miles up-river from the site of Mohave City, and only two miles below the present town of Bullhead City, there are a few old adobe ruins, all but crumbled away, and the tattered graves of a small cemetery.

In its issue of January 15, 1865, *Alta Californian* published a news letter datelined Hardyville, A.T., November 25, 1864. The new river town, stated the correspondent, was thriving steadily, with an excellent restaurant, saloon, stores, and other shops.

"Mr. Hardy, the enterprising founder of the town, is carrying on a regular trade between this and Prescott, the capital of Arizona; and also to Salt Lake City . . . Sam Todd, formerly of the Metropolitan Hotel in your city, has established a ferry here and is speculating largely in town lots. Mr. Kipp has a trading post here and does a lively business with the miners and Indians . . ."

Six months later, on June 1, 1865, Richard C. McCormick, then secretary of the Territory of Arizona, wrote a letter subsequently published in the *New York Tribune.*

"Hardyville," stated the secretary, "is a young, but active and hopeful settlement. It has a large trade from the quartz mining districts around it, and even from the Wauba-Yuma district, 40 miles in the interior, and from Prescott, the capital, 160 miles inland. Recently the Utah people have flocked to Hardyville for their annual supplies, finding it much easier than to go, as heretofore, to San Bernardino and Los Angeles . . ."

Despite its other endowments and virtues, Hardyville's chief claim to fame was its founder.

Even in that century-ago era when giants strode the land and sailed the seas, William Harrison Hardy was recognized as a man of uncommon stature and substance. On that day in January, 1864, when Hardy came riding overland to the river on horseback, he had already passed his 40th birthday—a goodly age in that day when man's life expectancy was short at best.

One of Hardy's intimate friends, John A. Riggs, in an unpublished manuscript of personal reminiscences in the files of Arizona Pioneers' Historical Society, at Tucson, pictures the founder of Hardyville as "an active, positive, energetic man, with wide vision and wonderful powers of endurance; in stature, a large man, big boned, angular, broad, square built, about six feet two inches in height and 200 pounds in weight; auburn hair, sandy beard, gray eyes, and heavy eyebrows. An educated man, whose life habit was cultural, genteel, moral, and well-bred. He neither drank, gambled, nor caroused in any manner. His language was without any taint of slang or profanity. For honesty, veracity, charity and general squareness, he had few peers . . . Graft and crookedness he despised; nor could he be bribed, bluffed or bullied into anything against his conscience . . ."

That such a paragon of rectitude could be thoroughly human and possessed of a well-developed sense of humor seems almost too much to expect; yet, the Riggs manuscript cites several incidents pointing up Hardy's general good nature. One of these is the tale of the "Colorado River Salmon."

Frontier trading posts of that era were invariably a gathering place for saddletramps, adventurers, pranksters, miners in from the diggings, and time-killers in general. One day when a large number of such hangers-on were decorating the front of the Hardy store its owner began explaining the merits of the new 70-per cent nitro-glycerine powder, familiarly known as "Giant Powder." To demonstrate the efficacy of the new explosive for taking fish in large quantities—a practice since outlawed but then still legal—Hardy rowed out into the river, dropped a "stick" overboard, and bent to his oars in an attempt to row upstream away from the spot. The water current on this particular day, however, was stronger than Hardy had anticipated and all his efforts were to no avail—the powder exploded directly under the boat, turning it bottom-side up, with Capt. Hardy beneath it. After the captain had crawled out and towed the boat ashore—to the great amusement of his audience—someone waded into the river and retrieved a fish, about six inches long, which had been killed in the blast. This finny casualty the jokesters then nailed to a post in the Hardy store and beneath it posted a large display card reading: "This huge fish, a Colorado River Salmon, was taken by W. H. Hardy, the great fisherman, after it had overturned his boat at great hazard to his life."

"Hardy," wrote the raconteur, John Riggs, "allowed the fish to hang in his store with the card until it was as dry as parchment."

Like all frontiersmen worthy of the name, Hardy had a well-developed bump of inventiveness and initiative. Evidence of this is found in the matter of the mail sacks.

"All mail sacks of that day were simply sewn, not riveted as they are today," wrote Riggs. "The pony mail routes were hard on sacks as they were carried behind the saddle of the rider, on horses or mules. The threads would soon cut out, leaving rips in the bag. Hardyville was a center for several of these routes and Hardy, as postmaster, had much experience with this defect in the construction. Whenever a bag came in ripped he would send it to his harness shop and have it mended with harness rivets. Later, on a trip to Washington D.C., Hardy had a personal talk with the postmaster general. He called attention to the defect in the mail sacks and recommended they be constructed with rivets. The suggestion was adopted, and Uncle Sam's mail bags have been riveted ever since . . ."

In the course of his years as storekeeper, postmaster, miner, county commissioner, and while serving in the upper house of the Territorial Legislature, Hardy's name became a household byword through Arizona. Scarcely less famous than the man himself was his favorite saddle-horse, Old Yaller.

Hardy had been riding the animal when he came to the river and the future site of Hardyville, in 1864—and he continued to ride him, through thick and thin and heat and cold, for another nine years. In view of Hardy's considerable weight and the advancing age of the horse, the animal's endurance seems almost phenomenal. On many occasions Hardy rode him to and from Camp Wallapai, on Walnut Creek, a distance of 120 miles, in a single day and without any ulterior effect. Upon arriving back at Hardyville, it is recalled, the animal would look almost as fresh as when he had left there early that morning.

As a means of pensioning the faithful animal, after he was 18 or 20 years of age, Hardy, in 1873, sold him to C. C. Bean, of Prescott. For several years thereafter, Old Yaller, still sleek and spry, could be seen drawing a light buggy through the streets of the city, recalled Riggs.

A little way above Hardyville, in the present vicinity of Davis Dam, was Polhamus Landing, named for Capt. Isaac Polhamus, another of the colorful and near-indestructible river boatmen. Arriving on the Colorado in 1856, when shipping on the river was still in its infancy and the present city of Yuma consisted of a single adobe building occupied by the Colorado Steam Navigation Company offices and storeroom, and by George H. Hooper and Company, Arizona's first "merchant princes," Capt. Polhamus continued to make his headquarters at that point for 66 years and died there in 1922 at the age of 94.[7]

Polhamus Landing was one of the later ports to be established on the Colorado.

"Wellton and Grounds, of Mineral Park, have established a new landing at a point about five miles above Hardyville and have named it Polhamus Landing," stated *Arizona Sentinel*, June 4, 1881. "They have cut away a large portion of the hill, on the bank of the river, and built a commodious warehouse and other buildings, and have graded a road

[7]Frank C. Lockwood, "Steamboat Captain on the Colorado," *Desert Magazine*, June 1941, p. 13.

from Polhamus to intersect the old road . . . shortening the distance to Mineral Park by about five miles."

Three issues later (June 25, 1881) the *Sentinel* remarked, "The Steamer *Gila*, Capt. Polhamus, arrived from the upper river on Tuesday, going up as far as the Polhamus Landing, which is 16 miles above Mohave. This is the new landing place . . . where more freight is now landed than at any other point on the river."

Some 50 miles above Polhamus, on the Nevada side of the river, was the landing at El Dorado Canyon. Next came the ominous Black Canyon, present site of Hoover Dam; and, finally, the Mormon settlement of Callville. Only when conditions on the river were near perfect was it possible for the steamers to ascend as far as Callville, and I have found no authenticated record of a paddlewheeler going above that point.

By 1900 the day of the Colorado River boats was waning; not that they had outlived their usefulness, but a new era was dawning in the Southwest. Railroads had come, highways were being built. Soon automobiles and trucks would begin to appear. Many mining camps which had supplied the lifeblood of the river shipping towns and the paddlewheelers already were ghost towns. Indian fighters had been succeeded by farmers. Farms were occasioning a need for assured water, and the only way that assurance might be given was by drawing upon Arizona's greatest water potential—the Colorado.

With the construction of Laguna Dam, a dozen miles above Yuma, it was no longer possible for steamers to pass up or down the river to Puerto Isabel, Sonora, for 50 years their meeting place with ocean-going vessels.

Thus was written finis to a way of life, a day, an era.

By 1910 most of the Colorado river boats had disappeared from shipping lists. Captains and crews scattered to other rivers. Ports, shorn of their livelihood, became ghost towns. Wharves were torn away by flood waters, or carried away piecemeal for the timbers they contained. Drydocks, no longer used, filled with sand and debris; and once-proud adobe walls crumbled back to dust.

Whenever a pleasant evening finds me camped along the Colorado—when the moon is full and the water spangled with diamonds and brushed by a cool whisper of breeze after the heat of day—I like to sit quietly alone on the river bank, my thoughts drifting back to those regal days before the broncho river was tamed and hobbled, when its mighty surging waters still tested man's mettle, and its course still formed the highroad to a vast storehouse of unknown wealth.

I think of the hazards endured, the heavy cargos moved, the grueling hours of labor, the loneliness of midnight watches; and, at last, I find myself mentally calling the role of the ghost ships. The queenly *Mohave*, the *Cocopah*, the *Colorado*, the *Esmeralda* and *Nina Tilden*, the *Gila*, the *Yuma*, the *General Jessup*, the *Cochran*, the *Aztec* . . . but though I call their names and listen, and strain my eyes peering into the dark of the night, there is no answering whistle, no sounding bell.

Arizona's ghost fleet of steamers has vanished like the wake from their own paddlewheels, the smoke from their own stacks.

Crumbling adobes, Cerro Colorado, 1960

"Whilst the early navigators were exploring the entrances to the bays, inlets, and river mouths of the Atlantic Coast to find places for the first colonists, the prospectors were searching Santa Cruz Valley and surrounding hills for silver, the precious metal of that age."

—*A Historical and Biographical Record of the Territory of Arizona, McFarland-Poole, 1890.*

Ghost of the Red Hill

FROM Kinsley's ranch on the Nogales highway, about 40 miles south of Tucson, a graveled road leads southwesterly into the desert foothills. Skirting the historic Sopori ranch, the road continues through rolling grasslands past the former mining camp of Cerro Colorado, and to Arivaca and points beyond.

My visit to Cerro Colorado found the old camp wholly deserted. As the year of its founding was then more than a century in the past, I was interested to see here more extensive ruins than mark the site of many ghost towns barely half as old. Investigation showed that some of the ruins remaining—possibly most of them—date from one of the camp's later periods of activity; but even this failed to diminish my interest. The Town of the Red Hill and the Heintzelman mine simply are so old that anything connected with them seems to wear the patina of age. Even the hard, rocky ground beneath seems steeped in history.

Wandering over the deserted townsite, so pleasantly warm and peaceful on that spring day in 1960, I located several dozen buildings and identified what appeared to have been the company offices, store, and schoolhouse, as well as numerous small dwellings. On a rise, not far from the main townsite is a communal grave holding the remains of 27 Indian laborers killed in a mine cave-in. Location of the mass burial was marked by an old wooden cross lying faded and neglected in the grass. Another small cemetery contains an unknown number of graves, including that of the mine's one-time superintendent, John Poston, brother of Charles Debrille Poston, assertedly slain by men in his employ.

Relative importance of the Cerro Colorado, like that of many Southwestern mines, is debatable. Some say it was a great producer with an abundance of rich ore from the grassroots down. To others the mention of its name brings—at the very best—a derisive smile, a shaken head.

Its ore, admittedly, was rich. For awhile there seemed to be a lot of it. Yet, some say that the mine produced no more than $200,000, all told. Others declare it had yielded $2,000,000 or more prior to 1861. Whether the mine ever gave forth as much wealth as was poured into it or whether, as Colonel Sam Colt, Hartford arms manufacturer, declared in 1859, it was "only a hole to bury money in," remains open to question.

In either case, the story of the Cerro Colorado is as fascinating as that of any mine in the state. Not only is it one of the state's oldest mines, but the roster of those connected wih its early operations—both financially and officially—reads like a directory of the men who played leading parts in the drama of Arizona's Beginnings.

Foremost among these were Charles D. Poston and Herman Ehrenberg. In 1854, with the ink barely dry on the Gadsden Purchase agreement, under the terms of which Mexico had ceded to the United States the southern portion of Arizona, (then New Mexico Territory), Poston and Ehrenberg made their way into the newly-acquired lands to investigate mining potentials. When they found near Tubac some old Spanish-Mexican mines which looked promising, Poston returned East and early in 1856 organized, at Cincinnati, the Sonora Exploring and Mining Company. General Samuel P. Heintzelman, on official leave from the U.S. Army, was named president of the new company. Directors were Samuel Colt, of Hartford; William T. Coleman, Major Hartley and Poston.[1]

After assertedly promoting a grant of $100,000 from the Texas & Pacific Railroad Company—same to be used in the opening up and development of mines in the southern part of the Territory—representatives of the Sonora Exploring and Mining Company established headquarters that autumn of 1856 in the little town of Tubac, then all but deserted. One of the first settlements made by Europeans in what is now Arizona, Tubac had been garrisoned by Spanish soldiers as early as 1752 and from this place in 1774 Juan Bautista de Anza had set forth on his history-making overland trek to California and the ultimate founding of San Francisco. The old adobe quarters at Tubac, evacuated by Mexican army troops following ratification of the Gadsden Purchase, were repaired and made comfortable for occupancy by the company's "field crew," by then expanding to include Frederick Brunckow, a young German mining engineer, who had joined the party at New Braunfels, Texas. With a store of provisions purchased for the winter, the group was ready to begin its exploration for potentially profitable mines.

"The mines in the Santa Rita mountains had been previously worked by the Spaniards and Mexicans, as was evident by the ruins of arrastres and smelters," wrote Colonel Poston. "Gold could be washed from the mountainsides, and silver veins could be traced by the discolored grass. When it became known in Mexico that an American company had arrived in Tubac, Mexicans came in great numbers to find employment, and skilled miners were obtained at from $15 to $25 a month and rations.

[1]Farish, *The History of Arizona* (1915-18).

Flour, beef, beans, sugar, barley, corn, and vegetables, were imported from Sonora at moderate prices."[2]

When ore in the Santa Ritas proved to be too low grade to be worked profitably under the conditions of that time, the company turned its attention to the west side of the Santa Cruz River. Here the outlook seemed more promising and early in 1857 there was entered in the *Official Records of Pima County* (Book A, Page 197), this claim:

"Know all men by these presents: That I, Frederick Brunckow, a citizen of the Demarcation of Tubac, County of Doña Ana, and Territory of New Mexico, by profession a mining engineer, have this day discovered a vein of silver ore in said Demarcation and County, near a noted landmark known in the vicinity as Cerro Colorado, and about one mile from the summit of said Cerro, on the Southeastern side, said vein running a course southwesterly from said Cerro; and I hereby publish and declare my intention to open and work said mine of silver ore, according to law, and denominate the same The Heintzelman Mine, and hereby claim the same to such extent as may be allowed by the Laws of Mexico or of the United States, for myself, my heirs, and assigns, and also for my associates in working the said mine, to-wit: Charles D. Poston, Herman Ehrenberg, Charles Schuchard, Theodore Moohrman, George W. Fuller, Samuel P. Heintzelman, Edgar Conkling, and William Wrightson. Such quantity and space for each as may be allowed by law, and the claim, possession or title of any other person or persons is hereby denounced. In testimony whereof, I hereunto set my hand and seal this First Day of February, 1857."

(Signed) F. BRUNCKOW.

The ore was rich. According to Farish it came in the form of silver-copper glance (a term used to designate certain minerals showing a splendent luster), and yielded from the grassroots $7000 per ton.

In the autumn of 1857 there arrived in the Santa Cruz Valley a detachment of the First Dragoons sent there to establish Fort Buchanan. Coincident with their arrival came an immense supply train of a dozen huge wagons each drawn by a dozen mules commanded by men in the employ of Santiago Hubbell, of Santa Fe. On their subsequent return east the wagons carried ore from the Heintzelman mine at Cerro Colorado the shipment packed in rawhide bags and loaded one ton to each wagon. Hauled to the steamboat landing at Kansas City at a cost to the mining company of 12½ cents a pound—$2500 a ton—the ore was distributed to cities throughout the Eastern United States where it served its intended purpose of attracting the favorable attention of Westward-looking investors.

The first run of bullion from the Heintzelman mine was made in 1858 from a small adobe furnace said to have cost about $250 to build and equip. Six hundred hours were required to smelt some 22,800 pounds

[2]Farish.

of ore, which yielded 2287 ounces of silver and 300 pounds of copper.[3] Freighted overland to Guaymas, Sonora, the bullion was shipped aboard sailing schooners to San Francisco where the silver was sold on the Asiatic market at $1.25 to $1.32 per ounce.[4]

With the founding of Arizona's first newspaper, the *Weekly Arizonan*, at Tubac, news of Cerro Colorado and the Heintzelman mine claimed priority.

"The Sonora Exploring and Mining Company, Major S. P. Heintzelman, president, under the superintendence of Mr. S. H. Lathrop, is obtaining from one furnace . . . between 300 and 400 ounces of silver per week," reported the *Arizonan* on March 3, 1859. "They have nearly completed extensive amalgamation works costing some $30,000. This week they have got the first silver from these works. When in full operation they will reduce about three tons of ore in 24 hours. The company are working over 100 men, most of them on the famous Heintzelman vein. Their main shaft is sunk 70 feet with the ore and vein continuing uniform. Some of the ore yields as high as $10,000 to the ton. None has been smelted yielding less than $900. The poorer ore is reduced by the barrel amalgamation process."

On the next week following, the Tubac news medium carried two quasi-legal notices from the office of the Sonora Exploring and Mining Company.

"John Streit, killed in a rencontre with one Antonio Ellsner on the 25th of December, 1858," set forth the first notice. "Both persons were in the employment of this company. Streit was a German by birth, aged about 30. He was a cabinet maker by trade and, at the time of his death, head carpenter of the company. His family yet reside in Germany, but he is said to have relatives residing in San Antonio, Texas. His friends may learn the particulars of his death, and his legal representatives can obtain the amount due him upon application to S. H. Lathrop, director of the mines, Tubac."

The second of the two notices carried virtually the same wording, except for the pertinent facts, and concerned one H. J. Olds who had died of consumption while in the employ of the company.

Even if the two legal notices served no other purpose, they informed the Territory and the world at large that mining camp "civilization" had come to Cerro Colorado.

Its cemetery was beginning to fill . . .

By the close of March, 1859, the Sonora Exploration and Mining Company was working 150 men and reducing some 1200 ounces of silver weekly. The seventh bar of bullion poured— the result of three days work of two amalgamation barrels— was shipped by Overland Mail to W. T. Coleman & Co., New York, and inspired the *Arizonan* to comment:

"People who are incredulous about silver mining in the Territory will doubtless be edified and comforted by the sight of a lump of pure,

[3]James H. McClintock, *Arizona*, S. J. Clarke Pub. Co., Chicago, 1916.
[4]Farish.

bright silver, weighing nearly 900 ounces, assayed and stamped with its value, ready for the mint or the silversmith. It was something worth looking at."[5]

In the weeks that followed, the *Arizonan* reported with regularity the weight and value of each successive bar poured and, by the end of May, the company's amalgamation works at Arivaca were yielding as much as $1700 for one four-day run of three amalgamation barrels.

Another scribe, lavishly enthusiastic concerning anything and everything that would advance the cause of Arizona, was Sylvester Mowry. Speaking before the Geographical Society in that same year of 1859, Mowry stated:

"The Heintzelman mine . . . bids fair to become more famous than any of the great mines of Old Mexico . . . At a cost of $39,000 the company has brought from San Francisco and erected dry amalgamating works from which they expect to obtain $3,000 per day, a million a year . . ."

Which was, of course, a prize example of counting chickens before the eggs are incubated. Production, however, continued good throughout 1860 and by 1861 the mine was yielding around $1000 per day and 250 men were employed. Sanguine as Lieut. Mowry's prediction had been, even it might have been realized except that on the Eastern seaboard, nearly 2000 miles away, states were preparing to fight states and brothers to fight brothers in a bloody and devastating Civil War.

In June, 1861, Lieut. Chapin, commanding officer at Fort Buchanan, received from Military Headquarters in Santa Fe an order to abandon and destroy his post, burn the commissary and quartermaster's stores, "and everything between the Colorado and Rio Grande that will feed an enemy."

Even if the war were to have no other ulterior effect on Cerro Colorado and the Heintzelman, this order to abandon Buchanan meant that any white settlements in the area formerly under protection of the post would be left henceforth to the dubious mercy of the jubilant Apaches and assorted ruffians. Feeling it would be impossible to defend the Heintzelman, once the troops had been withdrawn, those in charge of the mine decided to process the ore waiting for reduction, pay off all hands, and prepare for the worst—which was not long coming. Within a week after departure of the Boys in Blue, the Apaches staged a raid upon the mine and victoriously ran off nearly 150 horses and mules.

But the fact that Arizona had been abandoned by the military, and the mine presumably closed, did not mean that the mining camp of Cerro Colorado had ceased to produce silver.

"The Heintzelman mine, now known as the Colt's . . . is principally worked by gambusinos who are taking out a very rich metal which is sold at the mouth of the mine at half a dollar a pound," reported the *Alta Californian*, July 23, 1863. "This ore is reduced by the patio process

[5]Issue of April 14, 1859.

at Sarac, just over the line in Sonora, a town built up and increasing upon the stealings from the Heintzelman mine."

About this time the Sonora Exploring and Mining Company appears to have passed out of the picture, to be succeeded by the Arizona Mining Company.

"M. O. Davidson, director of the Arizona Mining Company, has arrived safely at the Heintzelman mines, Cerro Colorado, with a party of 80 men," reported the peripatetic mining scribe, J. Ross Browne, in the *San Francisco Bulletin* of March 18, 1864. "Mr. Davidson is an intelligent and skillful engineer, long superintendent of the Cumberland Coal Mines, in Maryland, and chief engineer of the Habana railroad, in Cuba. He speaks the Spanish language perfectly and is thoroughly qualified to control an extensive business of this kind."

Ross Browne's opinion of Director Davidson seems fully borne out in a letter written by John A. Clark of Santa Fe, then Surveyor General of Arizona and New Mexico, to I. M. Edmunds, Commissioner of the General Land Office, Washington, D. C., under date of May 24, 1865.

"In my letter to your office from Tubac, dated March 13 last, I gave you some account of my journey and the country through which I passed. . . . I visited the Cerro Colorado mines, about 21 miles westerly from Tubac. M. O. Davidson, Esq., superintendent, kindly furnished me with every facility for examining the mines and works . . . The Arizona Mining Company, the owners of these mines, are erecting reduction works here and will soon be in readiness to commence reducing rich ores of the Heintzelman and numerous other veins in that vicinity. The Enriguetta (*sic*) Milling Company, under the same superintendence, are also erecting works at the Enriguetta mines, about 13 miles southwest of Cerro Colorado. The superintendent has had a telegraph line erected, connecting these two mining haciendas, the first and only line of telegraph in that Territory. There are none in New Mexico."

This same John A. Clark, writing of his travels in the Hartford (Conn.) *Evening Press* of May 26, 1865, gives a graphic description of the remote mining camp of Cerro Colorado in that closing year of the Civil War.

"You ride into the plaza of the Cerro Colorado hacienda and a scene of apparently chaotic confusion, bustle and work, meets you. It was toward sundown. There was the last of this long string of huge Sonora wagons unloading; herds being driven in by Papago herders; little Mexican carts discharging corn into the granary; team animals, tied to a picket rope, being fed with nosebags . . . the noise of hammers and iron in the blacksmith shop; the creaking of the mule-power hoisting out of the old perpendicular shaft of the Heintzelman vein . . .

"We met Colonel D. and he promptly assigned a place for our escort to camp; took us to his room . . . It is a circular-roofed belltower at the corner of the great wall that is designed to enclose the whole hacienda. A tame Apache boy, his body servant, soon brought us our dinner and, afterward, by the blazing fire we sat into the wee small hours, the triangle outside striking the hours, showing the guards were on post.

"Colonel D. is one of those agreeable men that have the faculty of, at evening, throwing off business entirely. The vexatious cares, the weighty responsibilities of managing this gigantic undertaking, do not stalk like grim ghosts into the fireside circle to mar its general serenity. It is the rare habit of successfully throwing off care that enables men to accomplish great things. It enables the brain to rest itself and prepare for the morrow's work. I do not remember to have passed a more agreeable evening . . .

"The works contemplated are all to be enclosed in one wall to protect against external enemies, the danger of thieving, and so arranged are the towers to guard against a peon insurrection such as has occurred once before under the old management. A large reservoir has been made to hold the rains of the rainy season; a large engine is being put up in place of the small one used before. The mouth of the shaft is inside a high enclosure, and some 35 tons of superb silver rock, said by the assayer to average some $800 or $900 a ton, was being sacked up for shipment . . .

"As you move about here you see what is apparent disorder is really order, and that the appearance of confusion is because the buildings are not finished. Not half the things lately arrived are unpacked and, as will always be the case, those you get unpacked are useless until you get something that hasn't come yet. Among the things . . . that have arrived, is a telegraphic apparatus to enable the officer here to communicate with the Enrequita Mining Company's works, seven miles away, which are likewise under Colonel D.'s direction . . . There are some 100 Americans employed here, besides the Mexicans and Indians, making a large settlement."

From a perusal of the foregoing buoyant report, it must have appeared to Clark's readers that Colonel Davidson was happily married to the Heintzelman, and that the issue of their union could be nothing short of abundant prosperity and mounting cords of silver ingots. Unfortunately, an altogether different word picture was painted, less than three years later, by William A. Bell of the Royal Geographical and Ethnological Society.

"Eleven miles more brought us to the Colorado or Heintzelman mine," wrote Bell. "One or two hundred Mexicans still live here in huts which were built by the proprietors of the mine for their peons when they established their works. Systematic mining has, however, for some years, been discontinued, although the yield exceeded $200,000 in silver. The inhabitants now live by the pickings, and by extracting silver from the ore in the roughest possible manner. We inspected the square formed by the adobe houses of the gambosenos, [*sic*] poor Mexican miners who mine each on his own account and club together for mutual protection, and my guide tried to get a few eggs, and corn for our animals; but, failing in both and not liking the looks of the people, we continued on our way for four miles farther. It was then dark, and finding good grazing ground we picketed our horse and mule and went to sleep.

". . . four miles ride before breakfast brought us across a little dividing. ridge into another valley watered by a stream called the Aravaca. [*sic*] Here are the deserted furnaces of the Colorado mine, and a row of telegraph poles along the hills, uniting the places, shows very clearly that extravagance must have had something to do with the downfall of this mining enterprise . . . Nine miles from Aravaca is to be found Enviguetta [*sic*] another relic of mining enterprise, where a fine steam engine and a mill of, I believe, 20 stamps, with well-built houses for superintendent, employees, etc., now stand idle. One man takes care of this place . . ."[6]

Something in this dispiriting report must have stirred the interest of certain Britishers, for on January 14, 1871 the *Arizona Citizen* announced purchase of the Arivaca Ranch and Heintzelman mine by a General Ripley, of London, at a reported price of half a million dollars. As if fearful that its readers might hesitate to accept the story as true, the *Citizen* closed its report with the comment: "Evidence may be seen at the Recorder's office."

Unfortunately, however, neither Time's passage nor changes of ownership did much to improve the over-all picture at the Cerro Colorado. Five years after sale of the mine and ranch to General Ripley, a roaming representative of the *Citizen*, touring the region south of Tucson, reported in the issue of February 26, 1876:

". . . Returning down the valley we took the road which leads up a strip of rich valley land for several miles . . . to the once-famous mines and mining works of the Cerro Colorado Company. Here is ground on which was enacted as costly, foolish, and perhaps as bad a farce in the name of mining, as any of the many in our Western country. From all that . . . has been told us, we can but conclude that the projectors and local managers of this now effectually played-out and abandoned mining operation could safely enter themselves as competitors for a high premium in an exhibition of the hundreds of successful but senseless squanderers of money independent of that used in developing mines of possible value.

"Remnants of their stupendous folly, if nothing worse, may be seen at nearly every cabin and populated place in the country round-about. Here and there, on the public highway, may be seen large cast-iron wheels—in one place, near a dozen of a kind. Hundreds of pounds in weight and unlike anything we have ever seen in or about a mine or quartz mill. Pieces of engines, iron pipes, and iron in many and expensive forms, once belonging to that enterprise, may be seen for many miles about. All seems, for sometime past, to have become common property,

[6]*New Tracks in North America, a Journal of Travel and Adventure Whilst Engaged in the Survey for a Southern Railroad to the Pacific Ocean, During 1867-68* (p 102 Chapman and Hall, London, 1869).

to be taken and kept by anybody who wants it badly enough to appropriate it. The buildings are numerous and crumbling down.

"Some months ago, W. C. Ferguson, Alex McKay, and D. C. Thompson relocated the mines and reoccupied the Cerro Colorado village, for the old buildings constitute at least a deserted village . . . No one can say . . . that if the Cerro Colorado Company had wisely used its means, that it would not have been what its victims were led to believe it would be—a paying institution."

Charles D. Poston, of the old Sonora Exploring and Mining Company, evidently took exception to relocation of the Cerro Colorado by the aforementioned or other ambitious parties, and the *Citizen,* on October 18, 1877, published in its columns the following paid legal notice:

"CERRO COLORADO (Heintzelman Mine). All persons are hereby warned against negotiating any interest in the Heintzelman Mine, near Cerro Colorado, Pima County, Arizona, as the same was discovered by me in 1856 and after expenditures in development far in access of any requirements of the government, the said mine, together with Arivaca and other real and personal property, was conveyed to me by deed recorded in Pima County the 28th of August, 1871."

(Signed) CHARLES D. POSTON,
Florence, Arizona

Few mines in the history of Arizona have died a harder death than the poor old Cerro Colorado. Down through the years, other hopeful men, other optimistic companies and syndicates, tried their hand at making the old mine pay. None of these successive efforts, however, was rewarded by any great good fortune. Even mining men, copiously blessed with the hope that springs eternal, eventually came to regard the Cerro in much the same manner as tolerant villagers look upon the winebibbling town character—with sympathy and compassion, but not with the thought that he is to be taken seriously or the hope that he will ever amount to much.

In bringing to conclusion this more or less disjointed summary of the Heintzelman mine and Cerro Colorado, it seems but a simple act of justice to add a few words concerning Charles Debrille Poston, whose interest in the Heintzelman spanned a longer period of time than that of, probably, any other one man.

Though it was the hope of finding precious metals that brought him originally to northwestern Mexico and the Gadsden Purchase strip, mining was not the Be-All and End-All of Poston's life. Any worthwhile task that needed doing automatically became *his* task. Thus, it was Poston who first presented to President Abraham Lincoln, in 1862, an organized plan for partitioning New Mexico Territory to create the Territory of Arizona. With Territorial status granted by an Act of Congress in February, 1863, Poston was appointed Superintendent of Indian Affairs for Arizona—a post he subsequently resigned in order to represent Arizona as a delegate to Congress. Next came a federal appointment as Special Commissioner to Asia. Then followed a series of posts of gradually diminishing importance—chief of the government Land Office, in Florence . . . customs officer at various ports of entry on the U. S.-Mexico border . . . head of the agriculture experiment station, at Phoenix.

Like the old Heintzelman mine, whose fortunes and misfortunes he had shared and watched with interest over so many years, Poston's sun was sinking low in the west. When even token posts no longer were forthcoming, Arizona's 20th Legislature, in recognition of all he had done for the Territory, voted him a small pension for life.

On June 25, 1902, when death came to "The Father of Arizona," at the age of 77, it found the once-brilliant man almost forgotten. The closing years of his life had been spent in a small, dingy rooming house in Phoenix, amid filth and debris.[7]

Fifty-eight years after Poston's death, his name, and that of the Heintzelman mine, was again back in the newspapers and on the lips of thousands of Arizonans—all because of an inkstand!

Sometime in the early 1860s Poston had delivered to Tiffany's, world famous jewelry artisans of New York City, some 400 ounces of silver bullion from the Heintzelman, together with a sketch he had prepared as pattern for an elaborate inkstand to be fashioned by Tiffany's from the Arizona silver. Upon completion of the intricate piece of artwork, Poston, in 1865, presented the handsome stand to President Lincoln in gratitude for the help he had given in bringing Territorial status to Arizona. Following the president's assassination, the silver inkstand was inherited by Robert Todd Lincoln, the president's son, and in 1937 was presented to the Library of Congress by Todd Lincoln's daughter, Mrs. Charles Isham.

Early in 1960, through efforts of Senator Barry Goldwater, the Library of Congress made a temporary loan of the stand to Arizona Pioneers' Historical Society, which Poston had helped organize in 1884.

Gleaming as brightly and beautifully as on that day, 95 years earlier, when a war-saddened president had accepted it from the hands of a grateful Westerner, this creation of master silversmiths was given the choicest place of honor in the Historical Society's museum, in Tucson. There, for a full month, it was viewed and admired by thousands of persons—many of whom, I fear, had never heard of Charles Debrille Poston or the Heintzelman mine, and few of whom will ever lay eyes on the lonely, little gray ghost of Cerro Colorado.

[7]Ray Brandes, "Notes on Charles D. Poston," *Arizoniana,* 1960.

"They (the pioneers) were not conscious of dates and statistics, mileage and the march of destiny. They knew only darkness, fatigue, cold and sunburn, the insistent wind, the drenching downpour, the lonely danger of guard duty while wolves howled from the hills and the skulking Comanche fitted an arrow to his bowstring."

—STANLEY VESTAL,

The Old Santa Fe Trail, 1939.

Travel Was Seldom Tranquil

MORE THAN A CENTURY ago, when all the arid miles between Texas and California were included within the Territory of New Mexico, and the walled pueblo of Tucson was virtually the only white settlement between El Paso and California, stagecoach transportation was inaugurated in the region that is now southern Arizona.

This was a land that had come into possession of the United States through the Gadsden Purchase of only four years earlier. It was a land where the white man of that day rarely knew the privilege of growing old. Summer temperatures climbed to phenomenal heights. Ranches and mines were hundreds of miles apart. Known sources of water were almost nonexistent. Psychological wounds inflicted during the recent War with Mexico still festered and rankled. Merciless bands of ruffians and renegades—the scum of two nations—levied tribute upon the border country, and the dark menace of hostile redmen hovered over the land like a pestilential cloud.

In these perilous times of 1857 the San Antonio & San Diego Stage Company began shuttling passengers and mail across the Southwestern wilderness.

The company made no pretense that the trip was a pushover. In a paid advertisement appearing in the *San Diego Herald,* November 1, 1857, the stage line set forth that each man proposing to "go through" should provide himself with a Sharps rifle and 100 cartridges; a Colt's revolver and two pounds of balls; knife and sheath, a pair of thick boots and woolen pants, half a dozen pairs of thick cotton socks, three undershirts, three linen shirts, three warm overshirts, a Wide-awake hat, a cheap sack coat, soldier's overcoat, one pair of blankets in summer and two in winter, a piece of India rubber cloth for blankets, a pair of gauntlets, a small bag containing needles, thread, etc., a sponge, hairbrush, comb and soap in an oil-silk bag; two pairs of thick drawers and three or four towels.

"Such money as he takes," stated the tongue-in-cheek bid for patronage, "should be in silver or small gold."

Except for 180 miles over the "Great Sanddunes" and the near-roadless terrain between Fort Yuma and San Diego, where saddle stock was provided for the use of passengers, the entire run was made in six-mule coaches, and the company pledged itself to provide an armed escort through the Indian country "for the protection of Mail and Passengers." The fare from San Diego to Yuma was $35; to Tucson, $75; El Paso, $120, and to San Antonio, $150, including meals.

The San Antonio & San Diego Stage company was still in operation in September 1858, when John Butterfield's Overland Mail Company began carrying passengers and the United States mails between St. Louis and San Francisco, by way of Tucson and Los Angeles. Under terms of a six-year contract signed a year earlier, the mails were to be carried twice weekly over the 2759-mile route, with 25 days the maximum time allowed for each trip.

So smoothly and satisfactorily did the Butterfield line operate that the SA&SD bowed out of the picture at the close of 1858.

It is not the intent of this writer, nor within the province of this chapter, to review even briefly the history, politics, or operational procedure of the Butterfield line, nor of any other of the many stage lines that ultimately would operate within the boundaries of Arizona Territory. The intent, rather, is to consider as a whole this form of transportation that was responsible for bringing into being many of Arizona's most historical, least heralded, and almost wholly lost and forgotten Shrines of Yesterday—her stage stations.

Like the prehistoric Indian pueblos, the Spanish missions, river shipping ports, military outposts, and the later-to-come mining camps, the stage station, historically, stands alone—an entity apart. Although every frontier town on a stage route had its stage depot, as a matter of course, a way-station established in the wilderness for the sole purpose of serving the stage line and its passengers rarely, if ever, became the nucleus of a town. The stage station was established for a purpose. It served that purpose as long as the need endured. When the need ended—if the stage route was realigned to bypass that certain station, if a new station was established in a more fortuitous location, or if the line was discontinued altogether—the station was abandoned. Salvagable material was hauled away by its rightful owners, or by others. Walls crumbled, wells caved, corrals fell into disrepair. Desert brush reclaimed the site, and cartographers dropped the station's name from subsequent maps.

With few scars left in its wake and its location never so well-known or precisely defined as that of, say, a frontier Army post or mining camp, the site of such a station represents the ultimate in elusiveness. Having spent hours and days in an oftimes fruitless effort to locate even one of these old station sites, I have come to the conclusion that no phase of ghosttowning offers a greater challenge to prowlers into the past.

On all the established stage routes throughout Southern and Central Arizona stations were situated from 9 to 30 miles apart, the distance being determined by nature of the terrain, condition of the road, the presence or lack of water, and various other factors. Few of the station buildings made any pretense toward grace or refinement. Most of them were constructed of adobe, a few of stone. Floors, except in rare cases, were of the hard-packed desert earth. Roofs, of poles and grass and earth, turned the infrequent rains. Windows seldom knew the luxury of glass. Each station was staffed by two to four employes, up to eight or ten in bad Indian country. Upon the shoulders of these men rested a diverse lot of responsibilities. They fed and watered the stage animals—no sinecure in itself at those stations where water must be hauled 25 miles or more across the desert. They kept fresh relays of horses or mules harnessed and ready for the arrival of the incoming stage. They were able to shoe animals when occasion demanded, and at least one man at every station had a rudimentary knowledge of veterinary medicine—which often had to suffice for human ills, as well. Such station tenders had to be ready, day or night, to fight Indian attack, repel border ruffians, and stand off white desperados. If, in addition, someone at the station could whip up a reasonably tasty meal of beans, sowbelly, and saleratus biscuits, or "squaw bread," the traveler had reason to consider himself fortunate.

Life of the Butterfield Overland Mail was cut short in the spring of 1861 by outbreak of the Civil War. "This chapter in Western expansion, written in blood and sweat, was now brought to conclusion," wrote B. Ira Judd.[1]

"Apaches peered down from the grim Dragoons to behold those big leather carry-alls for the last time change horses and drive away. Freight wagons followed gathering up supplies. Two thousand horses and mules were called in from corrals and pastures. Stations grew silent. Doors creaked on idle hinges, windows stared vacantly at the dark blue sky. As the dust of the last coach and four had settled over the horseshoe route, and drifting sand filled its tracks, the watching Indians may be forgiven for exulting that the white man had left their land forever . . ."

But the white man, of course, had not left that land; nor did he intend to leave it.

Although a few horseback and stagecoach mail lines operated briefly and/or spasmodically in Southern Arizona during the War years, increased Indian depredations had the effect of discouraging staging. It was this pre-1870 period, however, that saw birth of the Prescott-to-Phoenix stage line through the Black Canyon, as well as the Ehrenberg-to-Wickenburg, and Ehrenberg-to-Prescott lines; and in 1874 the Texas and California Stage Company was organized to operate between Fort Worth and San Diego. Many of the old Butterfield stations, abandoned 13 years earlier, were rehabilitated and put back into use, and within a year daily mail service was established over the new line.

This decade of the 1870s was responsible for more of the historical incidents connected with staging in Arizona than was any other 10-year

[1] "The Overland Mail," *Arizona Highways*, Oct. 1958.

period in the history of the Territory or the State. By this time many mining camps were coming into production, Phoenix and other incipient cities had been founded, and the volume of travel to and from such points was increasing. Yet, the Indian and border bandit menace still was as great, or greater, than it had been in the earliest days of the Territory.

What stories the old stations could tell, if they could speak; and what a pity that even their crumbling remnants have vanished so completely.

What prowler into Arizona's past would not thrill to a sight of Burke's Station[2]—even a fragment of wall, or a heap of stones that might mark the one-time fireplace; even a mesquite stump which imagination might transmute into the "Hanging Tree."

Established as a station on the Butterfield Trail in 1858, Burke's was still seeing service well into the 1870s. At one time the station was owned by the celebrated Billy Fourr who sold it, in 1874, to one G. R. Whistler. On July 7, that same year, Whistler was murdered by his Mexican stable-hand, Ventura Nuñez, who grabbed such plunder as he could readily lay hands upon and headed south.

Next station to the west was Stanwix, then the headquarters of King S. Woolsey, terror of evildoers throughout Central and Southern Arizona. Woolsey and a small posse immediately took up the chase, pursued Nuñez 90 miles into Sonora, captured him, brought him back to the scene of his crime, and there hanged him to a mesquite tree near the station. Swaying in the wind, a grisly warning to all passersby contemplating evil deeds, the body is said to have decorated the "Hanging Tree" for a number of months. When the skeleton of the errant stableman at last fell apart, his bones were buried by Mexican freighters who erected over his grave a rude cross.

On another occasion, when Woolsey and two employees were cutting wild hay near Burke's Station, they were ambushed by a group of twenty-odd Apaches. Although the only weapon at hand was a shotgun loaded with buckshot, the haying party killed the Indian leader and succeeded in standing off the attack. At the first lull in the fighting the men sprang aboard the hay-wagon mules and raced to Burke's for reinforcements and arms. Upon returning to the scene of the attack, however, they found the Indians had fled, leaving behind the body of their slain leader. Hauling the corpse back to Burke's, Woolsey and his men strung it up to the old mesquite tree, where it dangled for years and was assertedly shot full of arrows by passing Pimas and Maricopas.[3]

On at least one occasion the rough-and-ready justice of the Stanwix-Burke's Station vicinity proved a trifle erroneous but nonetheless lethal.

In the *Arizona Citizen* of December 4, 1875, appeared two significant news items summarized from the *Yuma Sentinel* of the previous Saturday:

[2]Situated on the south side of the Gila River, in T. 10 W., R. 5 S., about 105 miles east of Yuma

[3]McClintock, 1916. *Arizona.*

STANWIX—Nov. 25: Harry Roberts will be in Yuma tomorrow with the dead body of Chavez, the noted highwayman of California. He was killed at Baker's Ranch, today. No doubt of his identity.

YUMA—Nov. 26: The body of the dead man was brought in and, terrible to relate, it was not the body of Chavez the Bandit.

Stanwix Station, about a dozen miles west of Burke's, must have been one of the very few oases along the line of the Butterfield. Writing in the *Daily Alta Californian* of June 24, 1860, a Mr. Wallace stated: "The Stanwix Station is a gem by the wayside. It is almost the first thrifty large place we have met out from Los Angeles. It is 85 miles up the Gila, over a very rough, dusty road. This is the residence of Mr. Wash Jacobs, the road agent on the section. There are farming operations here that would do credit to any country. The farm extends two or three miles down the river. Two hundred tons of hay have been cut and stored this season. Large fields of barley and alfalfa, and vegetables of all kinds . . ."

Next station east of Burke's was Oatman Flat, and the second station to the east—about 17 miles west of the present town of Gila Bend—was known in days of the Butterfield Trail as Murderer's Grave. The story is that sometime prior to inauguration of the SA & SD line—about 1854, I believe—a young man traveling this route in company with his guardian went berserk and killed the older man. Members of a wagon train party encamped at the spot immediately delegated to themselves the offices of judge and jury, and lynched the young man. With this dispensation of frontier justice a double funeral took place and the grisly spot became known to emigrants as Murderer's Grave. These burials of slayer and slain (and possibly others, by then), were augmented in 1873 with the burial of Ed Lumley, station keeper, who on August 18, that year, had been tortured to death by renegades in an effort to make him tell where his money was hidden.

"The problem of such attacks was one which all station keepers had to face," wrote Byrd Granger.[4] "Because there were no banks, there was nothing to do but hide money which came in to stations from travelers, and by other means. It was also a well known fact that money was secreted at stage stations. This led not only to robbery and murder but also to legends of buried treasure."

Another supposedly buried at this same station is Samuel A. Bishop, one of Lt. Edward Fitzgerald Beale's camel wranglers, who died of exposure and thirst in 1877.

Shortly after inauguration of the Butterfield Line, Murderer's Grave Station was renamed Kinyon's, in honor of Marcus L. Kinyon, then in charge of this division of the trail; and about 1872 confusion was compounded when the name was changed to Kenyon's, for Charles A. Kenyon, then superintendent of Moore & Carr's Stage Line between Yuma and Tucson.

[4]*Place Names of Arizona,* University of Arizona, 1960.

J. A. Moore, senior member of that staging partnership, had a step-son, John F. Crampton. In the sunset of his life this same Mr. Crampton, then a resident of Phoenix, gave to the Arizona Pioneers' Historical Society an unpublished manuscript covering many of his recollections of early life in Territorial Arizona. Some of the most fascinating of these anecdotes concern Mohawk Station, situated on the Overland Stage Road, fifth station east of Yuma and second west of Stanwix. In the early 1870s this way-stop was kept by a man named John Kilbright, together with his Mexican wife and two sons, John and Miguel.

"In 1872 my stepfather, my mother and myself, were driving over the road in a light spring wagon on our periodic inspection tour," recalled Mr. Crampton. "We stayed all night at Mohawk. Mrs. Kilbright, who had brought some hens with her from Hermosillo, Sonora, had saved about a dozen and a half eggs which she promised to give us for breakfast. Eggs, at that time, were considered a great delicacy in Arizona and we were all looking forward to the feast in store.

"Next morning Mrs. Kilbright fried bacon and a whole platterful of eggs, baked some nice saleratus biscuits, warmed a kettle of beans, boiled some black coffee, and set the meal on the table. Just as we were about to sit down, Kilbright's two large Siberian bloodhounds got under the table and started to fight. My stepfather, who was quite a large man, got on one corner of the table and old Kilbright on the other—but to no avail. The dogs upset the table and, of course, the big platter of eggs landed on the dirt floor, bottom side up . . ."

Some time after this episode John Kilbright took his own life by swallowing poison and jumping in the station well. On the day following, one of the stage teams got away from its driver and the runaway leaders likewise fell in the well. This remarkable duo of incidents Mr. Crampton dismissed with the succinct comment, "We had to dig a new well."

After Kilbright's death, the station was taken over by a man named Peck, from San Bernardino. One time a freight team knocked him down and the heavy wagon passed over his leg, mangling it badly. "Drs. DeCorse and Tagget, of Yuma, operated on him and put a couple of chicken breasts on his leg. After that everybody called him 'Chicken-leg Peck,'" recalled the old stager.

"Michael Traut was the blacksmith that shod the stage horses. We would set all the shoes in the shop at Maricopa Wells, and I would 'point' all the horseshoe nails, as those days they all come in 100-pound kegs from the factory and had to be straightened and pointed. Every month or six weeks Traut and myself would put our outfit in the spring wagon and drive over the line, shoeing horses. When we would reach Mohawk Station we would usually stay one day and two nights. Old Peck had quite a large bunch of Sonora goats, near 1000 head, and a couple of purebred Angora goats and was breeding them up. He was quite close and always fried his hotcakes in goat tallow. One could rake the tallow out of his mouth all day. When we arrived at Mohawk, Traut, the blacksmith, would always warn Peck, 'no flapjacks fried in goat tallow! This time, fried chicken!' Peck had about a hundred chickens, but if he'd had his way the flock would never have gotten any smaller . . ."

Mr. Crampton, his mother, and stepfather, J. A. Moore, for a number of years made their headquarters at Maricopa Wells Station, one of the most important stops on the entire line—or, for that matter, in the Territory.

It is impossible to know how far back into pre-history the pools or "wells" at Maricopa Station were of benefit to humankind. Some historians believe that Maricopa Wells may have been the watering place described by Father Kino in 1694, under the name "San Andres Coata," and may have been the "Las Lagunas del Hospital" of Father Garces, in 1775. The Mormon Battalion camped at these same pools in December 1846; and to covered-wagon emigrants and goldseekers of 1849 and the early '50s, "The Wells" constituted a popular camping place, affording the last surface water for some 40 miles to the west.

When the San Antonio & San Diego Stage line was established in 1857, Maricopa Wells—185 miles east of Fort Yuma and 100 miles west of Tucson—was designated an intermediate stop, where eastbound coaches met the westbound and mail was exchanged. At that time the station was but little more than a brush corral and a rude shelter.

In this same year of 1857, Maricopa Wells was scene of a great pitched battle between the Pimas, Maricopas, and whites, on one side, and a war party of some 75 marauding Yumas, Mohaves, and Hualpais, on the other. The beleaguered whites and friendly Indians at the post stood off recurring attacks until the Hualpais and Mohaves at last fled the field, leaving the Yumas to fight on alone until, it is said, "the sage-dotted plain was covered with their bodies" and only three survivors remained.[5]

Small bands of Apaches who occasionally attacked the post were invariably put to rout by the well-armed occupants, and one such warrior scout met a strange end. The incident is described by J. Ross Browne in his *Excerpts of a Tour Through Arizona*. "A few miles beyond the Maricopa village, on a rocky hill to the right of the road, our attention was attracted by a spectacle . . . " wrote Browne. "Looming up on the side of the hill, in bold outline against the sky, stood a rude cross upon which hung the dried body of an Apache, crucified about two years ago by the Maricopas. The legs and arms were fastened with cords and the head hung forward, showing a few tufts of long hair still swinging about the face. It was a strange and ghastly sight. The Maricopas do not profess the Christian faith but this much they had learned from the missionaries . . . that crucifixion was a species of torture practiced by the whites . . ."

After outbreak of the Civil War in 1861 brought discontinuance of service on the Butterfield Overland Stage line and abandonment of its stations, the Maricopa Wells site was pre-empted by General John B.

[5]*Pinal County Enterprise*, April 30, 1959.

"Pie" Allen, one of Arizona's all-time greats, whose name turns up again and again in published accounts of Territorial activities. A pioneer in the business life of early day Tucson, where he resided some 22 years, Pie Allen was to become in time, an inveterate follower of mining booms.

During his tenure of ownership at Maricopa Wells, Allen repaired existing buildings and corrals, added substantial improvements, and established a store where he carried on an extensive trade with Indian farmers of the surrounding area. In 1867 Allen sold his Maricopa holdings to George F. Hooper & Company, prominent traders of Arizona City, who erected still more extensive stone dwellings, stables, and workshops, until The Wells had become one of the most popular stopping points for travelers passing through the Territory.

Capt. C. C. C. Carr, writing in *The Journal of the United States Cavalry Association,* March 1889, recalled Maricopa Wells in the 1866-69 period, as "the only place between Fort Yuma and Tucson where the table fare ever rose above the sodden, greasy level of bacon and beans. Owned by the only firm of any importance in the southern part of the Territory, and controlling nearly all the government contracts and freighting business, they had their trains always on the road and, in this way, could supply themselves and their friends with all the necessaries and some of the luxuries of life . . . including the best brands of whiskey and champagne to be obtained in San Francisco. It was a veritable oasis in the desert."

During this same period, about 1868, Maricopa Wells became a station of the United States Customs Office, manned by none other than Charles DeBrille Poston; and, in 1870, the station and appurtenances were purchased by Mr. Crampton's stepfather, James A. Moore, and his partner, Larkin W. Carr. During the next eight years under this ownership, Maricopa Wells attained its greatest importance. Almost every day of the year found the well-stocked store, saloon, and eating house thronged by Indian farmers, roving miners, and cowboys. Dusty stages clattered up to the station, changed horses, satisfied the wants of their hungry or thirsty passengers, and clattered away. Weary trains of covered wagon emigrants plodded out of the east, rested briefly at the station, and plodded on toward the sunset; and all over the grassy, mesquite-shaded valley surrounding the station were encamped huge freight outfits of eight to sixteen animals each, their drivers and swampers shooting the breeze, repairing wagons, and making ready to resume travel.

In 1873, Maricopa Wells became a station on the telegraph system, then being established throughout Arizona, and its ascending star glowed all the more brightly. But so-called progress has a great way of playing havoc with the *status quo.*

By 1878 every citizen of Maricopa Wells old enough to talk was discussing jubilantly the Great Day when the nearing rails of the Southern Pacific would reach the station, thereby giving it world access and future unlimited. But the railroad survey passed ten miles to the south; nor did the projected branch line, to run north through Maricopa

Wells to Phoenix, ever materialize. Instead, a new town on the railroad, seven miles south of Maricopa Wells, was given its post office, its telegraph station—and even its name, with the second word deleted. With mail bags and passengers and freight traveling by rail instead of by stagecoach and wagon, there was no longer any excuse for Maricopa Wells, and the once-important way-stop ceased to operate.

All that remains of the old station today are a few low mounds of adobe walls grown over with mesquite and desert brush . . .

In view of the relative abundance of known incidents connected with pioneer life around Arizona's stage stations, it is saddening, almost staggering, to think how many other happenings must have taken place around these old stations—happenings amusing, terrifying, or pathetic, of which all knowledge has been lost.

Consider, for example, the station known as Nigger Well.

In the middle 1860s a bakery was being operated at Wickenburg by a man sometimes described as a Portuguese, but generally termed a Negro. If he had a name, I have never come across it. When this chap had saved quite a nice little nest egg he decided to establish a stage station about a dozen miles from Wickenburg, between the Salt and Hassayampa Rivers. As there was no permanent surface water for several miles in any direction, the first item on the agenda was to dig a well.

There are two stories—at least two—concerning subsequent events. Which, if either, is the true story, likely will never be known. One version is to the effect that after the Portuguese-Negro had hand-dug the well to a depth of about 30 feet, he disappeared. The excavation looked as if it might have been caved in from the top, and the man's camp showed evidence of having been robbed. But no one, apparently, was sufficiently interested to re-excavate the well to determine if the body of its owner might be lying at the bottom. A second version is that the Negro had taken in several partners. When water was not found at the anticipated depth, the partners became disgruntled and threw rocks in upon the Negro, who was working in the bottom of the hole, and killed him.

In either, or any case, the spot became known as Nigger Well and subsequently earned a bad reputation. A man by the name of Gus Swain is said to have been murdered there by Indians in 1881; and there is the story of a Chinaman robbed and killed at the same place.

Another Nigger Well anecdote concerns a freight wagon which is supposed to have broken down nearby. The freighter, who had aboard more than a ton of dynamite belonging to Goldman Brothers, of Phoenix, unloaded the cases at or near the well and covered them with a tarpaulin, expecting to come back and get them later. Before his return, "Frenchy" DeBaud came along. He noticed coyotes sniffing around the tarpaulin-covered pile and, without knowing the nature of its contents, halted his teams long enough to take a pot shot at the animals. It will

Fort Bowie, Cochise County

—Courtesy National Archives and Arizona Highways.

never be known if he hit a coyote or not, since his shot also hit the dynamite and effectually annihilated not only all the coyotes but the freighter's cache, as well.

Goldman's sued Frenchy for damages, but got nothing. Later, the freighter sued Goldman's for the freight bill—and likewise drew a goose-egg. J. M. Barney concludes the story thus:

" 'Did you kill the coyote?' Frenchy was asked.

" 'Not for cer-taine,' said the Frenchman. 'But I got one damn beeg hole to bury him een!' "

Another station that undoubtedly has its heritage of untold tales—as well as several that are matters of record—is Mission Camp, in Butterfield days the second station east of Fort Yuma, between Gila City and Filibusters' Camp. In December 1870, the station was purchased by one Peter Reed and his wife. Only a few nights later, on Christmas Eve, three border renegades murdered both the Reeds, as well as James Lytle, a stage driver, and Thomas Oliver, the station cook. According to McClintock, "The bandits were driven off by the opportune arrival of several soldiers in an ambulance." But just what is "opportune" about an arrival after everyone has been slain, isn't quite clear.

Another early-day tragedy is credited to Bluewater Station, about 45 miles northwest of Tucson. Established in 1859, this was one of Arizona's earlier way-stops and must have been quite fortuitously located. Captain Shinn, of the 3rd U. S. Artillery, in his report of June 2, 1862, wrote of Bluewater: 'The well here is 69 feet deep, with two and one-half feet of water standing in it. It will furnish 4000 gallons a day. There is good gramma and galleta grass nearby . . ."

The tragic happening did not occur until some years later, after the station had been purchased by a stage driver named Baker, who drove between Bluewater and Tucson. On one of his periods at home, Baker and his entire family were murdered there by bandits, who escaped into Sonora.

Blood also had a habit of flowing frequently at Picacho and La Cienega Stations.

Picacho, which Granger locates "approximately a mile southeast of the present Wynola Siding of the SPRR," was favored by having a well 200 feet in depth—the only water for miles around. On the debit side of the ledger was the fact that the station laid along the line of an old Apache trail to Sonora, thereby placing it in a highly lethal position. McClintock states that a burying ground near the station held the graves of at least 75 persons slain by the Indians.

La Cienega Station, established by the Butterfield line in 1858, was situated some 30 miles east of Tucson and three miles west of the present town of Pantano. One of La Cienega's notable battles occurred in 1867 when W. A. Smith and three companions were attacked here by Apaches. All of Smith's companions were slain, but Smith used his shotgun so skillfully that eight braves were left behind, killed or seriously wounded, and their slayer had earned the nickname, "Shotgun" Smith, which followed him to his grave. Three years later, a man named Scott Young, and the mail carrier, were slain at La Cienega by Apaches and the station again destroyed. Its former site now lies beneath the tracks of the Southern Pacific Railroad.

Next station to the east of La Cienega on the Butterfield Trail was San Pedro Station, at the crossing of the San Pedro River. This was about a mile below the present site of Benson, or a quarter of a mile below the railroad bridge. In 1871, William Ohnesorgen came to take charge of this station for his brother who had purchased it from Duncan, Renshaw and Fowler.

In files of the Arizona Pioneers' Historical Society, at Tucson, is an unpublished typescript dictated by Mr. Ohnesorgen in 1929. In it he relates many interesting incidents connected with his tenure at San Pedro.

"The station, like most of them, was made of adobe walls 18 inches thick, arranged in the form of a hollow square with portholes in each corner and no windows," recalled Mr. Ohnesorgen. " For light we had the doors; if we wanted *lots* of light, we went outside. . . . This (San Pedro) was an important point on account of the overland travel and the military. We kept supplies, forage, etc. I paid $3 a pound for butter at Las Cruces; bacon, $1.50 a pound; lard, 50 cents; flour, $10 a hundredweight. The flour came from Sonora, in ox-carts. . . . The government kept a picket of eight or ten soldiers at the station to protect us and the ranchers around, as well as the emigrants. Indians never attacked the station while I was there; but they had done so before I went there, and had killed two soldiers . . .

"In the winter of 1878-79 I built the toll bridge. There was no cut in the river, then. You could have stooped down and drunk out of it at any point. The bridge had only a 25-foot span. I bought four stringers from the sawmill in the Huachucas for $55 each. I built the bridge myself, so labor didn't cost anything. It did not pay until 1880 when they were bringing in machinery for Contention, Charleston and Fairbank. That year there was high water and one month I took in $400. How the people hated to pay toll! They declared it was a public highway. I told them the highway went through the river and they were welcome to use it, if they wished; but the bridge was on my land, and if they used *it* they would pay toll . . ."

The old station buildings, according to Mr. Ohnesorgen, were eventually washed away by a flood on the San Pedro—a hard thing to imagine in these days when the river, nine years out of ten, is a river in name only.

Also on the San Pedro, not far from San Pedro crossing, was the settlement of Tres Alamos. A stage station was maintained here for a time, also a school and telegraph station, and enterprising settlers of the vicinity banded together in the San Pedro Farmers Association.

"The first people to settle at Tres Alamos was a colony of 25 Americans," said Mr. Ohnesorgen. "They were all killed, in time. I remember that in 1871 four were killed in one day . . . Then the Mexicans settled there, but the water gave out. In 1876 or '77 William Gibson came and he stayed. He and a man named Roberts. They stayed by the cattle and made good . . . Gibson loved to play cards. He would come to my place and pay a man $2.50 to sit up with him all night and play cards . . ."

I have wondered if the Tres Alamos of Mr. Ohnesorgen is the same place advertised in the *Arizona Citizen* of May 5, 1877: "Tres Alamos House, on the San Pedro River. C. M. Hooker, proprietor. People traveling from Tucson to Camp Grant, Apache Pass, Pueblo Viejo, San Carlos, and Globe Mining Districts, will find good accommodations at a low price. Meals, 50 cents; beds, 75 cents. Horses, per day, $1.00. Travelers will find all kinds of groceries, liquors and cigars, which are needful to sustain life whilst the Apaches are out . . ."

So far as I have been able to determine, there was no such thing as a "typical" station keeper. They were good and bad, honest and dishonest, educated and illiterate. Certainly Darrel Duppa was far from being "typical." Even the station he kept—though not notably better or worse than the others—at least was different.

Duppa is reputed to have been a scion of English nobility. He is said to have been educated in European universities, he had traveled extensively and spoke several languages. What had brought him to Arizona Territory was not clear; possibly he was one of the British brotherhood of remittance men. Be that as it may, Duppa was one of the important men of early Arizona. He is credited with suggesting the mythical name "Phoenix" for the rag-a-tag settlement that has since grown to be one of the largest cities west of the Rocky Mountains; and in time, Duppa became owner of New River stage station on the Black Canyon route, about 35 miles north of the fledgling town he had named.

It must have been a curious sight to see this well-groomed, refined, educated, and sophisticated Britisher, holding forth in his New River station in a building described as having thin unplastered walls of ironwood boughs interlaced with rawhide, and a roof of willows.

"Unpainted furniture stood on the dirt floor upon which guests slept on blankets taken from the piles stacked against the walls. Guns, ammunition, saddles, whips and spurs, were suspended from the joists and crossbeams. Many dogs and mules were a part of the establishment, and a guttural-voiced gnome appeared only to prepare delicious and unusual meals in a corner of the room equipped as a kitchen . . ."[6]

Next station north of New River—if I have myself correctly oriented—was Calderwood Station, at the point where the Black Canyon Stage road crossed the Agua Fria River. For a quarter of a century this place was owned and operated by Captain M. H. Calderwood. Stationed at Tubac in 1865, with the California Volunteers, and at Camp Mason the following year, Calderwood rose to be Speaker of the Ninth Territorial Legislature, in 1877, and Sergeant-at-Arms of the 18th Legisla-

[6]*Arizona*, American Guide Series, Hastings House, 1940.

ture, in 1895. When not legislating Captain Calderwood ran his stage station, one of the best descriptions of which is that contained in a typescript by Frank T. Alkire, in the files of Arizona Pioneers' Historical Society.

"At Calderwood he built houses and barns, and dug a 90-foot well," recalled Mr. Alkire. "He erected a horse whim and had a whisky barrel for bucket, and put in watering troughs and barrels for water for the house and travelers' use. He served meals to stage passengers and others, and supplied grain and hay to animals. The stage company built a shed and corral here for their animals and changed teams going either way to or from Phoenix. During the years that Captain Calderwood owned the place there was much travel on the road—stages, freight outfits, miners, soldiers and cattlemen. At one time a large steam boiler for the Vulture mine was in transit, and to cross the river and pull the other grade there were between 30 and 40 horses hitched to the wagon.

"About 1884 Calderwood sold the station to M. M. Elder. In connection with the station, Elder ran a small general store and a bar. He furnished good meals to the traveling public, price four bits. Water for animals, two bits a head; canteen, ten cents. To fill a water barrel cost a dollar . . . This station and river crossing was about five or six miles north of the present Santa Fe bridge, on the Agua Fria River . . . Today, at site of the old station, there are a few stones, as if for a foundation; a deep sump, which was the well, and a big wash of boulders, which was the road. No one would guess that thousands of people and hundreds of thousands of pounds of freight, as well as several million dollars in bullion, once passed this way.

This same raconteur, Frank Alkire, also had some interesting recollections of Seymour Station, on the west bank of the Hassayampa River.

"Seymour Station was kept by a queer old couple, Danny Conger and his wife," recalled Mr. Alkire. "The station was run on much the same plan as any other, except when Danny and 'the old woman,' as he called her, had been too good customers of their own bar. It was difficult to get a meal, then. Danny, explaining that the old woman wasn't well, would open his little store of crackers, cheese, sardines, or some other like delicacy, for the traveler. He would give the visitor the key to the barn and corrals so he could get grain or hay for his horse or team, and always ended with the remark, 'You can sleep in the hay—but don't you smoke! The old woman is sick tonight, but in the morning she'll feel better and will give you a good breakfast!'

"In the morning, to the usual station menu would be added two fried eggs—a great luxury!—all for the usual price of four bits . . .

"A tough, rough muleskinner, in charge of a big outfit hauling to Vulture and other points on the Wickenburg road, had a little girl about 10 or 12 years old. The mother died; and as the father was on the road most of the time, the child was placed in the good hands of a family at Phoenix. The freighter occasionally took his daughter on trips with him in the big wagon. On one of these trips the child was taken sick and died at Seymour. With the help of some friends the freighter buried her on the high ridge across the Hassayampa from Seymour, near the main road. He built a small fence around the grave, and as long as he lived he repainted it every two years.

"The girl's death occurred quite a number of years before I came this way, but I met the old man numerous times," wrote Mr. Alkire. "He always mentioned the child. The little fence could be seen from Seymour Station, as well as from the road which ran at the foot of the ridge . . ."

In McMullen Valley, about 14 miles west of Aguila and four miles north of U.S. 60, a few low adobe walls mark the site of Culling's Station. Established about 1868 by an Englishman, Charles Culling, a former hotel-keeper, this station soon became one of the most famous on the Ehrenberg to Prescott run, and Culling grew to be highly regarded throughout the Territory. In May 1878, the *Weekly Arizona Miner*, of Prescott, termed him a "kind-hearted gentleman," and stated that during the ten years he had operated the station he had likely given food and shelter to more destitute persons than any other one man in Arizona Territory. Three months later, when Culling died at his station (August 4, 1878) virtually every newspaper in Arizona carried a news story concerning his death. The *Florence Enterprise* eulogized him as "a good, whole-souled, jovial man (whose) hearty welcome and pleasant countenance will be missed," and the *Miner* declared, "There is not a person who ever knew Charlie Culling but will mourn his loss . . ."

Culling was succeeded in ownership of the station by Joe Drew, a man slated to become almost as famous as his predecessor. An interesting story concerning Culling's Station, and Drew, was originally published in the old *Youth's Companion* and reprinted in *Arizona Graphic* of March 3, 1900. Writing of the well Culling had dug to a depth of 240 feet, thereby obtaining an inexhaustible supply of soft water, the reporter for the *Companion* noted:

"The water, sweet and cold, is raised in a great bucket made from a barrel. A revolving drum, above, is propelled by a blindfolded mule that knows to an inch just how many rounds he must make before the clanking bucket rises to the point where it tips itself into the trough. At that trough, and in the round tanks beyond, ever stand cattle come to drink. Other water is . . . at least 30 miles away in any direction. By the well passes the old Ehrenberg road, once the great highway in Arizona, and there concentrate roads leading to the great Harqua Hala and many other important mines of Western Arizona. It has been said that the road from the Hassayampa River to the well is blazed by the graves of those who have died of thirst upon their way.

"Two miners expired by the roadside two years ago, almost within sight of their goal; and, a fortnight ago, a prospector's body was found within rifle shot of the little station. A few days later a German lad came staggering to the tanks shortly after nightfall. He had lain down to die when he saw lamplight from a window in the distance. Then it was that Joe Drew, keeper of the well, determined to establish his 'Lighthouse.' Far up on a tall cottonwood pole, every night, there now swings a lantern to be seen for many miles across the level plain—a light of promise to the traveler, a lifesaving beacon not marked on the mariner's charts . . ."

Construction of the Parker Cutoff in 1905-6 eliminated the need for the old station at the crossroads and it was abandoned. "Since that time," wrote Will C. Barnes in the *Arizona Republic*, April 10, 1934, "the well has been utilized as a watering place for sheep and cattle outfits; but, today, the old adobe buildings are in ruins. The old whim lies on its side, near the well, broken and useless. A short distance above the well are several graves, only one of which is marked at the present time. This is of . . . Christian Berry, Company H, 7th California Infantry . . ."

Another of the graves is that of Culling, the Englishman, "one of Arizona's oldest and best citizens."[7] Who else may be buried at the old station is a secret known, presumably, only to God. Neither is it known—or ever will be known—how many lives were lost in the course of sixty-odd years of staging in Arizona Territory. Indians, bandits, accidents, heat and cold and thirst, each took its toll in the hard life of staging—a life that oftimes demanded the last ounce of courage from horses and mules, drivers and swampers, station tenders, and passengers.

In the closing years of the 1870s and the early '80s, Arizona's major east-west stagelines were replaced by completion of the Santa Fe and Southern Pacific Railroads. Connecting lines and feeder lines continued to operate well into the present century but, eventually, they too were put out of business by improved highways, automobiles, busses, and trucks. So gradual was their passing that Arizona scarcely realized they were going from the scene until they were gone.

Little remains as a memento of those days. The old stations, almost without exception, have vanished from the face of the earth. A crumbling adobe wall, a bit of stone foundation, a caved depression that was a well . . . only this, where once was life and bustling activity. In many instances, even the one-time location of a station's site is lost to man's knowledge. In museums and private collections repose a few stage coaches typical of their era, but with leather dry and cracked by age, their once-bright paint faded and peeled by time and sun and wind. There are a few timetables, a few pictures, the personal recollections of a few pioneer settlers. And there is at least one marker raised to commemorate a staging tragedy.

It is a white-stone monument in the cactus-grown foothills of the Vulture Mountains, a little way west of Wickenburg. Its location would be a peaceful spot were it not for the incessant traffic of U.S. 60-70 which passes so closely that the marker seems almost to tremble beneath the pounding of the everlasting wheels forever hurrying west toward Yuma and Los Angeles, east to Phoenix and Tucson, swallowing more miles in a single effortless hour than the stages of yesterday

[7] *Weekly Arizona Miner*, Aug. 9, 1878.

covered in a long and gruelling run. For every hundred drivers who speed past the monument with eyes fixed on the wide, white road ahead, there is possibly one who swings into the little pull-out there to gaze for a moment at the white stone marker. He reads the engraved plaque on the face of the monument, and his eyes rest for a bit on the bronze silhouette of a miniature Concord coach drawn by fear-frenzied horses, and driven by desperation.

When this one-motorist-in-a-hundred continues on his way, perhaps he drives a little more slowly, more thoughtfully. Perhaps he has a little better idea of stage coach days and the hardships and hazards faced by pioneer travelers through Arizona and the West . . .

The event commemorated by the white stone marker took place on a pleasant Sunday morning in November, 1871.

As the Concord coach jounced along the winding desert road between Wickenburg and Ehrenberg, a feeling of all-rightness coursed through the veins of its driver, "Dutch" John Lanz. Though the Dutchman had made but one previous trip on this run he was a veteran whip, who had handled the ribbons on everything from buckboards to borax wagons. With Arizona's winter sun lying softly on his shoulders like the warm hand of a friend, it is more than likely he found the run to his liking.

It was 'Pache country, sure; but it was Dutch John's boast that he could take anything the redskins could hand out—like the time the Injins had gone on the warpath and he had pushed a stage through 138 miles in 36 hours, without relief, and after an hour's rest had climbed back on deck and started the long return drive. "Apache country's no different than anywhere else," Dutch John always said. "Except a man's gotta be a little tougher, keep his eyes open wider—and have a helluva lot more luck!"

In addition to the Wells Fargo treasure box under the driver's seat and Uncle Sam's mail in the rear boot, the Ehrenberg stage that morning was carrying several passengers. Riding topside with Dutch John were P. M. Hamel and W. G. Salmon; within the coach rode C. S. Adams, Frederick Shoholm, F. W. Loring, Fred Kruger, and a Miss Mollie Shepherd.

Miss Mollie was soft and beruffled, gentle-spoken, and pretty as a picture in a magazine. Both Loring and Kruger were bidding for her favor, drawing her attention to coyotes trotting across the desert flats, to the grotesque saguaros towering beside the road, the dumpy barrel cacti, the wicked choyas, the long-tailed roadrunner birds which strove to outsprint the stage. As the son of a pioneer Arizonan, Kruger had a little the advantage over Loring; but the brilliant young Harvard scientist wasn't faring too badly. He had gained some knowledge of the West while serving with Lieutenant Wheeler's expedition, and what he lacked in this field was compensated by sophisticated charm and general attentiveness.

"And whatever is *that*?" breathed Miss Mollie, a slender gloved forefinger indicating a puff of dust on the far horizon.

"Could be an approaching sandstorm," said Kruger, the Westerner.

"How thrilling!" exclaimed the girl delightedly. "What if it overtakes us and buries us forever—coach and horses, and all?"

Both Kruger and Loring laughed gallantly; but an observer might have noted that the Westerner's laugh was cut a bit short and a serious look came into his eyes. Leaning out the door of the coach he called to Dutch John. Had he noticed the dust cloud? Yeah, the driver had noticed it. Probably just a windstorm . . . but he'd keep an eye on it.

In the rough terrain through which the coach was traveling, the little smudge of dust was soon lost to view and talk turned to other matters. But Dutch John, on his high seat, fingered his handful of leather ribbons, ran his faded blue eyes along the horizon, and wondered, uneasily. Just a dust storm, maybe; but in Injin country a man could never be sure.

Tossing their heads and snorting in the fresh crispness of the winter morning, the lead team swung around a blind curve in the road. Next instant, all hell had broken loose!

Squealing in terror the leaders reared and fell back upon the swing team in a frenzy of tangled harness and flailing hooves, and a blast of bullets and feathered barbs raked both sides of the coach.

Dutch John, the veteran stager, was slain in the first volley, as were both passengers on the seat beside him. Loring, Kruger, and Mollie Shepherd were wounded.

Leaping from the beleaguered stage, Adams made a headlong dash toward the hoped-for protection of roadside rocks, but shots cut him down within 50 yards. Despite their traveling companion's ill-starred attempt to escape, Kruger and the girl next sprang from the coach and started running up a sandy wash leading into the Vulture Mountains.

Hours later, and nearly ten miles from the scene of attack, passersby found the pair, their clothing torn by brush and bloodied by their wounds, their lips swollen and parched from fever. At Wickenburg, Kruger and Mollie Shepherd were given the best hospitalization afforded by that frontier community, and a volunteer posse of 25 men departed immediately for the scene of the massacre.

Their arrival found only grim silence and circling buzzards. The attackers were gone. The bodies of Dutch John and his passengers lay where they had fallen, sightless eyes staring at the desert sky. The coach, riddled by bullets and arrows, rested on its side where it had overturned. Mail pouches had been ripped open, their contents scattered. A large roll of currency lay undisturbed on the ground, and the Wells Fargo treasure box had not been opened. One of the lead horses had been shot and left lying in his harness. Tracks of the other five horses, as well as those of the attacking party, led off into the brown hills.

Responsibility for the deed was never definitely fixed, but its perpetrators were generally believed to have been "broncho" Indians from the Date Creek Reservation, presumably Apache-Mohaves.

With Mollie Shepherd's death from her wounds raising the total to six killed and one wounded, contemporary newspapers characterized the affair "the Wickenburg massacre," and called it "one of the most sensational" of more than 50 Indian raids taken place in Arizona Territory that year.

Such is the story back of a white stone monument in the desert foothills a little way west of Wickenburg . . . a monument that commemorates not only a tragedy of long ago, but also the travel and traffic, the drivers and stage stations, of an era now gone forever.

Picture Window, Charleston, 1959

"There is a class of men peculiar to our new mineral territories to whom the world has not yet done justice . . . I speak of those vagrant spirits commonly called "prospectors" who never make anything for themselves, but are always on the move to make fortunes for other people . . . The prospector is a man of imagination . . . Ragged and unshaved, he owns millions, yet seldom has two dimes to jingle in his pocket—for his wealth lies in the undeveloped wilds. The spirit of unrest burns in his blood . . . There is no desert too barren, no tribe of Indians too hostile, no climate too rigorous for his researches."

—J. Ross Browne,
Adventures in the Apache Country, 1863

St. Peter's River Was a Busy Place

Beside the evanescent San Pedro River, in the western part of Cochise County, drowse four little ghost towns. None of them is famous. Few persons have heard their names; fewer still have visited their sites. Yet, three members of this spectral quartet are direct by-products of nearby Tombstone, most widely publicized, most widely known, most written about and, probably, the most visited ex-mining camp in the world.

Through good, honest fiction, hypocritical fiction masquerading as fact, historical novel and documented tome, by radio, motion picture screen and TV, by magazines, newspapers, and word of mouth, the deathless story of Tombstone has traveled around the earth. In many of these presentations about all that has been historically accurate has been the magic name TOMBSTONE. From the standpoint of fame and notoriety, however, these fictitious "facts" have been equally as valuable to the town as the best authenticated history. Possibly even more so, since history, like some pills, seems to be accepted more readily if sugar-coated.

Tombstone, in my opinion, is the most fascinating ex-mining camp in Arizona, if not in the entire United States. I can't imagine any dyed-in-the-wool Western buff who wouldn't find pleasure in visiting these old streets and buildings were such an abundance of Western history was lived in the opening years of the 1880 decade. The Crystal Palace and Oriental saloons, the old Cochise County courthouse with its priceless collection of Tombstone trivia and trophies; the Can Can restaurant, the office of the historic Tombstone *Epitaph*, the Million Dollar Stope, the Birdcage Theatre with its rambling, almost-unbelievable museum; and Boothill, with its spurious but entertaining headboards and grave markers, are points of interest visited by tens of thousands of persons every year.

Tombstone—"The Town Too Tough to Die"—is a historical treasure, a showman's paradise, a tourist's delight. But it is not a ghost town. It has never been a ghost town and, very likely, never will be a ghost town. Consequently, Tombstone's place in this book must be limited to her connection with satellite camps which the passing years have transformed into genuine phantoms.

Throughout Tombstone's most active mining years—from 1879 through the early '80s—the town in the foothills had plenty of rich ore but no water for milling it. As a result, several of the large mining companies built mills along a ten-mile stretch of the San Pedro River. The ore, mined at Tombstone, was freighted downhill about nine miles to the pounding stamps, and a small town grew up around each mill or group of mills.

One of these towns was Charleston, situated on the west bank of the river about half a mile downstream from the San Pedro bridge on the road between Tombstone and Fort Huachuca. Across the river from Charleston, and a little way back toward the hills, was Millville. The prosperity, the very existence of Charleston and Millville, was contingent upon the bullion yield of their two great stamp mills. When those mills no longer had Tombstone ore on which to feed, the towns died—and stayed dead.

Today finds Charleston and Millville among the ghostliest of Arizona ghost towns. A few old adobe ruins, an abundance of miscellaneous rubble, the foundation and backwall of the Tombstone mill, the crumbling ruins of Dick Gird's "Big House" with its built-in treasure vault—little more remains of these two once-busy towns. Only this and a few printed references, a few photographs, and the recollections of a vanishing generation of oldsters who knew these towns as places where men lived and worked and died, and bars of silver bullion poured forth.

After its official laying-out in the winter of 1878-79, Charleston soon got itself organized as a maker-of-history. Beginning with Alfred Henry Lewis, feature writer for an Eastern newspaper, who visited Charleston in 1880 and subsequently wrote a series of stories in which the river town was fictionized under the name "Red Dog," pseudo-historians have had a delightful time glamourizing early-day Charleston. Here, if we could believe even half that has been written about the place, was a second Bodie, an embryo Dodge City, a short-lived Devil's Den where dead men littered the streets and gunsmoke drifted over the land like smog.

Although any mining or milling center of that era attracted some of the rougher element and a town had to show a reasonable amount of gumption or folks regarded it suspiciously, it seems doubtful if this Western namesake of South Carolina's lovely city was entirely abandoned to evil. Nor is that impression given by the recollections of Mrs. Mary E. Wood, who moved to Charleston with her husband in 1880 and knew the camp throughout its most active years.

At the time of Mrs. Wood's arrival at Charleston the town consisted of a single street—Main Street. Along this throughfare, according to Mrs. Wood, were located two good-sized adobe stores, McClure's &

Herrera's, and Zeckendorf's; a small wooden restaurant, two saloons, several shacks, and the residences of the storekeepers. At the west edge of town was the wooden schoolhouse and the ice house. Due to a lively smuggling trade between Charleston and Mexico the stores were always well stocked.

"Frequently a number of patient burros could be seen lined up on the street for packing or unpacking," recalled Mrs. Wood. "The Clanton boys and their friends sometimes had horse races south of town. Gamblers came down from Tombstone sometimes, but were not welcomed by either cowboys or millmen . . . The town was usually very quiet. Jerry Barton's reputation as a killer may have had a tendency to keep it quiet as he was constable as well as a saloonkeeper.

"Millville was noisy with the stamps of both the Gird mills going day and night unless shut down for necessary repairs. At first it worried the Eastern owners to have the men working for them on Sundays. Mr. Gird and Mr. Wood tried to tell them of the expense and time required to shut down the mills and to get up steam to start again. Besides, many of the men would get drunk and be unfit for work on Monday. The owners still insisted that it be tried. On Sunday they viewed the unsteady men with horror and, Monday, watched impatiently while the millmen were getting up steam to start the mill. On the next Saturday they said they really believed it was better to keep the men at work, even on Sunday . . ."

During August 1881, the mill managed by S. W. Wood produced the greatest amount of bullion of any month in its history—the total of $151,279.15 being recalled by Mrs. Wood with penny preciseness. This same month saw heavy rains destroy the mill dam and come close to drowning several townsmen who attempted to save it. Prior to this time the river had been spanned by a narrow footbridge but the flood widened the stream to such an extent that a more substantial bridge became necessary. As the town's first and only civic improvement, this bridge, according to Mrs. Wood, "was a matter of great pride."

Referring to the picture of "wicked Charleston" painted by modern historians, Mrs. Wood said: "If you came to Charleston looking for trouble, there were plenty of citizens who would gladly supply you with any amount of it . . . But the honest, law-abiding citizen went his way with little or any greater hazard than he faces, today, in any large city.

"Remember, too, that the stamps in Millville were grinding out a fortune. In twelve months time, from April 1881 to April 1882, a total of $1,380,336.97 in bullion was shipped from the mill—and not a dollar of it was molested by highwaymen. Tens of thousands of dollars were handled in the monthly payrolls without a single bandit ever trying to hold up the paymaster. And we lacked the 'civilized' convenience of such things as the armored cars that the city banks use today to transfer money. The bullion went out on the Charleston stage . . ."[1]

[1]From an interview published in *Tombstone Epitaph*, Oct. 15, 1929.

If the constable-saloonkeeper-alleged killer, Jerry Barton, was an important factor in the maintenance of law and order in Charleston, surely another factor equally important was Judge Jim Burnett who conducted his justice court affairs in the same arbitrary fashion favored by the celebrated Judge Roy Bean, of Pecos River fame.

Like other J. P.'s of that day, Burnett was supposed to turn in to the county board of supervisors all fines and fees collected by his office, whereupon the county would return to him as his salary a certain percentage of the money turned in. When Burnett failed to get as much in return as he figured was due him from his first three months in office he refused to make any further returns to the county, declaring that the Charleston justice of the peace office, henceforth, would be self-sustaining. Thereafter, he levied fines right and left and made no reports to anyone.

A drunken rancher, shooting up the town, was assessed 20 head of prime steers for disturbing the peace. Another cowboy was fined $50 for "reckless riding and endangering the lives of children," and a Mexican woodhauler stupid enough to come into town driving a horse stolen from a local livery stable was fined nine cords of wood—the exact amount he was carrying on his wagon. According to legend (which is how we historical writers tag any story we can't document), the owner reclaimed his horse, the Mexican went home pondering his misfortune in getting caught, and Burnett sold the wood to the mill and pocketed the proceeds.

Charleston's reputation for lawlessness is likely founded, in part, on an incident that occurred one evening in 1882.

M. R. Peel, popular young engineer employed by the Tombstone Mill and Mining Company, was sitting in the company office chatting with three good friends, when the office door was flung open by two masked riflemen.

Without any command or word of warning, both gunmen fired directly at the seated Peel. One shot took instant effect; the other slug buried itself in the adobe wall.

Still without speaking a word, or making any attempt to harm Peel's three companions or to rob the place, the two assailants dashed back to their horses, held by a confederate, and the three escaped into the night, leaving behind a dead man and three stunned witnesses.

Though not addicted to flights of rhetoric, the editor of Tombstone *Epitaph* apparently was deeply moved by the young man's murder and his subsequent funeral rites in that city.

"The simple but ever impressive ceremony for the burial of the dead by the Episcopal church, read by the youthful rector[2] standing at the head of the dead man, scarce older than himself, never before sounded more impressively solemn than on this occasion," reported the frontier scribe. "The storm of wind that had raged with great fury all day abated with the declining sun; the scurrying clouds had nearly all disap-

[2]The Rev. Endicott Peabody.

peared, and the sun shone forth in all its evening splendor, as the long cortege left the office bearing the remains to their final resting place in the city cemetery, where the concluding ceremonies of committing the earth to earth, dust to dust, ashes to ashes, were gone through with and he was left to his long untroubled sleep until the Resurrection Morn . . ."

The starkly plain headstone in Tombstone's famous Boothill cemetery is inscribed simply:

"In Memory of M. R. Peel, Born May 31, 1854. Died March 25, 1882."

About a month after the young engineer's mysterious and unsolved murder, the editor of Tombstone *Epitaph* made a personal visit to Charleston and reported himself agreeably astounded by the size and apparent prosperity of the place.

"It possesses excellent hotel accommodations, superior liquor establishments, a drug store, livery stables, and all the accommodations of a prosperous frontier town," stated the editor. "The buildings are all substantial and some of them are very creditable structures. What surprised us most is the amount of stock carried by the merchants. We had been led to believe that Charleston derived its existence entirely from the employes of the mill; but, in fact, it has a very extensive trade with the surrounding country and Sonora. Its Mexican business is daily becoming more important and will continue to increase until it reaches very large proportion. The town is well regulated and free from turmoil; in fact, it is one of the most peaceful places we were ever in . . ."

Possibly Charleston grew tougher as time went on, for James G. Wolf, who arrived in the river town in 1883, found it possessed of more impiety than was described by either Mrs. Wood or the editor. In a manuscript of his recollections of Charleston, now in the files of Arizona Pioneers' Historical Society, Mr. Wolf stated that the river town, in proportion to its size, was a far tougher and livelier camp than Tombstone.

"There were four saloons going 24 hours a day," noted Mr. Wolf. "All kinds of gambling games operated continually. There were lots of naughty girls living close to the saloons . . . When paydays occurred at Fort Huachuca, many of the soldiers came to Charleston to drink . . ."

But the days of Charleston and Millville were numbered. The great mines at Tombstone, whose treasure had nurtured the milling towns, began flooding with water in the middle 1880s; and before the close of that decade the new railroad town of Nogales, on the Mexican border, was capturing Charleston's lucrative trade with Sonora.

The double blow was more than the river town could weather. Her stores and hotels closed; her post office was discontinued in the autumn of 1888, and on June 14, 1889, the Tombstone *Prospector* published the town's obituary:

"Charleston, a city on the San Pedro River, which at one time was the liveliest place in Cochise County, possesses none of its old-time appearance now," it was stated. "It has been given over to the Mexican population who live in tents and houses claimed by no one. The sidewalks, awnings, roofs, doors and partitions in the large spacious stores that were once filled with thousands of dollars worth of goods, have been taken down and used for firewood. Not a store exists there now; whereas, several years since, several establishments, among them Sam Katzenstein, Springer and Hacks, Herrera and McClure, carried stocks valued at from $50,000 to $100,000 . . ."

Except for perennial rehashes of her turbulent past (which seems to have grown more turbulent with each rehashing), Charleston got little more attention from the newspapers of the land until the mid-war year of 1943 when the old ghost town was selected by the Untied States Army for use as a "practice location."

With part of the fallen debris cleared from her streets, a few new buildings erected and false fronts added to crumbling adobes that had once housed stores, hotels, and saloons, soldiers of the 93rd Infantry Division from nearby Fort Huachuca, using live ammunition, came here to practice street fighting under conditions similating those they might later encounter in the bombed-out cities of the worlds' war fronts.

With the war ended and the shooting and shouting past, battered Charleston returned to her briefly interrupted siesta and was still sleeping peacefully on the autumn day in 1959 when I made camp in the quiet, sun-warmed yard of her deserted railroad station. I had visited Charleston several years previously but, on that occasion, had failed to locate her sister ghost town of Millville, across the river. Now, I wondered how I had possibly overlooked it, as even from the depot yard I could glimpse the stone backwall of the old Gird mill looming above the thorn-scrub, about a mile to the north.

Shouldering a quart canteen of water and my camera, I set out in the general direction of the ruins.

It was a hot day. The sun, that October afternoon, was functioning with an enthusiasm quite out of keeping with the season, and not a breath of breeze was stirring. After making my way down an eroded gully and across the bottom of a wide, gravelly wash, I cut off into the flat that extends back from the river for close to half a mile on each side. It wasn't a pleasant place for a stroll. The thorn-scrub, which had looked so pleasantly green and carpet-like from the depot yard, turned into a tangled mass of botanical assailants, barbed, spined, spurred, and armed. With the strenuous exertion of floundering through the inhospitable brush—which seemed to delight in tearing at my clothing and planting long scratches on my anatomy—I was almost drowning in my own perspiration when I stumbled into an old wagon road. Though weed-grown and long unused, it still made an aisle of sorts through the thicket and I gratefully followed it the remainder of the way to the ruins.

The stone foundation and backwall of the old mill showed it to have been a building of considerable size, but I was more fascinated by the ruins of Dick Gird's "Big House," as it had been known in its day of glory.

Combining the functions of private residence and company offices, it had been a large and imposing structure. Though now roofless, and with all doors, windows, and usable lumber spirited away to serve other owners and other needs, several of the old walls were still standing to full height. Made of adobe bricks laid in mud mortar and plastered both inside and out, these walls were a full three feet in thickness and must have given the building an almost fortress-like quality.

Within one room of the Big House I was delighted to find a secondary room about the size of a clothes closet. Stoutly fabricated of fired brick laid in double-wall construction and originally fitted with an impregnable steel door, this had been the company's treasure vault where the great bars of silver bullion had been stored pending their shipment to the mint.

How much bullion, I wondered, had been stored in this old vault during the half-decade of its usage? I wondered, too, in which of these now roofless rooms the young engineer, Peel, had been chatting with his friends when the two unknown assailants burst upon the peaceful gathering and sent him to his death?

Both were interesting questions to ponder on a warm afternoon. But there was no one to answer, not even a vagrant ghost.

Less than three months after my visit to Millville the Charleston station house, where I had camped that night, was torn down for salvage. Every year finds the venerable adobes of the two old river towns a little more weathered, a little nearer the brink of oblivion, and the day is fast nearing when in all the world there will be no living man whose memory reaches back to the time when Charleston was in flower.

But even when that times comes, the Charleston story will still be told.

Down the river a few miles from Charleston and Millville is the old town of Fairbank. It isn't exactly a ghost town, even today. Last time I was there it still possessed a few inhabitants and a small country store—quite a come-down from eighty-odd years ago where there was a quartet of general mercantiles, half a dozen saloons, three restaurants, and a hotel.

Early-day Fairbank was scene of a shooting or two and one attempted train holdup; but likely the most exciting event in its history was The Night of the Flood.

The San Pedro River—like the Old Gray Mare, which ain't what she used to be—is today a desultory stream containing barely enough water to flow. Back when Charleston and Fairbank were feeling their oats, however, the San Pedro must have presented quite a different picture—at least on some occasions. One such occasion was in September 1890.

Later, in reporting the flood, Tombstone *Epitaph* pronounced the scene "a frightful one." Many persons had narrowly escaped being drowned. The onrushing waters had swallowed the town in the night when most of its residents were sleeping. Cabins were crushed and tumbled upside down, causing many persons to lose all their possessions, "even to the clothes they had laid off before retiring." Half the town was

Disappearing Walls of Brunckow Cabin, 1960

Ruins of Dick Gird's "Big House," Millville, 1959

thrown into panic, women fighting their way through the rising water trying to locate their children, and terrified children crying for their parents.

A Mr. Salcido, keeper of a lodging house, had dashed through his building, awakening each roomer and telling him to flee for his life.

"He had cleared the rooms and was leaving . . . when the flood struck the front door and filled the house with water before he could get to the back exit to open that door and let the water run through," reported the *Epitaph*. "The water was up to his neck in a moment and he struggled until help arrived and saved him. He was taken to Williams' drugstore where he recovered from his fright . . ."

Looking upon Fairbank today—high, dry, and complacent—it is almost harder to imagine it being ravaged by flood waters than it is to imagine the one-time presence of a great silver mill at the now-deserted site of Contention City, about three miles to the northwest of Fairbank.

Despite the town's strange name, it was no more quarrelsome than the average milling center of its era. Rather, it was named for the great Contention mine, of Tombstone, whose ore was milled at this point. The mine, in turn, had been named as result of a claim dispute between Dick Gird and Hank Williams.

"The town called Contention City has been laid out, surveyed and mapped, and the enterprising proprietors, D. T. Smith and John McDermott, are selling lots to all who desire to invest in the new city," reported *Arizona Daily Star*, September 23, 1879. "Although it is not one week since the survey of the town was made, a number of lots have been sold and preparations to build are going on as fast as material can be had. Already there are in operation one store, one hotel and restaurant, one saloon, one meat market, one Chinese laundry, and some dozen or more shanties which are buildings in embryo . . . In addition to the mill, which is now being erected, sites for two other mills have been selected and the prospects are that a third, for custom work, will be located here . . ."

The plat for the original townsite of Contention City, on file in the office of the Cochise County recorder, shows streets in one direction to have been named numerically from First to Ninth. These were crossed by Front Street, facing the river; Main, San Pedro, Mesa, and Grant.

In its issue of March 4, 1880, the *Star* was still singing the praises of Contention City, devoting more than half a page of space to a description of the new town, its mills, business houses, advantages, and assorted potential. The building housing the mill of Contention Mining and Milling Company is described as 170 feet in length, including the boiler, engine and recharge room, with a depth of 142 feet. "The fall of ore, from the time it leaves the dump floor till it reaches the settlers, is 60 feet. The foundation is in rock and solid cement, which has been cut out in terraces. The structure is frame, covered with tin . . ."

By 1881-82 Contention was becoming a well-rounded frontier town and an important center of activity, according to information supplied to Arizona Pioneers' Historical Society by H. L. Hayhurst. Merchants and other businessmen operating at Contention during this period included A. Castenada & Company, and E. E. Rupert & Company, general stores; J. M. Barney, drugs; D. Alexander, variety store; L. Larrieu, and L. Myers & Company, hotels; Simon Marks, clothier, and Shrewsbury and Baker, livery stables. Saloons and liquor emporiums were operated by John Dunn & Company, Romaldo Montoya, Stanton & Conley, Barela & Navarro, and Welch & Fenton, A. D. Kinnear operated the stage line, (later purchased by "Sandy Bob" Crouch), and also maintained a stable and blacksmith shop. During these years, said Mr. Hayhurst, Contention City was, for a time, railhead for the New Mexico, Arizona and Sonora Railroad, and contractors for building the road had their headquarters at Contention.

Among visitors to the San Pedro milling town in 1882 was William Henry Bishop, who briefly mentions that sojourn in his book, *Mexico and Her Lost Provinces.*

"We changed horses and lunched at Contention City," wrote Mr. Bishop. "One naturally expected a certain belligerency in such a place, but none appeared on the surface during our stay. There were plenty of saloons—the 'Dew Drop,' the 'Headlight' and others—and at the door of one of them a Spanish senorita smoked a cigarette and showed her white teeth.

"Contention City is the seat of stamp mills for crushing ore which is brought to it from Tombstone . . . The stamps are rows of heavy beams which drop upon the mineral on the mortar and pestle plan, with a continuous dull roar, by night as well as by day.

" 'That's the music I like to hear,' said our driver, gathering up his reins. 'Poundin' out the gold and silver. There ain't no brass bank ekils it!' "[3]

But the day-and-night thunder of the stamps in the great mills at Contention City was permanently stilled not too many years after those lines were penned.

By the late 1930s, when I first visited Contention City, the site of the formerly busy little settlement was almost inaccessible as result of encroaching catsclaw, mesquite thickets and barbed-wire fences. I found the walls of only four adobe buildings still partly standing, as well as two adobe columns that had flanked the main doorway of the big mill; and on the mesa, above the townsite, the old cemetery lay in peaceful neglect. Of the numerous headboards and crosses it contained, I was able to read the inscriptions on only three. One of these marked the grave of a Maud Burton, "daughter of E. C. and Kate Burton. Died Dec. 18, 1883." Another, only partly legible, had been erected to someone named Hughes, and the third marked the last bivouac of Colonel E. A. Rigg, Civil War veteran and Indian fighter, who later became justice of the peace at Contention and died there in 1882 at the age of 60.

[3]Quoted in *Arizona Writers Project Guide Book,* p 392.

Several years after visiting Colonel Rigg's grave on the lonely mesa above Contention City, I was researching some old files of the *Daily Alta Californian,* of San Francisco, and was interested to come across this "social note" in the issue of February 14, 1863:

"Lieut. Colonel E. A. Rigg, of the First California Infantry, is a benedict. He met his fate at Hart's Mill, near El Paso, in a sparkling, black-eyed Miss Cooper, a young lady from Rome, N.Y. . . . The Colonel is in command of Fort Craig and has placed that important post in excellent condition . . ."[4]

In addition to these four neighboring ghost towns of Charleston, Millville, Fairbank, and Contention City, there is another ghostly spot in this same part of Cochise County. It is not a ghost town. It is not even a mill or mine, or mission or stage station; yet, I consider its story quite as interesting as those of many more populous centers.

The Brunckow Cabin, situated about 200 yards south of the Tombstone-Huachuca road, at a point half a mile southeast of the San Pedro River bridge, is one of the oldest ruins in Arizona not built by Indians, Spaniards, or Mexicans.

Frederick Brunckow, born in Berlin about 1830, received a classical education at the University of Westphalia and went on to graduate from the School of Mines at Freiburg, Saxony. Emigrating to the United States in 1850 he accepted such employment as he could find. In 1856, while working near New Braunfels, Texas, as a shingle-maker at $2.50 a week, he gratefully accepted a chance to join the Sonora Exploring and Mining Company, then being outfitted for operations in the Gadsden Purchase which Uncle Sam had acquired from Mexico three years previously. Arriving at Tucson in August 1856, the company established headquarters at Tubac and began explorations in the surrounding hills.

Some 30 months later, on April 15, 1859, the following news item appeared in the *Weekly Arizonian,* of Tubac:

"Mr. Frederick Brunckow, of the Sonora Silver Mining Company, leaves this week on a visit to The States . . . Mr. Brunckow was one of the pioneers of the Sonora Mining Company. He came to the Territory with Colonel Poston and, in company with that gentleman, shared the many hardships and disagreeable circumstances which naturally environed the first mining company in this far off and isolated region. Now, having seen the company triumph over many difficulties, and having assisted in developing one of the richest silver veins in the world to a highly profitable condition, he proposes to pay a visit to the Atlantic States."

After reporting to the major stockholders of the company and being wined and dined royally at Delmonico's, Brunckow returned to Arizona and began work on his San Pedro silver mine, situated in the hills a

[4]Fort Craig's quite-extensive stone-and-adobe ruins are situated on the west bank of the Rio Grande, about 25 miles south of Socorro, N. M.

short distance east of the river. It was, presumably, about this same time that he erected several buildings, including the adobe cabin which is the subject of this sketch. He also established a store, took in a brace of partners, employed several Mexican laborers, and settled down to the business of mining silver some 20 years before production of ore was to get under way at Tombstone, less than nine miles distant. Had Brunckow lived to develop his San Pedro mine it is quite possible it would have been him, rather than Ed Schieffelin, who would have sparked the rush to that district. But poor Frederick Brunckow was not to be given that privilege.

On the morning of July 23, 1860, William Williams left the Brunckow cabin on horseback bound for Fort Buchanan, 35 miles distant, where he hoped to purchase a wagonload of flour. Behind, at the cabin, he left his cousin, James Williams, a mining engineer from Virginia; a Mr. Morse,[5] formerly professor of chemistry in a St. Louis high school; Mr. Brunckow, and David Brontrager, a German cook, who had come into the camp's employ only two days previously.

Due to the difficulty of procuring a wagon and teamster at the fort, Williams was unable to start back to Brunckow's with the flour until the morning of July 26, at which time he was accompanied by two young boys, the sons of a Mr. Acke, a Sonoita Valley rancher, who were to return the team and empty wagon to the fort.

It was about midnight before Williams and the boys neared the Brunckow establishment. As the premises were ordinarily overrun by a pack of noisy, mongrel dogs owned by the 11 Mexican peons in Brunckow's employ, Williams thought it strange that the sound of their approaching horses failed to trigger a bedlam of barking, and as he and the boys drew nearer the Brunckow residence they became aware of a peculiar stench that became almost unbearable as they entered the midnight darkness of the store.

Striking one of the two matches in his possession, Williams saw that the room was a chaos. Merchandise, torn from the shelves and counters was strewn and spilled everywhere, and upon the floor sprawled the bloated, putrifying body of his cousin, James.

Not knowing who had committed the murder and robbery, nor what possible assailants still might be lurking on the dark premises to serve him and the boys in the same fashion, Williams decided to make no further investigation but to return to the fort for help.

Except for the San Pedro ranch, 30 miles distant, in Sonora, the closest settlement to the Brunckow mine was Fort Buchanan. Under ordinary circumstances the 35 miles separating these two points was not regarded as especially far; but to the horrified Williams and the two young boys, urging the jaded wagon horses back over the long, rough trail traveled earlier that night, those 35 dark miles must have seemed almost interminable!

[5]In some contemporary accounts the man, Morse, is identified as "J. C. Moss," and as "Marsh."

When the trio arrived back at the fort during the morning of July 27 and related what they knew of the tragedy, Capt. Ewell, commanding officer, immediately dispatched a Sergeant Henderson and a file of soldiers to examine into the matter.

Late in the evening of that same day the German cook, David Brontrager, staggered into Camp Jecker, headquarters of the Sonora Survey Commission. Covered with bruises and scratches, barefooted and badly fatigued, the man related his version of the massacre which, he said, had taken place only a few hours after Williams' departure for the fort.

About two or three o'clock in the afternoon, he related, he had been sitting in the kitchen reading a book when two of the Mexican peons came in and requested a light for their little paper cigars. The German assented, and at almost the same moment heard several shots fired. When he attempted to go outside to see what was the matter, he found his way barred by the peons who told him he must remain inside the building and no harm would come to him. Other peons soon joined those holding him prisoner and he was told that his life had been spared because he had lived for some time in Sonora, was known to be a good Catholic, and that none of the peons had anything against him. However, said the spokesman, he must accompany them when they left and must give them his word he would not attempt to escape. This he promised.

After awhile, said the German, he was permitted to look around the premises. In the store he saw the body of James Williams, shot several times and with a deep gash on the head. Just outside the store, he found Mr. Morse dead from two gunshot wounds and, upon inquiry, learned that Frederick Brunckow had been stabbed and killed at the mine shaft.

Object of the attack was plunder. Only a short time previously there had arrived at the store from St. Louis some $2000 worth of goods. In addition, there was some money and a little jewelry, and seven horses and mules. Immediately after the massacre the peons and their families had begun selecting what goods they wanted, had packed the animals to the limit of their carrying ability, and had left before dark for Sonora. After traveling 30 miles without stopping, said the German, they had reached the San Pedro ranch and there had let him go free. Four days later he had wandered into the camp of the Sonora Survey.

Upon hearing the German's account of the tragedy, a messenger bearing that information was dispatched to Fort Buchanan, and the erstwhile cook was taken to the fort the following morning for examination. There, the commanding officer placed him in irons and put him in the guardhouse, partly to protect him from the enraged citizens, and to keep him as a material witness against the peons should they be apprehended.

Meanwhile, William Williams, Sergeant Henderson, and the soldiers, had been taking care of the gruesome details at the scene of the tragedy. The body of Mr. Morse had been found much eaten by animals; the body of Frederick Brunckow, at the mine shaft half a mile from the house, nearly all eaten; and the body of James Williams, in the store, badly decayed.

"The stench was so great that all hands were made sick, but by the help of whisky and camphor they gave the deceased a good burial, although they could not put them together," stated the San Francisco *Bulletin* in its report of the tragedy, on August 21, 1860. The account of the multiple murder, as presented in this chapter, is drawn from that report.

On August 2, the commanding officer at Fort Buchanan received word from the commandant of the Santa Cruz district, in Sonora, that a former employe of Brunckow, one Jesus Rodriguez, had been in Cananea the previous day and had boasted of killing Brunckow. At the same time he had sold some of the goods stolen from the San Pedro mine and had told virtually the same story related by the German, David Brontrager. Other members of the murderous band, it was believed, had made their escape to Hermosillo.

Upon receipt of this information the imprisoned cook was set at liberty, and one chapter in the history of the San Pedro silver mine was closed.

After the triple slaying in 1860 the Brunckow property apparently lay idle until its purchase by Major Milton B. Duffield, first United States marshal of Arizona Territory. Appointed to that post by President Abraham Lincoln in 1863, Duffield served until April 11, 1866. Six months later, on October 23, he bought the San Pedro.

To say that Major Duffield did not enjoy the wholehearted love and esteem of all his fellow citizens is about the most charitable thing that may be said for him. At least, it was no great surprise to the residents of Tucson and Pima County when Duffield was slain at the former Brunckow mine on June 5, 1874. His slayer, Joseph T. Holmes, surrendered himself to the authorities, pleaded self defense, was found guilty of manslaughter and sentenced to three years in the Territorial prison.

Reviewing the case a week after the murder, *Arizona Citizen* pointed out that the Duffield claim had lapsed; that the mine had been relocated by other persons more than a year previous to the slaying, and since that location Duffield had been very violent in his threats toward any and all who dared to interfere with his rights.

"Before leaving Tucson on his last trip, we understand, he made his will and frequently expressed his determination to drive off whoever he might find on the mine. From the fact of his making his will it is evident that he intended to succeed in his purpose or die in the attempt."

Elsewhere in that same issue of the *Citizen* (June 13, 1874), appeared the following editorial comment:

"He (Major Duffield) very frequently marched through the streets like an insane person, threatening violence to all those who had offended him. It is claimed by some good men that he had redeeming qualities. Such may be the case, but we are free to confess that we could never find them . . . But his stormy life has ended and he has gone to that unknown land whence none return. May his ashes rest in peace, and may his Maker deal more kindly and gently with him than he was disposed to deal with his fellow men."

After Duffield's death the one-time Brunckow property continued to bring bad luck to nearly everyone who had anything to do with it.

One exception was Ed Schieffelin, a former Army scout at Fort Huachuca. After deciding he'd rather be a prospector than a scout, Ed happened to wander up to the Brunckow cabin at a time when Bill Griffith and a friend were down from Tucson to do assessment work on the old San Pedro. The pair persuaded Ed to stand guard for them against chance red-skinned marauders while they completed their work.

One day, as Ed was sitting on a pile of rock at the Brunckow mine, a rifle across his knees, he was greeted by his former boss, Al Sieber, who rode up with a scouting party. According to Walter Noble Burns, Sieber asked:

"What're you doin,' Ed?"

"Prospecting, mostly," Schieffelin drawled.

"Whar?"

"Over yonder," Schieffelin waved his hand eastward toward the hills.

"Them hills?" scoffed Sieber . . . "All you'll ever find in them hills 'll be your tombstone . . ."[6]

But Al Sieber was mistaken.

When Griffith and his partner finished their assessment work and Ed Schieffelin continued on his prospecting way, it was in those same brown hills to the east that he made the discovery which was to give the West one of her greatest mining camps.

That was one exception. Generally speaking, however, the history of the Brunckow mine and cabin grew bloodier with each year that rolled around.

Writing in the *Arizona Democrat,* of Prescott, issue of May 20, 1881, the Honorable Pat Hamilton stated that the "dark and bloody history" of the Brunckow mine, with its "uninterrupted series of violence and murder," had given it an unenviable notoriety all over southern Arizona.

"Bronkow [*sic*] and three companions were the first victims, assassinated by Mexicans. Since then, no less than 17 men have been killed on this property. The graves lie thick around the old adobe house. Prospectors and miners avoid the spot as they would the plague, and many of them will tell you that the unquiet spirits of the departed are wont to revisit . . . and wander about the scene which witnessed their untimely taking off. The gloomy old building is unoccupied. The present owners of the mine, not caring to reside there, have put up a house some distance away. Although much blood has been spilt and many lives lost over this claim, it has not proven of much value. It it still in dispute and, as I passed, I noticed two men standing behind a low adobe wall, armed with Henry rifles . . ."

[6]*Tombstone,* Doubleday, Doran & Co., Inc., 1929, pp. 4-5.

As the story of the Brunckow mine and cabin had fascinated me for a number of years, I was both surprised and delighted to learn that remnants of the old buildings could still be seen. It was Jean DeVere, of Tombstone, Arizona historian and writer (and a completely charming person, as well!) who first gave me directions for locating the old adobe ruins.

"Stop at the last cattle-guard before crossing the San Pedro bridge, on the Tombstone-Fort Huachuca road," Jean had said. "The remains of the cabin set on a little rise about 200 yards to the south . . ."

Leaving my car at the designated point, I had barely started afoot down the dry hillside until my eyes fastened on a small earth-gray ruin looming just ahead, across a shallow ravine.

Moments later I was standing beside the weather-worn walls of Frederick Brunckow's more-than-a-century-old cabin, the scene of so many dark and despicable deeds, a breeding ground for legends and folklore, a place where phantom forms had every right to be.

Although the highest portion of the standing walls was not sufficiently old to have been built by Brunckow (the window and doorframes being fastened with wire nails, which did not come into use until the last quarter of the 19th century), this part was adjoined on the south by a series of adobe walls crumbled away to a maximum height of about three feet. Here the wooden-work was fastened with crudely made, square-cut iron nails—exactly the type that would have been available in Arizona in the 1850s and '60s. All considered, there seems no reason to doubt that this is the original cabin built by the young German engineer who found in this raw new land both the gratification of success and the tragedy of untimely death.

After spending nearly an hour around the cabin and its environs—where I searched in vain for any indication of graves—I started back across the hillside toward the road. About a quarter of a mile to the east set a small cluster of buildings which I assumed to be untenanted, and I decided to swing around past them and have a closer look.

When only a couple of steps separated me from the nearest cabin, I was startled to have its door suddenly flung open by a woman who seemed equally as surprised to find a stranger at her threshold as I was to have her appear before me. After staring at each other speechlessly for a moment, we both burst out laughing. She was a nice, friendly-looking woman, pleasantly plump, with snow-white hair and a blue gingham apron.

"Are those old adobe ruins across the wash the remains of the Brunckow cabin?" I asked, chiefly because I couldn't think of anything better to say.

"That's it!" said the woman. "Are you the owner?" Recalling the high mortality rate of folks who had owned the Brunckow cabin, I hastened to assure her that I was not.

"You fixin' to buy it, then?"

Again I shook my head. My sole interest in the place, I explained, was historical curiosity.

"Well, I must say, there's been plenty of history made at that old cabin," said the woman. "Guess you know about poor Mr. Brunckow being murdered there? He was a real fine gentlemen," she went on, almost as if the cabin's builder had been a personal acquaintence, or even a favorite uncle. "He always wore a tall silk hat and rode a donkey . . ."

The pleasant lady in the blue-gingham apron also told me of another man who had owned the property at one time.

"He was quite a hand to shoot first and ask questions afterward." The woman raised her eyebrows significantly. "He always said, 'Gunpowder's cheaper than lawsuits!' Then there was a Mr. Gallagher who owned the place quite a long while; but I don't know who owns it, now. That's why I thought you might be the owner . . .

"Guess maybe you've heard about all the blood that's been spilled around that old cabin?" she went on. "Twenty-six men, folks say, have been killed there and around the collar of the shaft. Workmen have murdered their employers; partners have killed partners. Outlaws, hiding out at the cabin, have fought over their loot and killed each other.

"I tell you—" the woman shook her head—"this is bad country! This whole San Pedro Valley has been soaked in blood. White men's blood, Mexican blood, Apache blood . . ."

For a moment neither of us spoke. Finally the desert woman pulled her eyes away from the indistinct smudge of gray adobe ruins on the nearby rise and turned to peer quizzically into my face.

"What would you think if I told you that sometimes, at night, it seems to me I can hear them?"

"Hear who?"

"*Them*!" said the woman, fiercely. "The ghosts of all the poor souls who have been killed around here! What would you think if I told you that?" Before I could frame a reply, she was speaking again.

"It seems to come from the sky. It's like the most beautiful music anyone ever heard. Like a heavenly chorus—high and thin, and miles away. Sometimes it sounds like white folks singing; other times, it's more like Apaches chanting . . .

"When I tell folks about it and try to get them to come out here at night and listen to it, they just laugh at me. They say I've lived on the the desert too long. They say all I hear is the wind whistling over the wires of the power line, yonder. Some of them even look at me as if they think I'm a little crazy, or something. But you don't think so, do you? I can tell by your face that you don't think so . . ."

"No," I said. "I don't think that—not at all!"

And I *don't* think so, either.

If ghosts can sing—and who is to say they can't?—surely the San Pedro Valley must have enough of them to form several phantom choirs.

Harvey and Mary Riggs at well drilling machine abandoned on desert near Mexican border, 1960

Weathered adobes at "New" Oro Blanco, 1953

"Guy has started to school (at Oro Blanco). A Mexican is teaching in a hovel with . . . a brush roof. The children sit on the ground, or on boxes from the store. If it is cold they have embers in an old bucket which is moved from one to another . . . Dr. Noon has three boys attending, so Guy has some company of his own kind. They are the only other white people here . . ."

—Diary of Mrs. Elihu (Alice Lee) Barker, 1881

Ghosts of the Border Country

AWAY DOWN in the southern part of Arizona, near the point where the U.S.-Mexico boundary leaves its east-west course and cants off toward Yuma, the ghost mining camps of Oro Blanco and Ruby drowse in the pleasant hills, and the old town of Arivaca goes its peaceful way.

Arivaca, too, was almost a ghost town when a few "outsiders" began discovering it as a delightful place to escape the complexities of city life. Since that time several new homes have been built there and some new business houses opened, and now it seems doubtful if the old town will ever pass from the picture. Which doesn't mean that it isn't still pleasantly remote from urban stress, and as lovely a place as almost anyone could ask.

Flanked by the rugged San Luis Mountains to the southwest and the Las Guijas to the northeast, the little settlement lies in a narrow green valley watered by Arivaca Creek, a pretty brook shaded by towering cottonwood trees. The town's short, quiet streets are bordered by picturesque old adobes, so mellowed by Time that the whole place looks like a "set" for a Western movie. So far as that goes, Arivaca could provide for such a picture not only the setting and props but also the drama—complete with Spanish friars, soldiers and Indians, mining booms and badmen.

How many years a town may have existed at this site is something no man can know with certainty. Hodge states there was a Jesuit mission at Arivaca in 1720.[1] Bancroft says: "Aribac or Arivaca appears on a doubtful map of 1733 as a Pueblo." Near the Pima Indian town of Ali Bac, situated at or very near the present site of Arivaca, the Spanish mine known as San Angeles de Quivavi is said to have been located in 1739,[2] and in the Pima Rebellion of 1771 Arivaca was scene of a decisive encounter when 80 Spanish soldiers, under Capt. Bernardino de Urrea,

[1]Hiram C. Hodge, *Arizona As It Is*, 1877. p 19.

[2]Christine Emery, *Westways*, May 1948.

engaged in battle a force of 2000 Indians, of which number some 200 were reportedly left dead on the battlefield.[3]

Application for a Spanish land grant covering "La Aribac" is said to have been filed in 1777 by a man named Ortiz. This grant, apparently, was not given final approval until early in the 1800s when the government of Spain sold "two square leagues of land"—about 11,500 acres—to Don Agustin Ortiz for the equivalent of $747, or six and one-half cents an acre.[4]

For a long number of years the Hacienda of La Aribac appears to have been a prosperous settlement with large herds of sheep and cattle and several productive mines. In 1828, however, when the hacienda was ravaged by the Indians, its inhabitants slain and the missions broken up, Arivaca was abandoned and still was untenanted in 1854, according to a report by Lieut. Michler of the U.S. Boundary Survey.

Two years later, in December 1856, Charles D. Poston, acting for the Sonora Exploring and Mining Company, bought "the estate of Arivaca" from Tomas and Ignacio Ortiz for $10,000 in gold. According to the company's report, issued in September, 1857, the property purchased included some 17,000 acres. The Sonora Exploring and Mining Company then established at Arivaca a smelter for reducing ore taken from the Cerro Colorado or Heintzelman Mine.

"The reduction works of the Heintzelman mine were located on the Arivaca ranch, eight miles from the mine, and connected with it by an excellent road," states Farish. "The process used was the European barrel amalgamation for argentiferous copper ores, and was introduced by Mr. Kustel, a German metalurgist, about the year 1859. About one and a half tons per day was reduced. Six dry stamps, a steam arrastra, one reverbatory roasting furnace, four barrels, a retort, and one refining furnace, together with a 10-horsepower engine, constituted the works. This was the first time where the barrel process for the treatment of argentiferous ores was used in the United States . . ."

Another "first" for the camp was chalked up in 1860 when Samuel P. Colt, Eastern arms manufacturer who had succeeded to the presidency of the Sonora Exploring and Mining Company, established at Arivaca the first public library in what soon would be Arizona.

By the middle 1870s men were speaking of Arivaca as "a coming mining camp." Several Anglo miners, including Barney Conly, Pat O'Mera, and Mr. and Mrs. T. C. Hayes, plus a few Mexican families, had established here a small settlement, and a Mr. Favorite, of Tucson, had opened a general store.[5] On April 29, 1877, miners working in the Cerro Colorado and Las Guijas mountains met at the Van Nostrand camp and organized the Arivaca Mining District.

After formation of the district and election of officers, 16 of the men in attendance at the meeting went on to form the Arivaca Anti-Apache

[3]Harvey Riggs, *Arivaca Briefs*, March 15, 1961.
[4]Bernice Cosulich, *Arizona Daily Star*, May 10, 1935.
[5]*Arizona Citizen*, Feb. 26, 1876.

Military Company. Reporting on their home-guard unit, the *Arizona Citizen* of May 5, 1877, noted that the guns issued to the company were found to have no "needles" to fire the cartridges, but that William Spiva, the Van Nostrand blacksmith, had put the weapons into condition for use. Capt. Poindexter, captain-elect, then was presented with a sword in a speech appropriate to the occasion.

Arivaca's first official post office was established April 10, 1878, with Noah Bernard as postmaster. The business section then embraced six saloons, two bakeries, two restaurants, one butcher shop, one blacksmith shop, a tonsorial parlor, a brewery, and a small factory where Mexican blankets were made.[6]

During the next 80 years Arivaca experienced the recurring ups and downs common to decadent mining camps. During this era one of the town's most prosperous periods was between 1913 and 1916 when it was the headquarters of a contingent of United States troops, all Negroes, who were stationed at this point to repel any possible invasion by the guerrilla warriors of Pancho Villa. After the threat of invasion had passed and the troopers had been removed to Fort Huachuca, Arivaca gradually declined. Some of its buildings were destroyed by fire; some of its citizens died, others removed to livelier and more prosperous centers.

Tucked away in the backroads country, 40-odd miles from Nogales and 65 from Tucson, the nearest larger towns, with no industry, no mining of importance, no tourists—almost, one might say, no excuse for being—there seemed every reason to suppose that Arivaca was going down the road to the Land of Ghosts.

But Arivaca didn't die.

Whenever business or recreation took me out that way in the decade following World War II, I was always pleased to find a few old timers still shuffling up and down the quiet streets, still gathering at the single store and the post office, still talking earnestly of lost mines and missions and mines still unfound, of big deals in the yet-to-come and even bigger deals in the fading past.

I always enjoyed going to Arivaca. It gave me a good feeling to spend even an hour in this homey little town where life seemed less bewildering and neighbors still had time for leisurely chats. Although my conscience never demanded an ironclad "excuse" for a jaunt to Arivaca, I felt especially justified in going there on a certain February day in 1960 that saw me driving the familiar road west from Kinsley's Ranch.

During the previous year one of my favorite Nevada friends, Hallie Jones, had married John Mansfelt, a mining man, and the two had set up housekeeping at Arivaca, where John had been wintering for 15 years or more. Besides wanting to visit Hallie and her new husband, I was starting work on this book and needed some first-hand information on several points of historical interest in the Arivaca region.

Hallie and John made me feel as welcome as spring's first meadowlark after a hard winter on the plains.

[6]*Arizona Sentinel*, April 3, 1880.

First of all, they wanted to show me the new home they were building. Even though the walls were then only two-thirds completed, I could see it was going to be a nice place—just about the nicest home in town. Hallie took me all through the building, identifying the big, cool living room, the bedrooms, the modern bathroom and kitchen, and John explained about the insulation and how the woodwork would be finished, and the materials being used in construction.

Kiln-fired adobe bricks, weighing 38 pounds each, formed the main component of the walls. Due to the fact that adobe turns both heat and cold much better than most other materials, it has long been highly popular among builders in the southwestern United States and Mexico, and nowhere more so than around Tucson. In order to meet the mounting demand for this material, said John, one of the large builders' supply concerns in Tucson undertook the manufacture of adobes on a mass production basis. After spending $225,000 for equipment and in fruitless trails and errors, the company was convinced that it couldn't produce adobes equal to those made under the most primitive conditions by Mexican craftsmen whose advantage lay in long generations of experience. Result was that the company had struck a truce with progress, scrapped its costly set-up, and gone back to patronizing the *paisanos*.

Seventeen Mexican outfits, in and around the little town of Sasabe, Sonora, a short distance southwest of Arivaca, were then working night and day, I was told, to fill a contract with one Tucson concern calling for the manufacture of 1,500,000 of the 38-pound adobes. In addition to paying the Mexican contractors $120 per thousand bricks, the Arizona purchaser was paying a small import duty, plus $60 per thousand bricks to have them hauled from Sasabe to Tucson, roughly 70 miles, and was reselling the adobes to building contractors at $200 per thousand.

When Hallie learned of my interest in Arivaca landmarks she asked if I had ever seen the old open-air jail?

"I've never even heard of it!" I said.

"Come on, then!" said Hallie. "We'll have time to visit it before supper . . ."

In the pleasant quiet of approaching evening we strolled up main street past a few occupied dwellings, past the empty hulk of the big hotel burned years ago, and the shells of the old Army barracks erected in the days of Villa.

These barracks have fascinated me since the first time I saw them. When building with unfired adobe it is common practice to retard weathering by covering the exterior walls with a coating of plaster, but it would be interesting to know whose idea it was to decorate the plaster of these buildings so that it has the appearance of skillfully-cut stone masonry! I had known the barracks several years before discovering that their apparent "masonry" is only plaster-deep, and I often have thought that the story behind that harmless but time-consuming camouflage would be an interesting one to hear.

A little way beyond the barracks Hallie introduced me to the old jail.

Two iron bars, to which heavy log chains still were attached, had been embedded in a slab of cement set flush with the ground. Securely handcuffed to the chains, malefactors had been left out in the open to meditate on their sins—or to study astronomy, as they chose—and to serve as horrible examples to the small-fry of the town. Although Hallie and I agreed that this type of *al fresco* jail would effect a neat saving in rent, utilities, and guard service, it was our opinion that wintertime tenants must have found it a chilly place to roost.

After supper that evening Hallie and John took me to the upper end of town to meet Harvey and Mary Riggs, and their Mexican foster daughter, Maria, from Caborca, Sonora.

Mary Riggs hailed from somewhere in the Back Bay country, and nine-year old Maria was my first experience with a Mexican child who had learned to speak English with a Boston accent. Harvey, who informed me he was half Cherokee Indian and could prove that claim if anyone questioned it, was employed by Arizona State Highway Department. Most of his free time was spent probing into the past history of Arivaca, the Cerro Colorado mine, and the Sonora Exploring and Mining Company. He also was the editor, publisher and star reporter of *Arivaca Briefs*, a breezy, mimeographed, four-page news letter, which he mailed to subscribers in every state in the union and quite a few foreign countries.

Naturally as some consonant groups fall into discussion of politics, fishing, or the World Series, our conversation throughout that evening was spiced and salted with tales of lost mines and lost missions, outlaw caves and hideouts and buried treasure. Some of the stories I had heard before, others were new to me. One that fell in the latter category was John's and Harvey's tale of the Oceanic mine and its rumored treasure cache of $70,000.

The Oceanic, I was told, was situated down near the Mexican border, about a dozen miles southwest of Arivaca. At one time the mine had been owned by Charles Tully, of the Tucson freighting partnership of Tully & Ochoa. In addition to the Oceanic, Mr. Tully also owned an important cattle ranch in the same vicinity and frequently kept about his premises large sums of money, a not uncommon practice among cattlemen and others who oftimes found it necessary to transact business far from any banking facilities. At the time of his death, the Oceanic's owner is reputed to have had on hand the aforementioned $70,000 in cash—which was nowhere to be found.

"First I heard of this rumored treasure was when one of Mr. Tully's daughters, then living in California, began coming back to the ranch each summer on her vacation," said John Mansfelt. "Since the ranch and mine property, by then, had fallen into ruin and disuse, folks around here had it figured that the only reason the woman would come back was to search for her dad's supposed cache of money.

"One summer, quite awhile ago, the woman appeared in Arivaca with a small house trailer. She bought some supplies and said she was going out to the ranch to stay a month. Three days later, however, she came tearing back through town—and nobody here has seen her since! The supposition, of course, is that she found the treasure cache she had been hunting all those years so had no reason to remain longer, or to return . . ."

When I learned there were adobe ruins on the ranch and at the mine, as well as an old cemetery, I tossed out a broad hint relative to visiting the place. Hallie, being an understanding hostess, made the proper response: "How about tomorrow?"

After driving out the Sasabe road nine or ten miles John turned left on a little-used trail that led off through a natural cactus garden where acres of chollas bristled luxuriantly. With every mile traveled the trail grew more indistinct and confusing and, at times, came close to disappearing altogether. Exercising the casual skill acquired by veteran desert prospectors, John maneuvered the Jeep around rocks and other obstructions, and back and forth across San Luis wash, and finally halted in the empty yard of what had been the ranch headquarters.

Although no buildings remained at the immediate site, a caved cellar showed where the old house had stood, and holes dug by treasure seekers gaped everywhere. A few of these excavations were quite extensive, two of them half as large as the cellar. Most, however, were no more than scrubby little pot-holes pecked out of the hard earth at the base of giant mesquite trees and under big rocks. Some of the holes were located in areas where there seemed no possible reason for digging—but treasure, I supposed, might be buried anywhere.

"Several years ago I brought a California fella out here," said John. "He was one of these folks who make a business hunting buried treasure and he had a carload of fancy metal detectors and doodlebugs. We stayed out here most of one day and he ran his 'scopes all over the place. All he found were some rusty horseshoes and scrap iron. I guess the little gal had already found her dad's money . . . if there ever was any to find."

After we had finished prowling over the well-pitted treasure site, John led the way up a low rise to the old ranch-mine cemetery, a small plot densely grown to thistles and weeds and half-heartedly surrounded by a sagging barbed wire fence. As well as we could determine by careful examination the plot contained 13 graves, nearly all unidentified. Four of the graves were headed by badly-weathered wooden crosses, and centering the cemetery was a crudely-contrived cement tomb bearing the family name of Ortiz. The baptismal name in the inscription was badly weathered but appeared to be Felistio; and traced in the cement of the tomb, evidently before it had hardened, was the date, in Spanish: *Septiembre 4 de 1914*.

What kinship, if any, we wondered, had existed between this Ortiz and that other Ortiz who, nearly two centuries earlier, had filed application with the government of Spain for the original land grant at La Arivaca?

From the little pioneer cemetery we returned to the treasure site and continued across the wash to an adobe ruin, windowless and doorless. Strewn about the yard were various pieces of placer mining equipment including two gold rockers, a length of sluice box and several sections of heavy, rust-corroded screen, and in searching around the old place Hallie and I found some fragments of willow-ware and sun-purpled glass.

Another half a mile along the rutted road brought us to a second adobe ruin. Beyond this point the trail rapidly grew worse. Since John had never been beyond the treasure site he had no idea which of the several dim wheel tracks might carry us nearest to the ruins of an old mine whose abandoned workings scarred the hillside some distance ahead—a mine he believed must be the Oceanic. Also claiming our attention was a large adobe ruin, dimly visible against the brown foothills to our left. To a dyed-in-the-wool desert prowler, such a building as this appeared to be holds forth more enticements than any castle in Spain.

After the road petered out altogether, John broke trail through the desert until we came, eventually, to an arroyo too precipitous for even the valiant Jeep. Here we transferred to the most primitive of all transportation systems—the two strong feet given each of us at birth.

The adobe ruin, about half a mile from the mine, had been a large and imposing structure, two stories in height, with a ground floor area 72 x 36 feet. When the mine was in operation, Harvey Riggs told us later, this had been a boarding and rooming house for the miners who had bunked in one large unpartitioned room occupying the entire second floor. In this same room, on special occasions, had been held gala all-night dances attended by settlers who came by horseback and buckboard from miles around. The lower floor of the building had been occupied by kitchen, dining room, and company offices, Harvey said.

After exploring the old ruin from stem to stern and top to bottom, we continued across the ravine to the mine and millsite. John prowled all over the site, studying the layout to see how it had operated; Hallie picked up some nice quartz crystals for her cactus garden, and I found a man's jackknife. Evidently it had lain out in the weather a long time. Its old-fashioned stag-horn handle was beginning to scale away and its blades were bedded in rust.

Late that afternoon—happy and tired and temporarily surfeited with exploring—we returned to Arivaca where we found an out-of-town Jeep parked in front of Hack Townsend's general store. Lashed across its hood was the carcass of a large javelina. In addition to a wicked-looking snout and a pair of formidable tusks, the beast was equipped with a coat of coarse brownish-gray hair, about four inches in length. The hunter who had bagged the creature said it would weigh about 40 pounds.

Next day Hallie packed a huge picnic lunch and we were off on another trail. This time there were three pickup-loads of us, including Harvey and Mary Riggs and Maria—the little Mexican girl with the Boston accent; and Mrs. Riggs' father and a friend of his, and Mr. and Mrs. Fred Noon and their daughter, Mary. The Noons, like the Riggs, are enjoyable folk to be around; and Fred, as a third generation resident in the Arivaca vicinity, would be a definite asset to any expedition interested in local history and landmarks. On this occasion, our target was a place known locally as Geronimo's Hideout, a promontory about ten miles northeast of Arivaca.

After a few miles on the Amado road we turned off to the right and followed a sinuous course easterly into the hills. From time to time we had to let ourselves through wire ranch-gates and close them behind us. The road grew tougher and steeper. We forded a bright little stream—an unusual sight in the desert; and, by and by, we reached the point where pickups could go no farther. Across a wash to the south a rocky knoll rose steeply to a granite rimrock likely 300 feet higher than the point where we were leaving the trucks. This, said Harvey, was "Geronimo's Hideout."

Crossing the wash we swarmed up the steep hillside, slick with matured grasses and bristling with sotols, bear grass, and cacti.

Whether the wily Apache, Geronimo, ever climbed that same hill is open to question. Perhaps he did. According to Harvey there's a local belief to that effect; and there seems little doubt that after Geronimo's escape from captivity, and prior to his final capture in 1886, he and a few members of his band did hide out somewhere in the vicinity of Arivaca.

Certainly the top of this hill gave every indication of having been designed for use as a stronghold. Encircling its summit was a series of man-built stone breastworks, some partially fallen, others in good condition. Crouching behind these primitive fortifications we found ourselves with an unobstructed view over tens of thousands of acres of desert hills, valleys and plains. Nowhere in the immediate vicinity is there another peak so well situated as a lookout point and back in those benighted war days, prior to bombing planes, it seems likely that even a handful of men—given ammunition, food, water, and the protection of these stone breastworks—could have held off half a dozen regiments.

After returning to the pickups, where we brewed campfire coffee and ate ravenously of the good lunches provided, Harvey asked if I'd care to see some bedrock *morteros* and an old well-drilling rig?

This happened to be one of southern Arizona's priceless February days with Spring just around the corner and every blade of grass and dormant leaf-bud waiting to welcome it. On such a day almost anything outside four walls sounds to me like high adventure. Just being alive and able to roam the untroubled hills, with the sun warming my blood and clear air flooding my lungs, is adventure enough in itself. If an old adobe ruin or a deserted townsite—or even a well-drilling rig and an Indian mortar—is thrown in for good measure, then my cup, indeed, runneth over.

Crossing a wide, sunny slope, where winter range grass stood tall and dense, we dropped down to a small stream of water. Pawprints of bobcats and coyotes—or maybe they were foxes—were imprinted in the moist sand at the water's edge, and soon after we started up the slope beyond, three big mule deer bounded off through the junipers.

The well-drilling rig was like nothing I had seen. About the size and general shape of a farm wagon, it was heavily built with an iron framework and timbers a foot square. Some of the original timbers had been replaced with poles handhewn from ponderous mesquite logs. The tongue of the contrivance was almost entirely rotted away, its fittings crumbled to rust beneath our fingers. The rig had been powered by horses, at least six of them at a time, said Harvey. Drilling, he said, had been accomplished by means of an eccentric wheel threaded with a cable which lifted a heavy drill bit and dropped it into place in the hole, over and over again, same as a modern churn drill pounds its way through solid rock.

Only a few steps distant from the old rig Harvey showed us what is, undoubtedly, the last water well ever drilled by this primitive piece of machinery. A pebble dropped into the six-inch casing fell freely a long way before the faint sound of a splash came back to our ears. Although he has inquired of all the local old timers, Harvey said he had never been able to learn anything about the history of the rig, and why it was abandoned at this particular place will probably never be known.

Back at the little stream we indulged in generous samples of its icy water, having no fear for its purity in this clean, unpeopled land.

With Harvey leading the way upstream, we soon entered the mouth of a narrow canyon where the sparkling water of the little creek filled a series of still pools, or *tinajas,* strung along the canyon's floor like a necklace of blue crystal beads. In the granite bedrock beside one of these pools, at the foot of a small waterfall, Harvey drew our attention to a group of Indian *morteros* ground from the solid stone. There were close to a dozen of the holes varying in size from six inches to nearly a foot across at the top, and from mere dimples, barely begun, to bucket-sized pits whose depth testified to long years of usage.

It was interesting to stand there in the quiet old canyon, thinking of these Indian women of no-one-knows-how-many years ago, sitting on this cool granite bedrock, beside this same little stream. Tirelessly, hour after long hour, they had pounded their heavy stone pestles into these stone *morteros,* pulverizing their mesquite beans or acorns, or dried jackrabbit meat, or whatever potential food that Fortune and their own hunger-spurred initiative had brought to hand.

Those who had worked here in that long-ago time had been happy and sad, by turns. They had gossiped and laughed, and played tricks on one another, and discussed problems and husbands. Mothers had taught daughters the skills of good mortaring, and brides had labored exhaustingly but in vain to achieve the effortless proficiency of mothers-in-law.

And all the while this had been going on the bedrock *morteros* had been growing deeper and this same little stream had gone chuckling on its way.

"What would you like to do, today?" asked Hallie Mansfelt as we sat at breakfast next morning.

I laughed. "What I'd really like to do is visit Ruby!"

"Okay!" said Hallie. "We'll do it!"

"But how?" I asked. When I had driven past this ghost mining camp a week earlier the gate across the road leading into the town had been wearing a padlock that seemed stout enough to serve the main entrance to Alcatraz, and the fence had been posted with a blustering Keep Out sign.

"The caretaker," said Hallie, "is a friend of ours—and I happen to know he's in town today!"

So Hallie and I walked down the Sasabe road a piece, and back through the mesquites to a small cabin where we found three men in serious consultation over an ailing Jeep. One of the trio was John Bieleki, then caretaker of the somnolent mining camp of Ruby, a dozen miles southeast of Arivaca.

"Sure!" said John, when Hallie told him of my wish to visit Ruby. "I'll be glad to give you the 'key to the city.' "

The key, in this case, was more practically the key to the padlocked gate and, minutes later, the Mansfelts and I were on our way.

Seven miles along the bumpy graded road that winds across the high mesa country to the southeast of Arivaca brought us to the adobe ruins of Oro Blanco.

As I understand it, this isn't the original town of Oro Blanco, but sort of a new Oro Blanco—and even it isn't especially new.

The first Oro Blanco came into being in the early 1870s when a party of Tucson prospectors made locations on a rich vein of gold-bearing ore upon which much work had been done in years past. In some of the old cuts large oak trees had grown to maturity "and none now living seems to know when the works were abandoned," reported *Arizona Citizen,* December 27, 1873. At the time that report was written, 22 arrastres already were in operation.

"Many thousands of dollars have been lately taken out and the yield is uniform—from $60 to $300 a ton," stated the *Citizen.* "Of the 300 feet developed, the vein runs from two to three feet in thickness and is regarded as a first class poor man's mine . . . The camp consists of about 40 miners, and ten homes have been built, five of which are stone and adobe. The miners have no doubts about the value of the mines, and from all appearances a very large mining town will soon grow up there . . ."

The fact that the Oro Blanco mine gave promise of becoming a first-class money-maker, plus its proximity to the international border, soon gave rise to an A-1 imbroglio: Was the mine in the United States, as its owners and operators asserted—or was it actually in Mexico?

With both Arizona and Sonora laying claim to the Oro Blanco, it was finally agreed that the Prefect of Altar District, and General Safford, of Arizona, each should appoint a surveyor to meet on the ground and settle, once and for all, the matter of the mine's location.

After careful observation the two surveyors found the mine to be in Arizona, a fraction more than two-and-one-half miles north of the international boundary.[7]

But Old Oro Blanco, and "New" Oro Blanco, as well, now are reduced to crumbled heaps of adobe, window openings without glass, and weathered walls without roofs.

"That's the Austerlitz," said John, indicating an old mine ruin close beside our road on the right. In appearance it differed little from thousands of other abandoned mines in the Western United States.

Who the original owners or operators of the Austerlitz may have been is, again, something no man knows. Certainly it had been worked many years before one in its series of owners—presumably an Austrian—named it for the Moravian town where Napoleon defeated the combined armies of Russia and Austria. Possibly this even may have been the old Tumacacori mine, one of the earliest worked by the Spaniards in what is now Arizona. When the old workings were rediscovered at the time of the Oro Blanco excitement in the 1870s, workmen found here many indications of earlier development, including drill holes of an archaic type four inches square, primitive tools, several human skeletons, and two rawhide bags filled with silver ore that assayed $7000 to the ton.

Reaching Ruby we employed our "key to the city," let ourselves through the locked gate, and rolled down the last hill into the silent streets of this Santa Cruz County town.

Ruby is not an especially old camp, not as Western mining camps go—at least, not under this name. Original locations were made here in the 1870s, but for 30-odd years after its founding the settlement was known simply as Montana Camp, after the Montana mine, its chief producer. Throughout these years the camp's nearest post office was at Oro Blanco.

After the Montana was patented by Louis Zeckendorf, of Tucson, in 1907, things began looking up a bit. More work was available, the camp began to expand, and on May 1, 1910, the town's first post office was opened under the name of Ruby, the family name of Mrs. J. S. Andrews, wife of the first postmaster.

The Andrews had come to Montana Camp in 1895 at the solicitation of their friend, J. Knox Corbett, who wanted them to take charge of the general store then being operated by J. B. "Pie" Allen, who had owned mining camp stores all over southern Arizona. Andrews eventually bought the mercantile establishment from the Zeckendorf interests, and he and Mrs. Andrews continued to operate it for 17 years. Selling out in 1913, they left Ruby while still enjoying good health—in which respect they were more fortunate than several of their successors were to be.

[7]*Arizona Citizen*, April 11, 1874; and June 6, 1874.

Ruby in 1940 —*Courtesy Erle D. Morton.*

Ruby's tragedy-haunted store and post office, 1960

First of Ruby's ill-fated postmaster-storekeepers were John and Alex Frasier, brothers, who had purchased the store from Phil M. Clarke, taking possession January 1, 1920.

On the morning of February 27, that year, after the Frasiers had breakfasted in their living quarters in the rear portion of the store building, Alex went out into the store and began price-marking and placing on the shelves some newly-received merchandise. John, meanwhile, busied himself washing the breakfast dishes and making their quarters generally shipshape.

As he went about his morning tasks, John Frasier heard the front door of the store open and close. Almost simultaneously, he heard the sharp bark of a pistol. Curious, but not sufficiently alarmed by the sound to arm himself, the man set down a dish he was drying and stepped through the connecting doorway into the store.

Standing by the cash register were two local Mexicans, laborers at the Twin Buttes mine south of Tucson, who were known casually to the Frasier brothers as Lara and Garcia.

Lara's right hand held a pistol, still smoking, and on the floor at his feet lay the body of Alex Frasier. A thin trickle of blood oozed from a bullet wound in his head.

For a stunned and speechless moment John Frasier eyed the still body of his brother; then he found himself looking into the muzzle of Lara's gun, and Lara was ordering him to open the safe.

It was the Ruby storekeeper's last earthly act. As the heavy vault door swung open, Lara's pistol barked again and the second of the brothers slumped to the floor, shot through the head.

After stuffing their pockets with currency and silver, and ransacking the store for cigarettes and a few other items they wanted, the cutthroat pair headed for the Sonora border, only five miles distant.

Next person to enter the Frasier store that morning was José Cuesta, a respected Mexican of the settlement. Finding the place a shambles, with merchandise strewn over the floor, Alex lying dead and John unconscious, Cuesta fled the scene in terror, his cries arousing the camp.

Justice of the Peace Maloney immediately dispatched a messenger to the nearest telephone, at Arivaca, from which point the Santa Cruz County sheriff was notified of the tragedy.

Taken to the military base hospital at Nogales, John Frasier regained consciousness only long enough to describe the double killing and to name Lara and Garcia as the murderers. On the fourth day after the shooting the Ruby storekeeper succumbed to his wound.

Phil Clarke, former owner of the Ruby mercantile, accompanied by several soldiers from the garrison at Arivaca and a pack of bloodhounds borrowed from Frank Bailey, chief of police at Tucson, took up the trail of the killers.

Clarke would have done better to have taken the hounds and left Uncle Sam's troopers in camp. The dogs, at least, would not have been bound by the protocol and redtape which prevented the soldiers from crossing the international border, even though the murderers were known to be camped less than a mile ahead.

By the time notification could be given to General Pina, commander of the Sonora troops, the pursued pair had effectually lost themselves in the rough, cactus-grown hills south of the line.

Like many murderers, however, Lara and Garcia were not satisfied to let well enough alone. In the mistaken belief that their identity was unknown, the pair returned to Twin Buttes as soon as they had squandered the bulk of their stolen gains.

As it had been surmised they would do exactly this, The Law was waiting for them.

While Deputy Sheriff Holliday and Garcia were shooting each other to pieces in true Wild West fashion—both being killed in a brief but furious exchange of shots—Lara again made his escape. He is presumed to have fled back into Sonora and, so far as known, was never seen again in Arizona.

Prior to the Frasier tragedy, Ruby—nee Montana Camp—had been a law-abiding settlement, unvisited by any major crime through all the years of its existence. But trouble seems to breed trouble, and scarcely 18 months after the slaying of John and Alex Frasier, death was to strike again in that same store building.

Ruby's second major tragedy took place August 26, 1921, when a raiding party of border bandits swept into Arizona from Sonora and attacked the store.

In the fight that followed, both the Ruby postmaster and his wife, Mr. and Mrs. Frank Pearson, were killed, and Mrs. Pearson's sister (later, Mrs. Woody Gatlin, of Patagonia) was wounded. Snatching the Pearson's three-year-old daughter, (now Mrs. William H. Anderson, Jr., of Nogales), the terrified young woman fled up a canyon to the rear of the store and so succeeded in eluding the killers.

Seven or eight bandits are said to have been involved in the raid. Two of this number—Placido Silvas and Manuel Martinez—were captured. Upon conclusion of the men's trial at Nogales, the jury deliberated only 15 minutes before bringing in a verdict of guilty. Martinez was sentenced to death; Silvas, to life imprisonment.

While the two prisoners were being transported to the state penitentiary at Florence, in July 1922, the automobile in which they were riding was involved in an accident and overturned near the small town of Continental, about 30 miles south of Tucson. Sheriff George J. White was killed in the crash and his deputy, L. A. Smith, suffered a fractured collar bone and several broken ribs, as well as internal injuries.[8]

Martinez survived the crash to keep his scheduled date with the executioner. Silvas subsequently escaped from prison and, presumably, returned to Mexico.

[8]*Arizona Citizen*, July 14, 1922.

Thus were concluded the two most spectacular events in the history of Ruby—events publicized far more widely by the national wire services, and remembered much more clearly today, than the production at Ruby of several million dollars worth of ore.

Upon entering the little ghost town, now drowsing so peacefully in the winter sun, our first stop was at the old building that had housed the Ruby Mercantile company and post office, where the two double tragedies had taken place nearly four decades earlier.

As in many small mining camps and other country towns, the post office had occupied one corner of the general store. It was still there, looking much the same as it must have looked throughout its active life. Over its window still hung the sign:

POST OFFICE, RUBY, ARIZ.

In the main portion of the store set several large pool tables, and on the rear wall we found where a portable screen had been hung when moving pictures were shown here on Saturday nights. Many of the bins and drawers and most of the shelving was still in place, but everything was befouled by packrat litter and seemed to carry the smell of neglect—and death.

Only a few yards from the store set the old jail—a small, stout building, of evil appearance. Although the single window was much too small to enable the escape of even a midget it was guarded with heavy wire screen, and the door was padlocked. Before this jail was built, I have been told, drunks were chained to a mesquite tree.

Exploring the three classrooms at the school house, we found a number of old desks, ink-stained and name-carved, and an assortment of textbooks cast aside at conclusion of the school's last term. Blackboards and primary-grade cutouts still lined the walls, but every nook and cranny of the rooms was littered by packrats and a drawer in one of the teachers' desks held a thriving family of bright-eyed baby mice.

From the school we crossed a shallow ravine to the hospital and company offices, all standing open and abandoned; and, finally, we climbed the hill to the hoisting works of the old Montana Mine.

Throughout the history of this camp, under both its names, the Montana always was the "big" mine. Although a small mill was erected and some production made between 1891 and '93, nothing especially important happened until around 1916 when the mine was leased by Goldfield Consolidated. During its three years of operation here, this company is said to have taken out some $265,000 in gold, silver, and lead. In 1926 the Montana was optioned to Eagle-Picher Company which built a 400-ton flotation mill and, prior to 1940, mined and processed more than 773,000 tons of ore, realizing from the operation an asserted $7,000,000. Between 1934 and 1940, when my friend, Erle D. Morton, of Tucson, was general manager of the Montana, it was recognized as the largest producer of lead and zinc in the state of Arizona.

These were Ruby's halcyon days. It was then that her population attained its peak of approximately 1200, with 300 men employed in mine and mill and close to 150 children enrolled in the local school.

During those same years the town knew the luxury of electric lights, had its own baseball team, and domestic water was piped 15 miles over the hills from the Santa Cruz River.

For six good years The Ruby glowed brightly.

Shortly before the onset of World War II, Eagle-Picher concluded its operations at Ruby and in 1942 sold the property—including the Montana Mine, mill, townsite, and most of the buildings—to Hugo Miller, mining engineer and assayer, of Nogales. At the time of my visit to Ruby, in 1960, Mr. Miller was still its owner.

From the mine we were afforded a good view of the entire town and its environs, including two small lakes nestled in the canyon almost at our feet. These had served originally as millponds, said John Mansfelt. Their water had been pumped through the mill, allowed to re-collect in the ponds, and pumped again. Every drop had been used and re-used until it was eventually lost through the natural processes of evaporation. In those busy times, with the mill handling an average of 390 tons of ore daily, seven days a week, the water in the ponds was always muddy and yellow, but now it was as clear and blue at that of any mountain lake.

On our way back to Arivaca we met Ruby's caretaker returning to his post. When we surrendered the key and thanked him for letting us visit the old camp, he said he was glad we had enjoyed ourselves.

"If you come out again you should bring your fishing tackle," he said. "The ponds are full of bluegills and crappies. You can catch a skilletful of them in only a few minutes . . ."

And then John Bieleki gave us a jaunty little salute and continued on toward his ghost town home, and John and Hallie Mansfelt and I headed back across the rolling hills to Arivaca and supper.

"The road leading to the (Mowry) mine from Fort Buchanan crosses a range . . . completely covered with oak, pine, sycamore, poplar, willow, and hazelnut. The . . . hills around the mine are covered with green oak, cedar, pine, and manzanitas. The whole country abounds with rabbits, quails, and wild turkeys. It is not a rare occurrence to meet droves of deer and antelopes . . . The amateur of more intense excitement may indulge in bear and Apache hunting."

—*Report of* F. Biertu, *Mining Engineer, 1861.*

Phantom Camps of the Patagonias

In the Coronado National Forest, south of Patagonia and almost within shouting distance of the Mexican border, are mining locations past all counting and not less than half a dozen ghost camps where mining or milling once was the Be-All and End-All.

One of the more interesting of these Santa Cruz County camps is Harshaw, founded in 1879 by David Tecumseh Harshaw, a former cattleman. Several years earlier, when Indian Agent Thomas J. Jeffords had demanded the removal of Harshaw's livestock from range assigned to the Chiricahua Apaches, Harshaw had transferred his ranching operations from the San Pedro Valley to this well-grassed and well-watered district then known to its Mexican residents as Durasno. Having some knowledge of mines and minerals—early newspaper accounts include his name among those persons who were mining successfully in the Santa Rita placer fields as early as 1875—Rancher Harshaw soon discovered the mine destined to provide the lifeblood of Harshaw City.

Development, started in the summer of 1879, was stepped up in 1880, and a lively mining camp began taking shape. On February 22, that year, *Arizona Star* reported 30 men were employed in grading a site for a new 20-stamp mill; also that the town already had seven saloons going full blast. John W. Davis and L. C. McBain had opened general stores; there were two butcher shops, a livery stable and two boarding houses, with several more grogshops and three additional stores, soon to open.

"New buildings," stated the Harshaw correspondent, "are going up in all directions, some with fine cut-stone fronts that would shame anything you have in Tucson."

By May, 1880, the camp had a lively newspaper, the *Arizona Bullion*, established by C. D. Reppy and Co., and midsummer found the town's main street three-quarters of a mile in length and built solidly on both sides with good buildings of adobe, frame, and stone—the latter predominating. Seventy-eight men, by that time, were on the payroll of the Hermosa Mine, situated on Hermosa Hill, two miles south of town. The ore was averaging $95 a ton in silver, the mill was operating with full satisfaction,[1] and in September, that year, Wells Fargo stage accepted responsibility for delivering ten bars of silver bullion, the first shipment made from Harshaw Camp.[2]

The metallurgical goose was still hanging high on that May day in 1881 when a reporter for the *Arizona Weekly Star* again visited Harshaw and was agreeably pleased to find there ten stores, several lodging houses and hotels, corrals, and blacksmith shops.

"Every other house around the town seems to be a saloon," noted the peripatetic newshawk, who also observed that the camp had a thriving and prosperous air, and that its trade with Sonora and neighboring mining camps was large. "New buildings are being erected," it was stated, "and a feeling of confidence pervades the community."[3]

Scarcely more than half a year later, strange to say, Harshaw was a ghost town.

Although several causes contributed to the town's untimely end, all might have survived if only the Hermosa and other mines of the vicinity had continued to put forth treasure.

The first blow came in the midsummer of 1881 when the Patagonia Mountains were subjected to a series of cloudbursts of torrential proportions. Plunging down the canyons at express train speed, each of these recurrent storms caused material damage to the town. The Harlow store was flooded, causing a stock loss of some $1500; the backwall of Goldberg & Sons big store was swept away and stock ruined, and many others sustained lesser losses.[4]

Undaunted by these events a second newspaper—D. L. Sayre's *Patagonia Sentinel*—made its debut in Harshaw only a week after the cloudburst.

With the devastating floods still painfully fresh in the memory of all, the town received another jolting blow when one section was swept by a major fire. Among the buildings destroyed was Henri Beauford's St. Charles Hotel, valued at $20,000; John Busher's livery stable, and several small shops and dwellings.[5]

And before the end of that same year of catastrophes, all the principal mines of Harshaw area began running out of ore and curtailing operations. Greatest of these mines, the Hermosa, had been in constant operation for 18 months with an average daily employment of 200 men. During that period the company had shipped $1,300,000 in bullion and its script had risen from a par value of five dollars to $49.50. After the

[1]*Arizona Weekly Star,* May 29, 1880.

[2]*Arizona Bullion,* quoted in *Pinal Drill,* Sept. 11, 1880.

[3]*Arizona Weekly Star,* May 5, 1881.

[4]*Tombstone Daily Nugget,* Aug. 6, 1881.

[5]*Tombstone Daily Nugget,* Oct. 30, 1881.

mine and mill closed and all employes were discharged (with the exception of about 25 men retained to prospect the mine in hope of finding additional ore), that same script dropped to $3.50, with no takers, and the town emptied like water pouring through a broken dike.

"Of 200 buildings, four-fifths are unoccupied and are almost valueless for any purpose whatever," wrote the Harshaw reporter of ***Tombstone Epitaph,*** January 16, 1882. "Windows smashed and doors standing open. One of the largest buildings in town, with fixtures complete and a No. 6 Herrings burglar-and-fireproof safe . . . was sold, lot and all, a few days ago, for the paltry sum of $350. The hills, adjacent to the town, have been denuded of the . . . beautiful trees by which they were adorned, and the birds that were wont to sing to us . . . have departed . . ."

Before the close of 1882 the only business concerns still operating in this town, where a population of 2000-plus had shriveled to only 70, were the Palace Restaurant and Jamison's Saloon. All other business places were closed and boarded up, and the camp was said to present "a most desolate appearance."[6]

More than 80 years have come and gone since Harshaw's obit was written. But, even though the camp is still a ghost town, I wouldn't say that it presents a "desolate appearance." Actually, it seemed a very pleasant and peaceful spot on that autumn day in 1957 when Mom and I made a pilgrimage to its ruins and talked with the few old timers who still called it home. We were shown the village church and a cut-stone building that had housed a saloon and dance hall, and we looked around some of the vacant adobe cabins, and visited the weedy little graveyard where most of the names are Spanish and the dates are of long ago.

With our respects paid to Harshaw and her named and nameless dead, we continued through the oak-forested canyons and hills toward Mowry and Washington.

About a mile from Harshaw we passed through the town site of Hardshell. Never very important as a town, it isn't, now, very important as a ghost. Only three small empty cabins and one fairly large adobe house remained at the site—but a heap of rubble and cast-offs behind one cabin rewarded us with a nice sun-purpled bottle and a brown-clay doorknob.

Another three miles put us in the ghost town of Mowry where we were delighted to find four large buildings of frame and adobe. One, evidently, had been a rather sumptuous residence; another looked as if it might have housed the company store and office, and the others appeared to have been sleeping and eating quarters for the men. That this many buildings remained was a pleasant surprise, since Mowry Camp is one of the oldest of the old . . .

History of the Mowry Mine—originally the Patagonia—is a complicated and controversial thing including mismanagement, Indian attack, politics, both clean and slightly-soiled; a duel fought with rifles; jealousy,

[6]*Tombstone Epitaph,* Dec. 16, 1882.

Former dancehall, Harshaw, 1957

Ruins, Mowry City, 1957

litigation, high-handed theft, and the arrest of the owner on charges of treason against the government of the United States.

Yet, withal, the Mowry Mine produced well over a million dollars worth of silver in barely four years time.

There is a persistent story, perhaps without foundation in truth, that the Patagonia Mine was first worked and abandoned during the days of the Jesuits then rediscovered about 1857 by two Mexican prospectors. These men sold or traded their prospect to a group of Army officers at nearby Fort Crittenden. The purchasers included Capt. R. S. Ewell, then in command of the United States troops at Crittenden and later to be Major General Ewell of the Confederate Army; Col. James W. Douglass, Lt. I. N. Moore, Lt. Richard M. Lord, and two men who have come down only as "Mr. Randal and Mr. Doss." Of the lot, only Doss and Col. Douglass were not connected with the Army.[7] Later, Lord and Doss sold their interest in the property to Elias Brevoort who resigned his postmastership at Fort Buchanan in order to become superintendent of mining operations.

Brevoort did not prove to be a good manager. With the mine operating unprofitably and its owners unable to agree on any major issue, the whole unhappy matter was resolved in 1859 by selling the property to Lieutenant Sylvester Mowry, a fellow officer, for $25,000.

Mowry immediately renamed the mine for himself and, resigning his commission, poured all his energies into developing his property and marketing its product. This, in itself, was no sinecure in that raw, sparsely-settled, and Indian-scourged land, where the first railroad was still 30 years in the future, and even wagon roads were scarcely more than pack trails through the wilderness.

Continuing the main shaft of the mine down to more than 200 feet, Mowry made connection with the old Mexican ladderways, ran many drifts and crosscuts, and began taking out amazing quantities of galena ore, a large part of it brought to the surface on the backs of Mexican peons.

In his book, *Arizona and Sonora*, sub-titled *The Geography, History and Resources of the Silver Region of North America*, published by Harper & Brothers in 1864, Mowry notes that ore from his mine, in the period 1860-61, was reduced in common Mexican blast furnaces.

"There are twelve of these furnaces at the reduction works, six of which are run alternate weeks," wrote Mowry. "The yield is, of course, lead and silver, which is shipped to Europe in bars weighing about 70 pounds each. These bars sell in England at from $200 per ton upward, giving a clear profit, over all expenses—mining, smelting, freight, insurance and commissions—of over $100 per ton. A portion of these bars are refined at the mines in the English cupel furnace (the Mexican *vaso*), to supply silver for the payment of current expenses. The silver is moulded into bars, from $2 up to $300, and is a ready and convenient

[7]*Report of F. Biertu, Mining Engineer, 1861.*

circulating medium in a country where coin exists only in the memory of some individual who has been to California. Twenty-five tons of the Mowry ores were sent to Europe as specimens in 1862. The result was an offer of £50 sterling ($250) per ton for the ore as it ran, properly cleaned . . .

"The yield of the mines with the present furnaces, when all are in operation, is about $4500 per week of silver," wrote Mowry.

The mine's yield of lead, at that time, could not be sold profitably on the world market inasmuch as the teaming freight from the smelter to Guaymas, Sonora, the nearest point where it could be loaded on vessels and shipped to San Francisco, was 12½ cents a pound—more than the material was worth. The lead, or litharge, however, was not a complete waste, being sold in Sonora to be used as a flux at such mines as the Bronces, Cruzecitas, Mina Prieta, and others containing refractory ores. Ox teams and mule teams came hundreds of miles to take delivery of the smelter refuse, and even at a penny a pound Mowry estimated that the sale of this product, alone, would pay all the expenses of operating the mine.

In addition to whatever he was doing at the time—soldiering, mining, or running for Congress—Sylvester Mowry was an indefatigible writer and speaker on the subject of Arizona, its resources and possibilities.

In the course of one such address, delivered before a distinguished audience in one of the Eastern cities, Mowry remarked that the streams of Arizona teemed with fish. The statement, in itself, was inconsequential enough, except that it backfired.

At the old town of Tubac, in Santa Cruz County, was then being published a weekly newspaper, *The Arizonian,* subsidized by the Sonora Exploring and Mining Co., and the Santa Rita Silver Mining Co., and edited by Edward E. Cross, who served in a dual capacity as Arizona correspondent of the *St. Louis Republican.* There had been no love lost between Cross and Mowry, and the frontier editor-reporter saw in Mowry's somewhat exaggerated assertion a marvelous chance for ridicule and revenge.

Writing in the *Republican* over the pseudo-signature "Gila," Cross scored Mowry unmercifully and concluded with the statement that only one kind of fish had as yet been discovered in the streams of the Territory. It was, he wrote, about as long as a man's finger and was generally known as The Mowry Trout.

Being called a liar, even in silky language, was something the lion-hearted sons of a century past refused to take kindly, and Mowry immediately challenged Cross to a duel.

Having opened his mouth and put his foot in it, so to speak, Editor Cross could do no less than accept the challenge, and it was arranged that the *passage d'armes* should take place at a location near Tubac, on July 8, 1859. Seconds to the principals were George D. Mercer, acting for Mowry, and John Donaldson, U.S. Custom's officer at Calabasas, for Cross; the weapons, Burnside rifles at 40 paces.

Day of the duel found a large gallery assembled for the event. Sporting men from Tucson, laborers from the nearby mines, and ranchers from the Santa Cruz and Sonoita Valleys, all had gathered to witness the affray and, perhaps, to lay a bet or two on the outcome.

Inasmuch as neither of the combatants was a gunman in any sense of the word, their first three shots were exchanged without effect. On the fourth shot Mowry's rifle failed to discharge. When it was decided that he was entitled to another shot, for free, as it were, Editor Cross stood with resolutely folded arms, waiting to receive the lethal ball.

Since firing at an unarmed man was regarded as terribly bad sportsmanship—almost on a par with shooting at a sitting duck—Mowry gallantly discharged his rifle in the air, shook hands with Cross, and declared himself satisfied.

Several leading newspapers of the Eastern United States, as well as the *Weekly Arizonian,* at Tubac—Cross's paper—subsequently carried in their columns a "card" of apology. After setting forth details of the duel, the notice stated:

"Mr. Edward E. Cross withdraws the offensive language used by him and disclaims any intention to reflect upon Mr. Mowry's veracity or upon his reputation as a gentleman in any publication he has made in reference to Arizona.

"Mr. Mowry, being satisfied from personal explanation that he has done injustice to Mr. Cross's character and motives, in his letter to the press of July 2nd., takes pleasure in withdrawing the imputations against Mr. Cross as a man of honor and veracity contained in that letter. Any difference of opinion which may exist between them, in reference to Arizona, is an honest one, to be determined by the weight of authority."

SYLVESTER MOWRY
EDWARD E. CROSS.[8]

Such copious apology should have seemed sufficient to close the matter, but the Mowry-Cross imbroglio evidently carried some benefits in favor of the former.

Only two weeks after the duel the duties of editing and publishing the *Weekly Arizonian* passed from Cross to Sylvester Mowry.

"With this number my connection with the *Weekly Arizonian* terminates," wrote Editor Cross in his swansong. "A change in its proprietorship involving a change in its character and tone, is one reason for this step. During the period I have edited the *Arizonian* it has been my object to labor for the best interests of Arizona, honestly and faithfully. Time will prove the right or wrong of my exertions. At present I would not alter a line that has been written. Some kind friends have cheered me,

[8]*The Arizonian,* July 15, 1859.

some people found fault. To the former I am thankful. To the latter I have no apologies to offer. Conducting a newspaper in a frontier country is always a perilous, precarious, and thankless task. It has been especially so in this case. Against many difficulties and discouraging circumstances I have sought to furnish a readable paper and, I trust, with some show of success. Those who succeed me have my good wishes for prosperity in the arduous enterprise they have undertaken . . ."[9]

During the next three years Sylvester Mowry continued to operate and further develop his highly-successful silver mine at Mowry City, soon running its average yield up to $1000 per day. He also found time to dip into politics and to make occasional trips back to "the States" where he lectured before geographical societies, explorers' clubs, and other select groups—always on the subject of Arizona and Sonora and their wonderful potential.

Meanwhile, the Civil War had got under way, and in that conflict lay the beginning of the end for Mowry.

It has been said that the Arizona publisher-miner was sympathetic toward the Confederacy and the Southern Cause. Yet, he was not a Southerner—he was born in Rhode Island. Furthermore, my considerable research into the matter has not turned up one piece of actual proof that he was not a loyal Unionist. Be that as it may, Sylvester Mowry was arrested June 8, 1862, under orders of General Carleton, on charges of treason against the United States government. The specific charge: That he had given aid and comfort to the enemy by furnishing it with lead from the Mowry Mine for use in the manufacture of bullets.

Taken to Fort Yuma, in California, Mowry was confined as a prisoner of war, and all his property, including the Mowry Mine—then producing $1000 per day in silver—smelter, workmen's quarters, et cetera, was confiscated by the federal government.

After being closely confined at Fort Yuma until November 4, 1862, Mowry was then unconditionally released, by order of the United States War Department, there being no evidence against him, either oral or documentary.[10]

With Mowry's complete exoneration and release, it would have seemed only a matter of common justice that the valuable Mowry silver mine should have been restored immediately to his ownership and indemnity paid for the time it was out of his possession.

But the mine was not restored. In the spring of 1864 Mowry brought suit against Generals Carleton and West for $1,000,000 for damages sustained in his arrest and the confiscation of his property; yet, even with the suit pending, the following notice appeared in the June 28, 1864, issue of *Rio Abajo Weekly Press*, of Albuquerque, New Mexico:

[9] *Arizonian*, July 21, 1859.

[10] *Journal of the Senate of the United States*, June 13, 1864.

U.S. MARSHAL'S SALE OF CONFISCATED PROPERTY

Condemnation and sale made by the United States District Court of the Second Judicial District of the Territory of New Mexico at the May term thereof, 1864 . . . I will on Monday, the 18th day of July 1864, in the public plaza of the town of Mesilla, county of Dona Ana, offer for sale to the highest bidder . . . the real estate known as the Mowry, or Patagonia, silver mine, together with all buildings, machinery, and lands thereto belonging, situated in the county of Dona Ana. Also sundry chattels, goods, effects, wares, and merchandise . . . The above mining property is considered as rich as any in the United States, and is fully supplied with all the machinery necessary to work it. It is now being worked with good results . . .

The sale took place as scheduled, and the mine was sold—to United States Marshal Cutler.

The price paid for this extremely valuable property, oddly enough, was only $4000.

Following the forced sale of his mine, Mowry launched a long, costly, and heartbreaking battle to regain title of his property, or to secure reparation. Whether Mowry succeeded in those efforts is a disputed point. Farish states that after exerting every possible effort to gain reimbursement for his seized property, the former Arizonan died in London, England, in 1871, in poverty. Historian McClintock, on the other hand, declares that Mowry received from the federal government, in 1868, a token payment of $40,000 indemnity.

If McClintock, rather than Farish, is correct in the matter, payment of even $40,000 still would have covered but a bare fraction of the true worth of the rich mine and smelter property.

Soon after Sylvester Mowry's death the mine fell into disuse, eventually being relocated by a group of Tucson merchants. Through trading and purchase the "Old Mowry," in time, came into possession of S. Silverberg, San Francisco financier, and A. Steinfelt, pioneer merchant of Tucson. More years passed when little or no work was done, neither production, development, nor maintenance. Shaft and manway caved in, drifts became choked with waste, buildings fell to ruin.

During the course of more than 100 years since its seizure, the Mowry has been worked from time to time by owners or leasers, and by various methods. Never again, however, has it produced so abundantly or held forth such wonderful promise, as during the few great years when it was owned and operated by Sylvester Mowry. During those years an estimated $1,500,000 in silver was taken from its vaults.

On a low knoll, not far removed from the old adobe buildings and other ruins that mark the site of Mowry City, is a weedy little graveyard, untended, unkempt, unhallowed. All but a couple of the two dozen-or-so white men assertedly buried there are said to have succumbed to violent death—chiefly at the hands of Apache Indians.

Though his mouldering bones do not rest in the rocky soil of this mining camp version of God's Acre, it does not seem too far-fetched to believe that if ghosts ever walk these precincts, there stands tall among them the phantom of persecuted Sylvester Mowry—one of the most brilliant and able men of Arizona's formative years, a loyal, energetic, and persevering worker for everything he believed would benefit his adopted state.

From the ruins of Mowry City, Mom and I drove south another four miles through the oaks and junipers of the Patagonia Mountains to the ghost town of Washington. Here we found about as many vacant buildings—and as few occupied—as we had left at Mowry. In addition, there was a small country store, obviously in operation although closed at the time of our visit which fell on a Sunday. Across the ravine back of the store we prowled around what appeared to be an old machine shop, and investigated a large dump of jet black smelter slag. While in its molten state the slag had flowed down the side of the ravine, like black lava, hardening in all manner of intricate loops, bubbles, and bumps.

Mowry City had been a ghost town more than 20 years before Washington came into existence. Although several prospects had been discovered in this vicinity as early as 1862, the region was plagued with recurrent Apache raids and development of the property had lagged until 1880 when a mining location, ambitiously named The Bonanza, was staked by Thomas Shane and N. H. Capan. Washington Camp began taking form but still did not prosper too well until around 1889 when the Duquesne Mining and Reduction Co., of Pittsburgh, acquired several promising local properties and founded the town of Duquesne, a mile or so to the southwest.

One of the greatest of the Washington mines was the Pride of the West, a copper producer, located in 1880 and owned for a time by Senator W. A. Clark, of Montana and Jerome fame. In only eight years (1899-1907) the mine is said to have produced some $1,400,000 in copper and lead.[11]

At their peak, around 1900, Washington and Duquesne each had a claimed population of close to 1000—but not for long. Though considerable mining was done in the vicinity, even as late as the early 1950s (chiefly by W. C. Humphrey and his son, and C. O. Byrd and A. R. Byrd, Jr., who still owned the property at time of our visit to the old camps) both towns are long-since ghosts—Washington having given up its post office in 1890, and Duquesne in 1920.

Our pilgrimage to the ghost town of Duquesne revealed to us a pleasantly-wooded site scattered with a few vacant adobe cabins, many empty foundations, a stone chimney or two, and an abundance of cast-offs from 80 years of intermittent mining.

There is a persistent story that George Westinghouse, of the famous Westinghouse Electric Co., once was a resident at Duquesne where

[11]Dunning.

some of his people assertedly owned mines. According to the story, young George created quite a furor when he equipped his Duquesne home with fancy bathrooms complete with hot and cold running water—an effete touch almost unheard in that land of little rain and fewer streams.

Unfortunately, I have not succeeded in finding any documentary proof that George Westinghouse ever lived in Duquesne—but I'm still hoping to find it.

To separate the history of Washington and Duquesne is almost impossible. Children of both towns attended the same school, built midway between. The two camps fought together and laughed together; they played ball together, and whatever was the fortune of one was the fortune of both. In contemporary news accounts, the two names, Washington and Duquesne, are linked almost as inseparably as ham and eggs.

When I mentioned this characteristic to an old timer at Patagonia, he chuckled.

"Yep!" he agreed. "That's how it was! The two camps was so close together folks used to say that when Duquesne's tail was stepped on, it was Washington that barked!"

About 20 miles east of Duquesne lie the Huachuca Mountains . . . and somewhere in the Huachucas, according to legend, is the site of the General Booth mine.

The story of the Booth mine fascinates me, despite the fact that I can offer no bona fide proof that such a mine ever existed. The only time it has come to my notice was by way of a worn newspaper clipping in the files of Sharlot Hall Memorial Museum, at Prescott. Dated November 1894, but otherwise unidentified, the clipping reads:

"Way up in one of the most inaccessible portions of the Huachuca range of mountains there is situated, perhaps, the most curious mining camp in the world. The 16 men who daily toil in the General Booth mine are all members of the Salvation Army, and the profits of the mine all go into the treasury of that organization. Old Dick Taylor, the discoverer, was one of the best known and most successful prospectors in the Territory. He was, perhaps, the most profane man in the Southwest. His vocabulary of invective was something phenomenal and was brought into use on the slightest provocation. It is said that after each sale, Taylor would invest in a year's grubstake, then spend the remainder of the money in the wildest debauchery. At one time he ran through with $10,000 in two weeks at Phoenix, throwing the money away in the most reckless manner. His money gone, he returned to his prospecting and touched no liquor until his next sale.

"It was immediately after the sale of the Apache mine, for $15,000, that he started on his wildest debauch in Phoenix and at last, reduced again to poverty and weakened mentally and physically, he professed religion at an open air meeting of the Salvation Army in Phoenix. For over two months he marched with the Army at its nightly meetings; but, finding such routine life too irksome, he again started for the hills. Nothing was heard of him for over two months, when, one day, he again turned up in Phoenix with a burro-load of rich ore which set the town ablaze with excitement. Some of the nuggets which he brought were valued at $50. Everyone was surprised when he announced his intention of turning his latest find over to the Salvation Army. It was accepted, and 15 men volunteered to develop it.

"Work was commenced over a year ago, under the direction of Taylor, and has progressed steadily since. The ore is treated in two crushers erected near the mine, and it is proposed soon to bring in a 10-stamp mill. Strict religious discipline is maintained in the camp, and the profits, after paying the necessary expenses, go into the treasury of the Army, to aid in the work of that organization. Two shifts, of six men each, work underground, while the remainder attend to treatment of the ore and transportation of necessary supplies. Supplies are freighted 100 miles."

Keeping the foregoing story in mind, it is interesting to note the general similarity—and, yet, the specific differences—between it and Byrd Granger's story of the mining camp of Sunnyside, also in the Huachuca Mountains, of Cochise County.

"In the 1880s, a Scotsman named Sam Donnelly was one of the better patrons of the tough waterfront bars in San Francisco," wrote Mrs. Granger. "His regeneration began when he overheard men plotting skullduggery and suggesting that Scotty accomplish the deed at the cost to them of a drink of whiskey. Scotty saw through his alcoholic haze clearly enough to give the men a tongue lashing, after which he wandered out to stumble into a Salvation Army meeting. He became a Salvation Army officer and street preacher whose path finally took him to Tombstone. There he ran into a man named Sinclair, the locator of the Copper Glance (sometimes incorrectly referred to as the Glance) Mine in the Huachuca Mountains. Plans developed for a religious colony, to be christened Sunnyside, at the mining prospects. The mining locations paid off and the community prospered. All went well until the late 1890s when Brother Sam died. He was irreplaceable. Added to that, the mines played out, debts accumulated, and the honest people of Sunnyside gave the creditors the mines."

So there it is—were there two Salvation Army colonies in the Huachuca Mountains, both working mines and both active in the 1890s . . . Or was there only one colony, and some slightly garbled "facts?"

"The sun never shone on a finer grazing country than upon the 300 miles west of the Rio Grande. The traveler has before him, throughout the entire distance a sea of grass, whose nutritious qualities have no equal, and the stock raiser in January sees his cattle in better condition than our Eastern farmer his stall-fed ox.

—SYLVESTER MOWRY, *Arizona and Sonora, 1863.*

Spectres of Spur and Plow

IF GHOSTS of the past are fancied to inhabit abandoned stage stations and moribund mining camps, surely other phantoms, no less spectral, must haunt the sites of deserted ranches.

Where today finds only corrals reduced to bleak ruin and once-fruitful acres reverted to desert or trodden underfoot by the advancing battalions of progress, yesterday's tillers of the soil gave their blood and sweat and sacrificed even life itself to establish in Arizona the fields and herds of permanence. It is no sinecure to husband the land where the simplest acts of everyday living are fraught with hardship and hazard and attended, all too often, by violent death,

Hostile Indians, bitterly resentful of white man's intrusion and determined to drive him from their lands at any cost, shared the picture with cattle rustlers and border bandits. Even Nature, over much of the Territory, seemed aligned with the opposition, dealing out scorching summer temperatures and, oftimes, withholding the rains until crops had shriveled and thirst-tortured cattle ringed desert waterholes dried and cracked by the sun and heat.

Man soon learned the lesson of the arid Southwest: He who would sow with any hope of consistent reaping must have the wherewithal to supply water to his fields.

Which pioneer Arizona rancher actually was first to adopt the practice of irrigation has been the subject of much historical hair-splitting. Actually, that honor does not belong to Don José Redondo, or the Mormons; even the Spanish mission fathers must take a back seat.

The true pioneers of water husbandry in Arizona were members of a prehistoric race which had vanished from the scene long before any white man set foot in these United States.

The name archeologists have given these original farmers is Hohokam—"The People Who are Gone."

To modern engineers, bolstered by scientific knowledge, powerful machinery, and the best precision tools and instruments money can buy, it is no monumental achievement to plan and bring into use an irrigation project. But what modern man would have the courage to undertake such a project, or the fortitude to carry it through to completion, were he completely untrained in engineering skills, if he had no metal tools or wheeled vehicles, no beast of burden?

Spurred by existing need, with the priceless ability to think and reason as his sole guide and blueprint and with only his own human strength to meet the needs of labor force and machine, the Hohokam farmer of possibly 2000 years ago took the first steps toward bringing water to the thirsty valleys of Southern Arizona.

Using roughly-chipped stone hoes and axes lashed to short handles by means of skin thongs, these Neolithic engineers cut through the roots of giant cacti and mesquite and laboriously loosened the dirt to be removed. Using their own bare hands in lieu of a shovel, the Hohokam scooped the loosened material into a woven basket. When the basket had been filled it was carried by human strength to some point on the line of the canal where fill-dirt was needed. There it was emptied, and tramped firmly into place, and the task of filling another basket was begun.

Working in this painfully slow manner, basket after basket and day after day, these incredible Arizonans eventually excavated irrigation canals as much as 25 feet in width, 15 feet deep, and 25 miles long!

According to findings of the water resources policy commission of the United States government, these Stone Age canals formed a virtual network over New Mexico and Arizona and in the present vicinity of Phoenix, alone, carried water to an estimated quarter of a million acres of farmland!

Profiting by the labors of their vanished predecessors, the later-arrived Pueblo Indians developed an irrigation system even more extensive. As a consequence, when other North American tribes were still largely nomadic, drifting with the seasons, dependent for their livelihood upon the wild game, seeds, and fruits provided by Nature, Indians of the Pueblo country already knew a high degree of security. Living in well-built apartment houses of stone or adobe construction, they cultivated their irrigated fields of maize, squashes, beans, melons, and cotton, and stored the yield of their labors in thick-walled granaries.

Were we to conjure in imagination a parade of Arizona's ghostly ranchers, surely the leaders in that parade, the vanguard, the flag-bearers, must be these Neolithic men, these first tillers of the soil in the valleys of the Salt River and the Gila.

Next in line of march would be herdsmen from the Spanish missions.

The first permanent breeding cattle in what is now Arizona are believed to have come in with the cross of Father Kino in the last decade of the 17th century. Numbering about 700 head, these breeders had been driven north from the Jesuit mission of Nuestra Senora de los Dolores, in Sonora, to Kino's new mission outpost of San Xavier del Bac, a little way southwest of the present city of Tucson. During the next several years until his death Father Kino imported additional herds of breeding cattle and other livestock to new missions established in the fertile valleys of the Santa Cruz, San Pedro, and Sonoita Rivers. At these same missions also was planted, for the first time in Arizona, the fruits, vegetables, and grains of Europe.

Following the discovery of precious metals in Pimeria Alta—Arizona—great haciendas were established by wealthy Spaniards and the middle 1700s found nearly every arable valley in the present counties of Pima, Pinal, Santa Cruz, and Cochise green and luxuriant with tilled fields of wheat, corn, beans, and various fruits.

"Thousands of enterprising Spaniards cultivated the rich valley of the San Pedro and scattered settlements flourished at every suitable stream and spring," wrote Sylvester Mowry.[1]

The Spanish hacienda of San Bernardino, near the present Arizona-New Mexico boundary about 17 miles east of Douglas, is said to have embraced, at its peak, around 100,000 cattle, 10,000 horses, and 5000 mules. Its *patron grande* lived in almost regal splendor in an elegant adobe mansion 100 feet square, surrounded by flower-filled gardens and orchards of oranges, limes, pomegranates, figs, grapes, apricots, peaches, and other fruits suited to this mild climate.

Yet, when John R. Bartlett visited this once-splendid hacienda in 1851 he found its field and orchards in ruin, and only gutted walls remaining to tell the story of slaves in revolt.[2] In reducing their Indian laborers to serfdom, the Spanish grandees had failed to allow for the fact that the enslavement of a proud people seldom is marked by permanent success. Aroused to fighting pitch by the ever-increasing demands and cruelties of their tyrannical masters, the "tame" Indians eventually had rebelled and with the help of tribesmen still unsubdued had driven out or massacred their oppressors.

This story of revolt and ruin, with only minor differences, was repeated at Babocomari (where 40,000 cattle are said to have roamed the range); at the Sopori, Reventon, Calabasas, Sonoita, Tubac, Sierra Bonita, Agua Prieta, San Simon, and Pueblo Viejo. Once beautiful haciendas where luxury and opulence had reined, all had been reduced to blackened timbers and ghosts, their fields laid waste, their great herds slaughtered and scattered. Between 1828 and 1843 hundreds of Spanish settlers in Arizona had lost everything they possessed, including their lives. By the mid-Civil War years, Mowry found here hardly a trace of their earlier civilization "except scarcely visible ruins."

After the acquisition of Arizona by the United States in 1848, and the Gadsden Purchase of 1853 some effort was made to reestablish herds. Due largely to depredations of the Apaches—who are said to have stolen or slaughtered some $100,000 worth of Arizona livestock in the year

[1]*Arizona and Sonora*, Harper and Brothers, New York 1864.

[2]*Personal Narrative and Incidents in Texas, New Mexico, California, Sonora, and Chihuahua.* Vol. I, p 255.

Old cabin, Cochise County, 1959

1854, alone—none of these efforts was crowned by much success.[3]

Except for draft animals used by the '49ers and trail herds driven from Texas to California, the first cattle brought into Arizona by an Anglo-American are said to have been 200 head of stock imported by William Kirkland in 1857 and placed on the Canoa ranch, south of Tucson. Most of these animals eventually fell to the Apaches.

Already bad for close to a century, Indian depredations worsened soon as Army troops were withdrawn from Arizona at beginning of the Civil War, and the local situation, between 1861 and 1874, has been called "the most sanguinary and violent in the annals of the Southwest."[4]

Prominent among the few pioneer ranchers who courageously defied the redskins during this period, were Pete Kitchen and Elias Pennington. Each was successful, to a degree—but the cost was great. Year after year they saw their fields destroyed, their buildings burned and cattle stolen. Pete Kitchen's 11-year-old son was shot and killed no more than 200 yards from the family residence,[5] and one of the daughters of Elias Pennington was kidnapped by the Apaches, forced to accompany them until she fell from exhaustion, then lanced through the body, thrown over a cliff, and left for dead. Upon recovering consciousness the girl dressed her wounds as best she was able and, after several days' effort, succeeded in crawling back home. Later, her father and only brother were slain by Indians.

Such were the calculated risks faced by cattlemen in Arizona during the Civil War years and the half decade immediately following. It was this same period, however, that brought into being one of the Territory's greatest pioneer ranches. This was the domain of the Redondo brothers, Don José and Don Jesus, who had located in the Yuma Valley when only mesquite and catsclaw jungles blanketed the land.

Employing at a single time as many as 300 Mexicans and Indians, the Redondos cleared the land, broke the virgin sod, leveled the ground, dug 27 miles of gravity-fed irrigation canals at a cost of $25,000, and planted crops. At its peak of production the Redondo ranch—known as the Hacienda de San Ysidro—embraced some 2000 acres planted to grain, corn, beans, alfalfa, melons, vegetables, sugar cane, and peanuts, as well as fruitful orchards of apricots, dates, peaches, pears, and pomegranates, and extensive vineyards. The *hacienda* included a two-story adobe flour mill, a winery, store, cookhouse, blacksmith shop and stables, as well as an elaborate two-story adobe residence with tiled floors and ceilings beamed with peeled cottonwood logs. Here the owner, Don José, lived with his wife and six children and his servants, while cattle bearing the Redondo brand numbered into the thousands and ranged for 100 miles along the Gila and Colorado Rivers.

When United States government land surveyors divided the Gila Valley into townships and sections in 1874, the *hacienda* fell under a

[3]*Southwestern Historical Quarterly*, Vol. 36, No. 1, p 51.
[4]*Arizona Historical Review*, Vol. VI, No. 4, p 12.
[5]*Arizona Citizen*, June 27, 1872.

cloud that curtailed further development. Following Redondo's death, four years later, his heirs sued to obtain title, but were granted only the one section of land on which the main buildings were situated. Family members gradually drifted away and the corrosive forces of disuse and neglect began gnawing at the big house, the flouring mill, the winery, the groves, and fields. Windows, doors and timbers of the well-built structure were spirited away by neighboring ranchers and miners. Wind and weather began wearing down the adobe walls, and treasure seekers began their insidious probing in search of buried treasure.

These combined forces of devastation and destruction have done their work so thoroughly that today finds only a few ruins to mark the site of the once-great agricultural empire of Hacienda de San Ysidro.

Another early-day rancher in this same part of the Territory was William B. Rhodes, or Rood, said to have been a member of the Jayhawker Party of Death Valley fame.

After passing four years on the Reventon ranch, between Tucson and Tubac, Rhodes sold this property to Elias Brevoort, about 1860, and subsequently located another ranch on the east bank of the Colorado River between La Paz and Fort Yuma. Extending some 16 miles along the river, these holdings were known as Rancho de las Yumas.

Rhodes married and became the father of two daughters. His fields spread wide and green and luxuriant; his herds waxed fat and multiplied.

But tragedy, all the while, was stalking Rhodes' trail.

While crossing the Colorado River during a period of high water in April 1870, Rhodes was drowned when the light rowboat carrying him and his foreman, Alex Poindexter, struck a snag and capsized. By clinging to the upturned boat, Poindexter succeeded in making his way to shore.

Judged by the standards of his day Rhodes had been considered a wealthy man, and when the public administrator was unable to locate more than a few hundred dollars on deposit with merchants of nearby towns it was generally assumed that the bulk of the rancher's supposed fortune must be buried somewhere on his property.

"The story of Rhodes' gold has brought treasure hunters from all over Mexico and the Southwest," wrote Harold O. Weight, probably the leading researchist into the life history of William B. Rhodes—or Rood, as Weight prefers to call him. "No year in the past 82 but has seen some attempt to uncover it. The adobe walls are gouged and undermined. The adobe floors are pitted. The land surrounding has been trenched and opened to depths of 30 feet. In recent times treasure detectors have been used.

". . . under the assaults of treasure seekers and weather, Los Yumas is vanishing. Each year legends and traditions gather more thickly about its broken walls. Seen by moonlight through scudding clouds, the black and lonely wall fragments above the silver sheen of the Colorado seem a likely place for a ghost.

"Before many more years have passed, the last of the adobe walls will be level with the dark mesa."[6]

Another of the notable ranchers of Arizona's Territorial days, possibly the most widely noted of all, was King S. Woolsey.

Beginning about 1860 King Woolsey ranched at various locations, including Stanwix Station, a few miles west of Sentinel; at Agua Caliente, 18 miles northwest of Sentinel; near Phoenix, and on the Agua Fria River, about 18 miles east of Prescott. Many are the tales told of the old fire-eater who feared neither consequences, force of superior numbers, or divine wrath, and was quite uninhibited by either international boundaries or ethics.

If I were to learn that King Woolsey's ghost occasionally comes back to prowl around a crumbling adobe cabin on the former Woolsey ranch, at Agua Caliente, I wouldn't be at all surprised. A stout palo verde tree beside that cabin became known locally as "The Hangman's Tree" after the old King had assertedly supervised the hanging there of not less than nine outlaws, broncho Indians, and assorted renegades.

Among the first Anglo-Arizonans to turn their efforts toward field crops rather than livestock were colonists sent out by the Church of Jesus Christ of Latter Day Saints, more commonly known as "the Mormons."

In the spring of 1877 nine Mormon families from St. George, Utah, comprising 71 persons, established the settlement of Camp Utah, or Utahville, in the Salt River Valley east of Phoenix. A letter from Daniel W. Jones, president of the settlement, originally published in the *Deseret News*, of Salt Lake City, and subsequently reprinted in the *Arizona Miner*, gives an interesting insight into farming practices in the Arizona of that day.

Under date of April 26, 1878, Jones wrote:

"The land here is level and easy to pulverize . . . If you will call down here soon we can show you some 300 acres of good wheat put in by our colony with a crooked-stick plow, without so much as a ram's-horn point. This country is so productive and easy of cultivation that, notwithstanding this colony was too poor at seeding time to buy a common plow, from present prospects we hope to be able to square up and have enough for bread and seed for the coming season. All is well with us . . ."

By the 1880s, a majority of Arizona's ranchers could echo the "All's well" remark of President Jones. The dove of peace was beginning to sprout pinfeathers. Most of the renegade Indians had been sent down the long trail "where the pony tracks all point one way," or were growing fat on government reservations and Uncle Sam's beef. Conditions finally reached such a state that about all the cattlemen had to worry about was drought.

During the scorching dry year of 1885, according to a story originally published, I believe, in the *Range News*, of Willcox, a meeting of south-

[6] *William B. Rood*, a Calico Print Folio, 1952.

eastern Arizona stockmen was held at that place. In view of the long-sustained drought some wag moved that the meeting be opened with prayer. The motion was seconded (probably by another wag) and Big Dan Ming, owner of a ranch in Graham County, was nominated to ask the invocation.

Big Dan, six feet four inches in his stocking feet, was undaunted by the assignment.

"Oh Lord," he began, "I'm about to round you up for a good plain talk . . .

"Now, Lord, I ain't like these fellows who come bothering you every day. This is the first time I ever tackled you for anything, and if you'll only grant this I'll promise never to bother you again.

"We want rain, Good Lord—and we want it bad. We ask you to send us some. But if you can't or don't want to send us any, for Christ's sake don't make it rain around Hooker's or Leitch's ranges, but treat us all alike.

"Amen!"

General Mercantile, Mineral Park, Mohave County, 1951

"After some years residence here the writer is more than ever confirmed in the belief that while there is much to contend with in Arizona, there is much to contend for, and that despite all the drawbacks and discouragements the Territory will yet command a large and prosperous population, and abundantly repay the government for the outlay required to save it from the savage.

—ROSSITER W. RAYMOND, *Resources of States and Territories West of the Rocky Mountains, 1871.*

Three Old Towns in Mohave County

APRIL AND MAY are wonderful months for adventuring in Western Arizona. Winter's chill is past; summer's heat is in the yet-to-come. If the season has yielded even a little moisture, April finds the hills painted with wild flowers—yellow poppies and orange mariposas, yuccas and cacti and ocotillos—and every rocky wash shimmers with the spun gold of palo verde. Desert quail and cactus wrens are nesting, and the soft voice of wild doves makes every canyon a place of sweet tranquility.

On such a day in 1951 I explored the ghost mining camps of Mineral Park and Cerbat. My friend, Marion Johnson, of Las Vegas, Nevada, helped me do it.

Marion and I had been told of these two old towns by Charles Labbe, a retired mining engineer, who had shared with us a few fascinating bits of their history and had sketched a rough map to help us locate them.

Sunrise had found us on our way. By six o'clock we had traveled from Marion's home in Las Vegas to Hoover Dam, had crossed the Colorado River, and were climbing into the topaz-and golden hills of Arizona. With our hearts singing in the joy of the morning, and my little old red car skimming along the highway as if it had sprouted wings, we passed the turn-offs to Temple Bar and the ghost town of White Hills—long one of our favorites. By the time most honest folk were bestirring themselves for breakfast, we had reached the sideroad to Chloride.

Situated in the foothills four miles east of the main highway, Chloride is a rather interesting old town where a hundred-or-so citizens still piece out a living. There are a few small business houses, a few nice dwellings, and more than a few picturesque old homes standing in various stages of disrepair, with woodbine and morning-glories scrambling over their porches and big cottonwoods and Chinaberry trees throwing heavy circles of shade in their yards.

At the head of Chloride's principal street and a little above the level of the town, looms the dingy surface works of the great Tennessee-Schuykill mine. Back in 1898 the Tennessee held forth sufficient promise to spark construction of the now-defunct Arizona & Utah Railroad, a 24-mile line that gave Chloride direct connection with the A. T. & S. F. In the course of intermittent operations between 1898 and 1948 the Tennessee is said to have produced around $7,500,000 in gold, silver, copper, lead, and zinc and there are plenty of Chloriders who maintain that the old mine still has a lot to offer.

Guided by Charlie's pencil-sketched map we picked up a dirt road in the center of town and headed east along its sinuous course. Six miles of bobbing over hills and hollows, through a desert land sparsely dotted with yuccas, brought us to a division of the ways. After another map consultation, we chose the lefthand road fork. The desert began pushing together to form wrinkled hills, the hills narrowed into a canyon—and soon after entering the canyon's mouth we began seeing ruins of the old town of Mineral Park where the ghosts had taken over long years before either Marion or I was born.

First building of consequence was a big roofless adobe, floorless and doorless, its window frames barren of glass. We remembered Charlie telling us that this first large ruin formerly had housed a general store and cafe. Across the road, closer to the wash, we found the ruins of a small adobe ore mill. While I was taking pictures of the two places Marion collected some attractive samples of turquoise, several bits of sun-purpled glass, and quite a handful of square-cut iron nails.

Leaving my car parked beside the old store, we wandered on up the road, higher into the canyon. We soon came to a small stream of water—barely a trickle by Eastern standards, but a pleasing and delightful sight in this arid land. In the bushes fringing the little water-course birds were singing and chattering—something about nest-building, I believe—and wild bees were humming drowsily. Crossing the stream we continued along the old road, halting occasionally to peer into shallow prospect holes or to examine some crumbling foundation or rock wall.

It was Marion who discovered the old stone jail. Although roofless and with its cell-block missing, one of the building's two windows was still in place and securely barred. The barred frame of its second window lay on the ground nearby where it had been discarded by some earlier-comer who had used it as a grate for his campfire.

One of the larger rock dwellings had walls nearly three feet in thickness, plastered inside and out with adobe. After the exterior of the wall had been smoothed nicely, white paint had been used to block its surface into small rectangles so that the effect, from a little distance, was that of a brick house. At the windows, gray with dust and grime and cobwebs, hung tattered lace curtains, and the walls of living-room and bedrooms were covered with wallpaper bearing the ponderous patterns of Victorian days—maroon and poisonous green, with an abundance of great pink roses and fancy gilt curley-cues.

Near the head of the canyon we came upon a large inoperative mill. Circling the building, our road narrowed and steepened and canted off toward a cabin visible on the ridge above. A trespass sign guarded the foot of the grade. Gambling that its warning was intended for motor transit and night prowlers rather than for honest folk who come wandering afoot in the full light of day, we continued upward along the trail.

The woman who came down from the cabin to meet us didn't seem to mind that we had ignored the posted notice—and the little black and white dog who accompanied her was wagging his tail in high-gear and seemed to be laughing all over. The woman invited us up to the cabin and asked us to sit down. She offered us hot coffee and gave us an introduction to her husband, a disabled veteran of World War I. She said the two of them were caretakers of the mill and 32 mining claims in the vicinity.

Sitting on the cabin steps, in the mid-morning sunshine, we scratched the floppy ears of the little dog and listened to this desert couple's tales of Mohave County. They told us how the first gold had been discovered here, and how Mineral Park had been founded; how it had captured the county seat from Cerbat, in 1877, and, through sundry shenanigans, had lost it to Kingman nine years later. They told of the gallant river steamers and long mule trains, and of the dark days of Indian terrorism when a man's life depended on his gun and his ability to sleep with one eye open.

"That low mountain yonder is Cherum's Peak," said the woman. "Cherum was an Indian chief. When he died, about 1912, the Indians cremated him according to their custom, then took his ashes and his herd of horses and rode up that canyon to the east. At a point part way up the canyon they stationed guards to see that no white man followed them. After riding several miles farther to a secret spot, they buried the old chief's remains. When the funeral party rode back down the canyon only half of Cherum's horses came back. The others had been killed and left for Cherum's spirit to ride in the Happy Hunting Ground . . ."

When Marion asked if there was any good turquoise to be had in that vicinity, the woman nodded.

"There's quite a lot of it over in Todd Basin, about three miles southeast of here," she said. "The Oro Plato mine there was worked for years, and the Montezuma mine was worked quite extensively long before white folks came to this region. When they were mining in the Montezuma, about 75 years ago, they found a shaft containing many stone hammers and other mining tools like those found in the old Aztec mines in Mexico . . ."

After we had visited an hour or so I told Marion I thought we should be getting along our way. The woman was quick to protest.

"You just got here!" she exclaimed. "Stick around awhile. Pop and I get lonesome out here. We don't have much chance to talk to anybody.

Folks down at Chloride don't like us," she confided. "They say we're too tough. You know why they say that? Just because when I see somebody prowlin' around the mill at night, lookin' for anything that ain't nailed down or chained to a post, I shoot first and ask questions afterwards. That's why they don't like us. I've got a good .30-30 rifle—and I'll throw down on the Old Nick himself if I ever see him prowlin' around that mill at night!"

After we had explored the ruins some more and had eaten our lunch in the shade of the old adobe store building, Marion and I drove back down the canyon and off on another sideroad to the Gross Ranch. Charlie Labbe had told us we should visit this place.

"Tell them you want to see the old apricot tree," Charlie had said. "I'll bet it's the biggest one in the state of Arizona . . ."

We located the ranch house easily enough but were disappointed to find there no one but a Mexican caretaker. Although the old fellow was the soul of politeness and showed us the tree with obvious pride, there wasn't much he could tell us about it. He said the tree was at least 50 or 60 years old and that it bore a big crop of fruit every season. But when Marion asked him who had planted the tree, he shook his head and shrugged.

"*Quien sabe?*" he said. "Who knows?"

Returning to our main sideroad we followed it to a sign reading "O'Brien Mines." The woman at the mill had said we should turn off here a mile and a half to the R. C. Ferguson place.

"Mr. Ferguson is a nice old fellow," she had said. "He can tell you a lot about Cerbat and Chloride . . ."

Instead of the sun-scorched miner's shack we expected to find, the Ferguson place was like an oasis! An attractive little house was almost hidden beneath luxuriant trees and vines. There were flowers galore, and a flourishing vegetable garden.

When Mr. Ferguson answered our rap at the door and we made known our quest, he nodded eagerly.

"Yes, yes!" he nodded. "Of course! Come in!"

Mrs. Ferguson, pleasant faced and white haired, hustled a couple of chairs for us and two refreshing glasses of cold water, and we sat down to talk.

Mr. Ferguson had come to Cerbat in 1896.

"It was never a very large town," he said. "Both White Hills and Mineral Park were larger; but, so far as that goes, they were larger than Kingman, too. They were good, law-abiding towns. Chloride was sort of wild. It had 16 saloons when I came here, and there were quite a few shooting scrapes . . ."

Before Marion and I said goodbye to the Fergusons, about an hour later, Mrs. Ferguson showed us through their home in which every room was spic-and-span as a showcase. With obvious delight she exhibited the big, white-enameled, kerosene-powered refrigerator in her kitchen, and their gleaming bathroom with its white porcelain tub and flush toilet operated by water piped from a spring. The same source supplied irrigation water to the small orchard and garden where we were shown peach and fig trees laden with swelling fruits, and vines bearing long bunches of green grapes.

"If you'll come back later in the summer when this stuff is ripe, we'll give you all you can haul away," said Mr. Ferguson.

Guided by the old man's explicit directions we easily located Cerbat and its one remaining house. Until a short while before, Mr. Ferguson had told us, this place had been occupied by a Mrs. O'Brien whose husband had come to the district many years earlier.

"The old girl finally grew so feeble her friends didn't think she should be living alone, so they practically kidnapped her and packed her off to Kingman where she would be nearer doctors and a hospital—if the need should arise," our informant had said.

In our explorations of the old townsite, Marion and I came upon a small house trailer occupied by a pleasantly-plump lady from Los Angeles. She said she and her husband owned a mining claim in the vicinity and were up there to do the annual assessment work. When she confessed—a little too happily, I thought—that they had been doing assessment work on this same claim 20 years and in all that while had never taken out one penny's worth of ore, I made some remark concerning 'the hope that springs eternal.'

"Or maybe these folks just want a good excuse to get out in the desert each spring," said Marion.

The trailer lady laughed—a nice, friendly, wholesome laugh. "It's about six of one and half a dozen of the other," she said.

That was my introduction to Mineral Park and Cerbat, about a dozen years ago.

Later, when I started running down their history in earnest, I learned that these two towns are among the oldest Anglo-American settlements in Mohave County.

In 1863, when white men in western Arizona were few and far between, a prospecting party headed by John Moss and further composed of William Furlong, William France, and two others named Walton and McCall, had made the first mining location in the county—a gold ledge subsequently known as "the Moss mine." Encouraged by the discovery and early sale of this ledge for $90,000, a second prospecting party—this consisting of San Francisco mining men under the guidance of Moss—in 1864 prospected through the Union Pass, Cerbat, and Wallapai Mountains, and on the western slope of the Wallapais formed the Wauba-Yuma mining district. After making several locations, the party was driven out by hostile Indians. Returning two years later they again made a desperate attempt to work two of their most promising claims, the Pride of the Mountains and the Florence. Indian hostilities again forced their departure.

During the next four years Indian depredations made it impossible for a white man to live in Mohave County except at the several small river ports. By 1870 the situation was improving a bit and a few prospectors were beginning to congregate in small settlements which ultimately developed into the towns of Mineral Park, Cerbat, and Chloride. These were mining camps in every sense of the word—but not boomtowns. No fanfare attended their founding. Cerbat soon became the third seat of Mohave County—taking that honor away from Hardyville which, in turn, had captured the official seat from Mohave City; and on the same December day in 1872 post offices were established at both Cerbat and Mineral Park. Despite these several signs of progress and urbanity, little news of either town reached the outside world before 1876. On June 1, that year, Mohave County's first newspaper, *The Wallapai Enterprise,* edited by J. W. Leonard and C. F. Mitchell, made its bow at Mineral Park—and bowed out again, six months later. Another five years passed during which the only news of Mineral Park and Cerbat is that contained in chance paragraphs that filtered in to the journals of Prescott and Yuma.

"A Letter from Cerbat," published in the *Arizona Weekly Miner* July 30, 1875, is unnoteworthy except for one brief item that affords a fleeting insight into the perils of early-day travel in western Arizona.

A Negro man, together with his Chinese bride and the latter's mother, had left Cerbat in a horse-drawn wagon bound for Pioche, Nevada. The day had been a scorcher. The water containers carried by the trio had proven defective and their contents had leaked away. When the team gave out, four miles from the Colorado River, the Negro had started on foot for water.

"Before he could return," stated the succinct item, "both women were dead."

Another news item which leaked out (and probably was not hailed with delight by residents of Mohave County) appeared in the *Arizona Citizen,* of Tucson, February 5, 1876:

"Michael Dekay was hung in Cerbat on the 14th for the murder of his wife. This was the first execution in Mohave county, although there are others who should have been hung long ago if justice were done them."

In more serious and constructive vein is a letter written by one W. M. Ewing and published in *The Mining Record,* September 10, 1881.

Within eight miles of Mineral Park, stated Mr. Ewing, were more than 2000 mining locations, many of them showing large bodies of ore, gold, silver, argentiferous galena, and copper carrying both silver and gold.

"Embraced within this area are the camps of Chloride, Cerbat, Todd Basin, and Stockton, all within 15 miles of the railroad survey," wrote Mr. Ewing. "Each of these camps has supported since 1871 a goodly population of miners who have held their claims year after year in the firm belief of their future value. During this time prices of provisions and material have been higher than in any other mining district in the United States, and the cost of shipping or milling ore so great that nothing less than $200 per ton would leave a margin to the miner; and yet, these men have lived comfortably on the profits of their mines and in many instances made a handsome surplus . . ."

Such was the situation in October 1881, when the *Alta Arizona,* a weekly newspaper, was established at Mineral Park with W. H. Cureton as editor and publisher.

The *Alta* wasn't around to celebrate its second birthday. Before that time it had been absorbed by the still-more-youthful and likewise more robust *Mohave County Miner.* Under the journalistic guidance of Anson W. Smith the *Miner* continued to publish at Mineral Park until Jan. 1, 1887, when it followed the county courthouse to the new town of Kingman, on the railroad. There the *Miner* is still going strong. Most of the research for this chapter was done in its basement warehouse where are preserved its files (and those of the *Alta*) for more than 80 years back.

Scanning those priceless journals, that week-by-week saga of life in a small mining camp weeks removed from anything resembling urban civilization, I was deeply impressed by the splendid self-sufficiency of a people who could so well adapt their lives to circumstances. They made and enforced their own laws, they established and financed their own schools. They cared for their sick and buried their dead; they contrived their own amusements—all without TV, radio, or movies, and with but few books, magazines, or outside newspapers. That they were aware of their isolation is certain, but they refused to let the situation dominate or intimidate them. Rather, they embraced it and made light of it.

Even though these citizens paid as much as $200 a trip to have mail brought from Prescott and felt lucky if they received it in less than 15 days; even though a letter mailed from Mineral Park to the mining camp of Signal, in the same county, was forced to travel 1000 miles by way of Maricopa and Yuma to reach its destination, still the people of western Mohave County never lost their feeling of union, of oneness with the Nation. Problems faced by the Nation were their problems; its holidays were their holidays. Furthermore, and insofar as possible, those days were celebrated exactly as the forbears of these men and women had celebrated them back in Indiana and Vermont and all the other long-established places from which these people were sprung.

On Fourths of July all business houses of the town were decorated with flags and bunting. The day was passed firing "salutes" and shooting firecrackers. In the afternoon, citizens of the town gathered at the courthouse where Mr. Hyde, or W. H. Krider, or Anson Smith, or some other solid and respected burgher, read the Declaration of Independence. Someone else recited a patriotic poem or two; the *Star Spangled Banner* was sung, and the day concluded with fireworks and a Grand Ball which always brought out "a large and orderly crowd."

Washington's Birthday, in the Nation's Centennial year of 1876, was celebrated in Mineral Park with great verve and eclat. With that same day also seeing the first bar of bullion cast at the Mineral Park mill, there was a double reason for jubilation and the old newspapers inform us that eggnog and champagne "flowed like water, and all hands were happy."

Strangely enough, though Indian-scourged miles of torrid desert stretched between their tables and the nearest outside sources of provender, Mineral Park's menus were not bare of delectables.

Scanning the columns of the *Miner* we learn that the "finest peaches and grapes raised in Arizona are to be seen in Mr. Krider's garden in Mineral Park," and that Thomas Logan's four-horse team and wagon "came into town on Thursday loaded with live turkeys, sausage, fresh pork lard, raisins and various other luxuries" grown on the St. Joseph farm where Mr. Logan formerly resided. We read that A. F. Simons "comes over from Beale Springs about twice a week with watermelons, grapes, cucumbers, etc., and takes away about all the loose change there is in town." A paid advertisement notifies readers of the *Miner* that W. H. Taggart and Company "will receive regular shipments of fresh vegetables, oysters, Brooker's pork sausage, bologna sausage, etc., during the cold season," and a news item sets forth the interesting fact that John Burns was in from his mountain ranch with a load of fresh watercress. "He reports his peach orchard in fine condition and the prospects good for an immense crop this summer. He will also have a fine crop of grapes," it was stated.

Life in Mineral Park, quite naturally, was not all peaches and cream and fresh watercress. Wherever there are human beings sin is almost certain to rear its ugly head, but The Park's malfeasances seldom reached the point where arguments were settled with six-shooters. Instead, a "paid reader," inserted in the news columns of the *Miner* might point an accusing finger at some townsman generally unidentified by name:

"Any person having any of my carpenter tools is requested to return them forthwith and save trouble," warned W. H. Krider on one occasion. "Especially the parties who have my moulding plane marked H.C.C., and the party who borrowed my saw last week."

Even the *Miner* once took occasion to remark: "Somebody along the line of the Mineral Park and St. Thomas mail route makes a business of appropriating copies of the *Miner* addressed to Wooley, Lund and Judd, St. George, Utah. Now this business," cautioned the frontier journal, "has got to stop or someone will get into trouble."

The advertising columns of early-day mining camp newspapers are my special delight.

To be told that Mineral Park had stores and shops is interesting—but to read the actual paid messages of those mercantilers and shopkeepers is infinitely better.

Scanning the 80-years-ago advertisements in the *Alta Arizona* we learn that the partnership of Lamson S. Welton and W. F. Grounds, who styled themselves "The Pioneer Mineral Park Merchants," were the heaviest advertisers, as they continued to be throughout most of their tenure at Mineral Park. The charms of "white space" were then still undiscovered by advertising men and the columns devoted to their weekly announcements are filled like a bag of grain, from top to bottom and side to side.

"We have on hand, just received by the last steamer, an elegant assortment of clothing and gents' furnishings," set forth Messrs. Welton and Grounds in one of their typical display advertisements, November 5, 1881. "Also drygoods and fancy goods consisting in part of red flannel, brown drill, ginghams, duck, shawls, overalls, shirts and drawers, perfumery, toilet soap, combs and brushes, thread, needles, boots and shoes, miner's boots a specialty, hats, a large assortment of all styles; hardware, iron, steel, horseshoes, muleshoes, saws, locks, files, shovels, hammers, rasps, staples, axes, rope, hatchets, screws, picks, nails, bolts, pitchforks, etc., etc., crockery and cutlery, tobacco and cigars, tinware and woodenware, every article desired by miners or families. Liquor and wines, giant powder, gunpowder, caps and fuse at reduced rates and in quantities to suit. Paint, oils, and glass, miners goods of all kinds. Give us a call and obtain goods at prices lower than ever before . . ."

How easy, reading that advertisement, to conjure in imagination this all-inclusive frontier store, its hustle and bustle, the mingled smell of new leather and dried apples and sauerkraut, clerks hurrying about their duties measuring coal oil, fitting boots, cutting glass, explaining the merits of saws and muleshoes, selling "perfumery" and "fancy goods" for coming balls, and white flannel for pioneer babies to be.

Welton and Grounds . . . how often their names appear in the columns of the *Alta* and the *Miner*. What wide awake merchants they must have been, and how vitally important to citizens of Mineral Park and Cerbat in their isolated position.

From issues chosen at random through the 1880s we learn that Welton and Grounds were building an addition to their store to hold the immense stock of goods being received by every steamer from California . . . Welton and Grounds were piling up goods to meet the wants of that rapidly growing section of the Territory . . . The next steamer would land 100 tons of freight for Welton and Grounds . . . A load of hides from Sam Crozier's stock ranch had been sold to Welton and Grounds and would be sent to San Francisco by the next steamer, and Indians of the vicinity were bringing in large quantities of fresh green hay which they disposed of to Welton and Grounds, receiving one cent a pound, which they took out in calico, flour, and other staples . . .

So it went, issue after issue. Nor were Welton and Grounds the only frogs in Mineral Park's puddle.

In most of the early issues of the *Alta* Dr. A. L. Burdick recommended himself as a physician and surgeon, and James J. Hyde called attention to himself as an apothecary and dealer in patent medicines and drugs. Krider Brothers (W. H. and J. P.) advertised their Post Office Store as headquarters for groceries and provisions, livery and feed. O. H. Fehr was a boot and shoemaker on Main Street; John Huges, a blacksmith and wagonmaker.

"DID YOU EVER GO TO MOHAVE?" inquires one display advertisement. "Better go. Charlie Bussey will take you through on time. Leave Mineral Park at 3 p.m. Get to Union Pass a little after 5. When you reach that place you will find that liberal-hearted gentleman, Ben Spears. If you are dry, walk in . . ."

Comstock and Lawson were advertising their meat market and family grocery store "with cigars and tobacco on the side." M. W. Henkle was proprietor of the Mineral Park Waterworks, and Mrs. G. W. Brooks did dressmaking and millinery.

"Do you want your washing, ironing and mending done cheap? Take your linen to the undersigned professional. He will give you such work as will cause you to come again and keep coming. Prices bedrock," advertised Sam Long. Another advertiser we may suspect of Oriental extraction was Jack Wong Ping, owner of the Mineral Park Restaurant. "The proprietor of this well-known restaurant is an artist in his profession," stated Jack without any trace of false modesty. The Palace Hotel, of Mineral Park, was being operated by one Kelvin Cuzinno, who announced himself "prepared to accommodate the public with as good fare as northern Arizona affords," and the advertisements of H. White and T. A. Christie commend to the public, respectively, the merits of White's Pioneer Saloon, and Christie's Place.

How strange to realize so much business activity, so much buying and selling and trading, in that quiet canyon where, 70 years later, Marion Johnson and I had found only sun-warmed ruins and quail, and a caretaker who wanted to visit.

Though Mineral Park and Cerbat had their period of greatest activity during the time when the Apaches were warring on other sections of Arizona Territory, there is virtually no mention of local Indian depredations in either the *Alta* or the *Miner*. Rather, the Indians at Mineral Park seem to have fallen in the nuisance category along with other minor annoyances, such as mice and mosquitoes, which neither enhance life nor materially shorten it. Indicative of this attitude is a brief news item in the *Miner* of January 17, 1886: "A number of Wallapai Indians have taken possession of an old cabin up the wash near town where night is made hideous by the war songs of the braves, and a savage can-can is danced to the discordant music of an inverted tin bucket."

For some reason, however—possibly as a matter of preparedness, more likely because other wide awake towns in Arizona were doing it at the same time—Mineral Park, in June, 1886, organized a troop of home militia, The Mohave Rifles. Fifty-six stalwarts joined the organization as charter members. Bond to insure the issuance of arms and other equipment was provided by Dan Smith, T. J. Christie, E. L. Burdick, and J. F. Ball, and a citizen with the unusual name of Robert D'Yhr was elected captain. So far as I have been able to learn from the old files, the well-drilled Mohave Rifles were never called upon for any purpose more serious than that of lending spice to patriotic parades.

Most tragic incident in the usually placid life of The Park appears to have been the collapse of the Moses Steen home.

Brother of Robert Steen, then deputy U.S. marshal for this region, Moses Steen lived with his wife and three children in a large adobe house having a full basement walled in stone. The winter of 1884 was extremely wet, for Arizona, and there had been an especially heavy storm the night of the tragedy, April 1. On the morning following, as the Chinese cook at the Palace Hotel was on his way to work, he discovered the Steen home collapsed in a mass of rubble.

The Steen's youngest child, Mabel Grace, about four years of age, had been sleeping in one corner of the house in a double bed with her sister, Nellie, aged 14. That corner of the house was partially standing and Mabel was still alive—though her sister, who had been sleeping at her side, was dead. The girls' eight-year-old brother, Madison, was pinned in the wreckage of beams and brick, injured seriously but still alive. Mr. Steen, apparently, had been killed instantly, since Madison had not heard him or Nellie cry out, but had listened to his mother moaning and crying until death finally released her from her pain. Though suffering intensely from his injuries, the little boy insisted that rescuers save his baby sister, Mabel, before attempting his own removal from the wreckage.

It was Mrs. Ballard Hadman, of Seattle, a daughter of Mabel Grace, who gave me the story of the Steen tragedy.

"It was an extremely hazardous undertaking to get mother out of that insecure, unsupported corner," said Mrs. Hadman. "An old timer in his eighties volunteered to go in after her. I believe he was called 'Old Man' Haas. The miners formed a human chain and began passing out adobe bricks until they had cleared sort of a tunnel into the wreckage. Mr. Haas then crawled in until he could get hold of one foot, whereupon he dragged my mother free. She was unconscious and bleeding from the ears.

"Not until his little sister was safely outside the wreckage would Uncle Madison allow the men to effect his own rescue. He had been pinned down by a fallen beam, which lay on one shoulder. His collar bone was broken, also several ribs."

Mrs. Hadman's grandfather, grandmother, and Aunt Nellie, are buried in a single large grave in the old cemetery at Mineral Park. According to the *Miner* of April 13, 1884, their funeral "brought the largest number of people into town of any occasion in its history."

The mines of Mineral Park and Cerbat were numerous but not phenomenal. The Cerbat mine reportedly had a production of about $250,000 in gold and silver, and the locator of the Metallic Accident mine, situated about a mile north of The Park, is said to have taken out in only a short time some $60,000 worth of ore from a cut 20 feet deep and 100 feet long. Other mines prospered briefly, faltered, and passed out of the picture. Throughout the active life of the two camps their citizens clung to the belief that eventual completion of the Atlantic and Pacific Railroad would prove the panacea to cure all their problems, even as it ameliorated their isolation.

That hope, unfortunately, was ill-founded. The railroad's long awaited coming to Mohave County served only to lure trade away from the older towns to newer upstart towns directly on the rail line. Several of these towns immediately began clamoring for the county seat, and in November 1886, electors of Mohave County were asked to choose a permanent location for their courthouse. They could vote either for Mineral Park—where the county's seat of government had been located since its removal there from Cerbat in 1877—or they could vote for any town on the railroad.

Counting of the ballots revealed that Kingman had won the decision with 271 votes. Hackberry had been favored by 132 voters, and poor old Mineral Park by only 99. The county records were moved forthwith and in January 1887, Anson Smith packed the type cases and imposing stone of the *Miner* and followed the county seat to its new location.

Shorn of its two greatest mainstays—the county courthouse and its newspaper—Mineral Park withered and died. That, in brief, is its history.

The story of Cerbat is a more nebulous thing, largely because the town seems never to have had a journal of its own. As a consequence, anyone assembling its history is limited to the scattered bits of informaton that appeared in the *Alta* or the *Miner*, or crept into the journals of other Arizona towns.

One of the earliest mentions I have found of Cerbat, as a settlement, is that contained in the *Arizona Citizen* of May 3, 1873:

"We learn that the location of the county seat of Mohave county will have to be settled by a lawsuit. The contest is between Cerbat and Chloride City, the former having been the more fortunate in the number of votes cast, but the latter claims that the way it was done was 'dark and mysterious.'"

The *Alta Arizona* of November 5, 1881, contains the advertisements of two attorneys at Cerbat: Davis and Stephenson, counselors at law, and W. G. Blakely, who announced himself "ready to practice in all the courts of Arizona Territory." In its next issue following, the *Alta* reported that the Cerbat mill was working "in a manner to please everybody," and in another issue we read: "J. W. Blakely has opened a blacksmith shop at Cerbat and vows he can shoe anything that wears hair."

Elliott's History of Arizona Territory, 1884, credits Cerbat with possession of "several mercantile houses, saloons, blacksmith shops, and one furnace for the smelting of gold and silver." It was further stated that two physicians and two lawyers were in practice at Cerbat, and that a free school, providing a six months term, was in operation.

Actually, the only graphic picture I have found of life in Cerbat is contained in a letter published in the *Mohave County Miner*, January 14, 1888. Mining law, at that time, provided that anyone wishing to retain possession of an unpatented claim must restake it on the first day of January each year. To minimize the danger of claims being jumped, miners customarily made their re-locations soon after midnight, December 31. It is this activity of which the letter treats.

"New Years Eve I spent at Cerbat," wrote the anonymous newshawk. "From the amount of claim jumping one would have imagined that neighborhood to be rich as the diamondfields. There were lawyers, assayers, carpenters, saloon keepers, miners, mine owners, and superintendents, all on the leap. Their headquarters was at Frenchie's Saloon where the time and toddies were compared up to 10 p.m., when each slipped out and took a circuitous route to the property he had in view, in order to mislead the others, and many of them met on the same claim. High up on the cliff back of town, old Bill Raywood had a 60,000 candlepower bonfire burning on the dump of The Wrecker, warning all parties that property was represented, and occasionally rolling a huge boulder down the mountainside warning them it was especially dangerous to come up."

When I last revisited Mineral Park in 1961, ten years after the visit made there by Marion Johnson and myself, I found but little change. The old graveyard looked a trifle more neglected; desert brush had grown a bit higher around the old foundations, adobe walls had crumbled a bit lower.

It was very quiet in the canyon. Once again the magic of April was on the land. New generations of desert quail and doves were setting up housekeeping; somewhere, a mockingbird was singing. A tiny trickle of sweet water coursed down the rocky stream bed, and wild bees hummed drowsily. All this was unchanged—yet the place seemed immeasurably lonelier than I had remembered it. It took me a little while to understand why.

On my first visit here, with Marion, I had known virtually nothing of Mineral Park's history and had seen it only as a place where nameless ruin merged with nameless ruin—just an old townsite where we could hunt little mementoes of the past and savor the glory of an April morning. But now that I had read the story of Mineral Park in the old newspapers I was no longer seeing the deserted canyon as only a place to hunt souvenirs and pass a pleasant day. Now I was viewing it as a place where men and women had come and briefly conquered; had lived and loved and hoped and died, had sipped from the cup of joy and drunk the draught of tragedy.

I didn't stay long at Mineral Park—and I didn't go to Cerbat, at all. I didn't wish to disturb its weary ghosts.

Mill at Mammoth, Pinal County, c. 1890

—Courtesy Percy Leaverton.

Ore Freighting

—Courtesy Arizona Pioneers' Historical Society.

"They (the prospectors) have done more to open up our vast interior territories to settlement and civilization than all the scientific expeditions ever sent across the Rocky Mountains. The indomitable courage, the powers of endurance, the spirit of enterprise, the self-reliance and the fertility of resources exhibited by this class of men . . . have no parallel in the annals of daring adventure."

—J. Ross Browne, 1865.

The King Wore a Silver Crown

Silver King and Pinal City had their beginning in the days of Apache warfare when any man foolhardy enough to venture alone into the mountains of Arizona Territory was lucky to keep his scalp, whether or not he discovered a mine.

As an aid in combatting Apache depredations, General George Stoneman had established Camp Picketpost, about three miles west and a little south of the present site of Superior. Here, in 1872, he began construction of a road leading into the more inaccessible portion of the Pinals.[1]

Among soldiers assigned to this road-building detail was a chap named Sullivan. One evening, as he was returning to camp after a day's work on the grade, Road Builder Sullivan sat down to rest—so runs the tale. As men are prone to do in a mining country, Sullivan idly examined the loose pieces of rock scattered over the ground at his feet. Among them he noticed some heavy black fragments which aroused his interest—especially so when he discovered that these pieces, upon being pounded, would flatten metallically but would not shatter. Collecting several of the samples Sullivan went on to camp where he made no mention of the discovery to his comrades. Soon afterward, when his term of service expired, Sullivan made his way to the ranch of Charles G. Mason, on Salt River. Here he exhibited the black ore on numerous occasions but never revealed exactly where he had found it. When the man subsequently disappeared it was presumed that he had been slain by Apaches, or had perished while attempting to return to the place where he had found the rich silver sulphide.

[1]James M. Barney, writing in *Arizona Municipalities*, August 1940, states that this pack trail was, for years, the main traveled route into the mining districts of Globe, and vicinity. Ray Brandes, historian of Arizona Pioneers' Historical Society, tells me that the old road still may be followed, in part, and that it roughly parallels the present Superior-Miami-Globe highway (U.S. 60-70).

During the next two years Mason and others, who had seen the samples exhibited by Sullivan, organized several prospecting parties to search the Pinals in hope of finding the source of the black nuggets. On several occasions such parties were almost on top of the spot without knowing it—one location, the Silver Queen, being made only a mile and a half from the place where Sullivan had collected the samples. In 1874 a prospecting party headed by Mason located the Globe mine, in what later would be known as Globe District. The following spring, Mason and four of his friends—Benjamin W. Regan, William H. Long, Isaac Copeland, and another man whose identity apparently is lost to history—took a train of pack animals to the Globe location with the idea of bringing out a shipment of ore.

While returning from the Globe mine, March 21, 1875, this party was attacked by Apaches and its now-nameless member was slain. Taken to General Stoneman's original campsite, about a mile southwest of the present Craig farmhouse, the body of the unfortunate miner was interred in one of the abandoned stone ovens formerly used by the soldiers for baking bread. Soon after completion of the simple obsequies, the four surviving members of the party made camp and Copeland went to bring back one of their mules which had strayed. When he found the animal standing on some croppings he paused to examine them and moments later came rushing back to camp shouting, "I have struck it!"[2]

Gathering around Copeland, the excited and hopeful prospectors passed the pieces of croppings from hand to hand and all agreed that this was the same black silver ore Sullivan had found some three years earlier. The Silver King location was made the following day, and two weeks later (April 3, 1875) the *Arizona Citizen* reported:

"Pinal county stands head now in the matter of a silver deposit, according to all acounts . . . The discovery is about 32 miles by trail and 35 by wagon road from Florence, at the foot of the Stoneman Grade. It consists of a deposit of ore worth thousands of dollars per ton; in fact, some pieces of pure silver weighing pounds are said to have been found . . . Mr. Reagan [*sic*] . . . brought a ton of the ore to town, a sample of which was assayed here showing it worth $4340 per ton . . ."

Convinced that the mine was "too good to last," Copeland and Long soon sold to Mason and Regan for $80,000 each, and in January, 1877, Mason sold his half interest to Colonel James M. Barney, of Yuma, for a quarter of a million dollars.

Despite the dire predictions of sundry calamity howlers, news from The King continued good.

"Two trains, one drawn by 18 and the other by 20 mules, bringing 36,050 pounds of rich silver ore from the famous Silver King mine of Arizona, arrived in town yesterday," reported the *San Diego Union*,

[2]This version of the discovery of the Silver King mine is taken from *The Prospector*, Feb. 16, 1884, a weekly journal published at Quijotoa, A. T., by Harry Brook, former San Francisco newspaperman and a writer on the *Pinal Drill*, at Pinal City, during the early days of the boom at nearby Silver King.

January 19, 1877. "This ore is consigned to W. W. Stewart & Co. by Colonel Barney, and will be forwarded to San Francisco for reduction. We predict a first-class sensation among mining men when it arrives there. The freight trains presented quite an imposing appearance as they came into town. The immense wagons . . . are the biggest we have ever seen here. The hind wheels are seven feet high and look as if they might weigh a ton each."

San Francisco Stock Report of April 30, 1877, listed a sale of 33 tons of Silver King ore, best of which brought $4650 per ton, and the lowest grade $1230; and two months later (June 30, 1877) *Arizona Citizen* noted that the Silver King had been shipping 30 tons per month for about a year, no ore being shipped that yielded less than $1000 per ton.

"Actual receipts from sales of ore in San Francisco have averaged $1400 per ton, a few selected lots going as high as $15,000 to $20,000 per ton," it was stated.

Benjamin W. Regan, last remaining owner of the four original locators, sold his half interest in the mine in May, 1877, for a reported $300,000. The Silver King Mining Company, organized in San Francisco that summer, was capitalized at $10,000,000, and James M. Barney was named as manager.

By the close of 1880 the Silver King Company had paid a dozen dividends, aggregating $525,000, without levying one assessment.[3]

Meanwhile, in the vicinity of the King mine, had grown up a small, well-behaved mining camp; and five miles downgrade on Queen Creek, at the northeasterly foot of Picket Post Mountain, flourished Pinal City, milling center for the King ore.

Perry Wildman, one of the leading merchants and citizens of Silver King and Pinal City, in a manuscript of reminiscences written for the Arizona Pioneers' Historical Society under date of October 30, 1926, recalled many historical and humorous incidents in the lives of the two camps.

"Silver King was quiet and orderly as one could wish," asserted Mr. Wildman. "We were a peaceful, contented, happy people. Each man was a law unto himself and anything unruly was not tolerated. We never locked our doors, and robbery was a thing unknown. Gambling, of course, was wide open but always on the square . . ."

Water was very scarce. Bill O'Boyle, who had control of the Silver King water system, brought in on pack animals and delivered to the homes a supply sufficient for drinking purposes and cooking. This was sold at five cents a gallon.

"When we wanted a bath we either went up to the mine and caught the condensed water from the boilers, or waited until it rained and caught the water as it ran off the roof of the house," recalled Mr. Wildman.

Although rainfall in the Pinal foothills ordinarily is light, one cloudburst came close to putting the Silver King camp out of business.

[3]*Pinal Drill,* Dec. 18, 1880.

About four o'clock in the afternoon of a scorching hot day in August, 1883, two small clouds had appeared on the horizon. Growing rapidly in bulk, they had assumed tremendous proportions within half an hour when they met at the top of the Stoneman Grade.

"In less time than it takes to describe it, water poured out of the clouds, creating a perfect torrent that came down the side of the mountain with a roar," wrote the pioneer mercantiler of Silver King. "Large boulders, five or more feet in diameter, rolled down the ravine like pebbles. A miner in his cabin was caught by the torrent before he could escape to higher ground and his body was afterward recovered about ten miles below.

"My store building was flooded with water and many sacks of grain carried off; also platform scales and many boxes and barrels of merchandise that were in storage awaiting transportation to Globe by pack-train. Among the articles in the warehouse were about 20 rolls of barbed-wire fencing, all of which were carried out of the building. When we went down the ravine, after the flood had subsided, we found the wire a tangled mass stretching from bush to bush along the banks for miles. We gave up trying to save any of it . . ."

When the Reverend G. H. Adams, a Methodist minister, asked Mr. Wildman if there wasn't some building in the town where religious services might be held, the storekeeper had said he didn't think so—but hastened to add, "Let's see what we can do about it!"

"We first went to Thompson and Bowen's saloon where four or five faro games were running, as well as other games, and several were patronizing the bar. I suggested to the boys in the saloon that we needed money to build a church, which we could also use as a school. It was wonderful the way the men at the tables and bar threw the coin into my hat as I passed it around. From there we went to Bill O'Boyle's and Bob Williams' place with like results."

Donations from other townspeople and the mine swelled the fund to $1200, a sum sufficient to buy all the lumber and material needed. With labor voluntarily supplied, a nice building was soon erected on the hillside opposite Williams' hotel. In addition to providing a meeting place for roving ministers of any denomination, the "social center" was used for dances and other entertainments, and for school classes. As it took a long while to get an appropriation from the Territorial School Fund, residents of the camp defrayed the costs of school and maintenance by holding dances, basket socials, and other fund-raising affairs. Miss Fannie Jordan was remembered by Mr. Wildman as the first teacher at Silver King. Later, he said, Miss Angie Doran was employed.

"The Silver King Company, as a rule, paid their men off monthly in drafts on the Anglo-California bank at San Francisco," recalled the former storekeeper. "In order to keep the money from leaving town as much as possible, I made it my buisness to bring cash into the camp and had to use all sorts of ways to accomplish my purposes. I very seldom had it shipped by express as there were so many holdups of stages between Florence and the mining camps. Most of the time I would have my friend, Don Carlos Hayden, at Tempe, put a bag of silver inside a sack of grain mixed in a load of several like sacks brought to me by the Mormon freighters. Although I pursued this course a long time and brought into the camp many thousands of silver dollars, I never lost one single dollar. Sometimes I would get the coin from Los Angeles by freight, packed in boxes with merchandise. It was somewhat risky but I felt it was safer than by express.

"One time I was rather late providing for payday, and I wrote to Union Hardware and Metal Co., in Los Angeles, to ship me by express five kegs of nails, each keg to contain about half its weight in silver dollars and the other half in nails. As nails were worth only 12½ cents a pound it was a foolish thing for me to do, because the express charges would be more than that per pound and most anyone would think it queer to see nails going through by express. After I mailed my order I regretted doing so, but could not cancel it and had to await results.

"The nails came through all right as far as Florence; but when the stage came in to Silver King the day I expected them I asked the stage driver if he had any nails aboard for me? He replied that there were five kegs of nails on the sidewalk in front of the express office, but he could not bring them up that day as he had such a heavy load of passengers and baggage. So the nails laid out on the walk all day and night, sat upon, and rolled about; but they came up on the stage next day and it turned out all right. No one was the wiser but myself—and I never tried that method of bringing in money again!

"We had no ice and water was always scarce. I had great difficulty in supplying butter during the summer months and would ship it in from Los Angeles in two pound rolls, in tight casks, pickled in salt brine. The butter and brine would amalgamate, and it had to be sold by the dipper-full."

On one occasion a group of the boys at Silver King developed a craving for a fish dinner and asked their accommodating storeman to procure the main ingredients. The fish was duly ordered from Kansas City, and since it had to be packed in ice and repacked several times en route the ultimate cost was $60 for only 25 pounds of meat.

"But it was a treat," recalled Mr. Wildman. "We didn't care how much it cost!"

As milling town for the King ore, the fate and fortune of Pinal City was inescapably bound to that of Silver King. Pinal, however, was the larger and more lively of the two towns.

"A year ago we were almost without hotels, drug stores, assay offices, mechanics and professionals," reported the *Pinal Drill*, July 9, 1881. "We have now two large hotels, half a dozen excellent restaurants, nearly a dozen elegant saloons, two large blacksmith and wagoners' shops, two drug stores, groceries, watchmaker, photograph gallery, brickyard, lime kiln, and seven well assorted stores, some of them with immense stocks of goods of every description. We have a large lumberyard, two large livery stables and corrals, bath rooms, two barber shops, six lawyers, four doctors and a preacher, several immense freight teams. We've got a bank, Wells Fargo office, our post office has become a money order office. We have built a new church and improved our schoolhouse. Large, elegant stone buildings have been erected. There are two large halls and an Oddfellows lodge . . . All is life, energy and progress . . . Our population has more than quadrupled. Whole streets have been built and suburbs are becoming covered with houses . . . The *Pinal Drill* has printed, issued, and scattered over 30,000 copies within that time; that is, one year. They have gone all over the country. We know it has been quoted in over 700 different papers in America and some in Europe . . ."[4]

Among the components contributing to Pinal's life and energy, if not its progress, was a sometimes obstreperous Oriental colony. One shooting affray in the "China Block" elicited the following comment from the *Drill*:

"Qui Gee shot Go Ghu. Dang Fook and Wang Wy shot Sue Gee, all in general row at the Chinese gambling house . . . Dang Fook and Wang Wy ran away and the police are after them. Sue Gee was shot in the mouth and shoulder. He cannot eat and will probably die. Go Ghu is dead. Qui Gee is under arrest. Dang Fook and Wang Wy are likely to be caught . . . Dr. Bluett says that Sue Gee shall not die. He is fixing a wire, ironclad, fireproof jaw on him so as to secure the safety of Sue's masticating machinery hereafter. The funeral of Go Ghu took place Thursday afternoon . . . in a very orderly and respectable manner."

The laying-away of Go Ghu, in July 1881, may have been the Chinese funeral described by Perry Wildman in his early-day recollections of Pinal in the Arizona Pioneers' Historical Society archives.

"The Chinese friends of the deceased wanted their comrade buried 'allee same 'Melican man,'" wrote the boomcamp storekeeper. "They were told it would cost them about $40 to have the ceremony performed, to which they agreed. I do not recall who it was that officiated as master of ceremonies, but he had dug up a copy of Shakespeare somewhere, and when the procession reached the grave he happened to open the book at the play of Romeo and Juliet. This he read, with all the solemnity and reverence he could command, and the funeral was a pronounced success. The Chinese were satisfied and paid the $40, which was soon distributed over the bars of the various saloons . . ."

One of the prominent citizens of Pinal's Chinese colony was Jim Sam, owner of the Bank Exchange Restaurant and Chop House. Although he was one of the best-known and most highly esteemed Orientals in the Territory, Jim refused to acknowledge his Chinese ancestry.

"I no Chinaman!" he would say contemptuously. "I born Hongkong. I Blitish subject!"

[4]For some reason unknown this news story failed to mention the fact that Pinal, at this same time, also had two breweries—the United States Brewery, on Pearl Street, located in a building 143x38 feet in area, and the Pinal City Brewery, also on Pearl.

Jim took to wife a Mexican woman. With the birth of their first child he said to an acquaintence, "Me Blitish subject; my wife she Mexican. What you think him little boy is?"

"Half Chinaman and half Mexican, ain't he, Jim?" said the friend.

"No," grinned the proud father. "Him little sonny-gun!"

One day a self-asserted bad man known as Shoot-em-up-Dick, wearing fringed buckskin, two guns, and a Bowie knife, swaggered into Jim's restaurant and ordered the best meal the house provided. After eating, he demanded the best cigar available and started leisurely toward the street door, ignoring the cashier's stand. Withdrawing a big six-shooter from under the counter, Jim sung out cheerfully, "Hey, fliend—you forgettum somesing, yes?"

"No, you damned yellow heathen," snarled the departing badman. "I didn't *forget* nawthing! I was your *guest*. I'm Shoot-em-up-Dick!"

"Oh, so?" said Jim Sam blandly, meanwhile directing the muzzle of his gun toward the other's belt buckle. "So—you Shoot-em-up-Dick, I Shoot-em-down-Sam! You pay pletty damn soon, or Shoot-em-up-Dick be pletty damn dead!"

According to the twice-told tale, Dick paid.

Another storied character in Pinal's past was Jack Keating who drove mail stage from Silver King and Pinal to Florence. Holdup men had waylaid him so many times that when Jack would see a stranger approaching on the road he would halt his team, clasp the lines between his knees, and raise his hands high over his head. If the stranger did not happen to a holdup man, Jack would act as if Fate had betrayed him.

"All right!" he would growl. "But, by God, I'll bet th' next one of ye'll ask for the box!"

What was likely the least boresome day in the history of Pinal City—definitely so, if one is to set aside the Day of the Flood when the town was all but wiped off the face of the earth—was the day of Bill O'Boyle's murder and the lynching of his slayer.

O'Boyle, the highly respected owner of a saloon and hotel at Silver King and a second saloon at Pinal, was shot and killed in this last mentioned place of business as result of an argument over a card game. His slayer, a man named Fales, alias Goode, was said to be the leader of a gang of five desperadoes who had been terrorizing the two towns, "making threats against citizens and shooting recklessly on the streets."

Following the killing, Fales was arrested and taken to Lewis' Corral where he was left in charge of a deputy sheriff. About 10 o'clock that night a party of armed men, said to number 25 or 30, stormed the corral, demanded and received the prisoner, and summarily suspended him by the neck from a pole behind the corral.

Reporting the lynching in its issue of August 30, 1884, *Arizona Silver Belt* described the victim as 23 years of age, "a very intelligent man and evidently well educated."

In at least part of that statement the frontier scribe was in error. If the desperado had been "very intelligent" he would have thought twice before murdering a man as wholly popular as Bill O'Boyle.

Reposing in the files of Arizona Pioneers' Historical Society are several fascinating manuscripts written by an anonymous pioneer who assertedly arrived at Pinal City in 1881.

Included among these recollections are several anecdotes concerning Maggie O'Boyle, widow of the murdered man.

"Mrs. O'Boyle was a hard-working, good-hearted Irish woman who loved to go to all the dances in both towns," wrote this nameless raconteur. "At the dedication of the first large adobe schoolhouse erected in Pinal, the dedication was opened by prayer by the Rev. Mr. Williams of the local Methodist mission. After the services the dancing started, and Brother Williams stood watching the dancers. Suddenly Mrs. O'Boyle grabbed him by the arm and said, 'Come on, damn ye! Don'tcha dance?'

"After leaving Pinal I did not see Maggie for perhaps 25 years," wrote the former Pinaleno. "Then, one day, while in Casa Grande talking to Charlie Bennett, I noticed a little old dried-up woman crossing the street. Charlie said, 'Do you know who that is?' 'I do not,' I answered. Then Charlie called to her, 'Oh Maggie! Come over here.' When she came, he asked 'Do you know who this is?' She peered up into my eyes and said, 'Oy don't know ye, and Oy don't want to—but who in hell are ye, anyhow?' "

It was this same anonymous pioneer who recalled a system of highgrading commonly practiced by teamsters between Silver King and Pinal City.

"In hauling the ore from the mine to the mill," it was recalled, "the drivers frequently had occasion to speak forcibly to their animals. Rocks were the most forcible arguments at hand—and, usually, it was the very finest silver specimens which were thrown at the teams. This was especially apt to be true if some friend of the driver happened to be passing at that time; and the friend was almost always curious enough to see what kind of a rock had hit or missed the mule. It is thought that considerable highgrade ore, which was intended to be milled, passed into other channels by the mule-skinner route."

With the Silver King mine pouring forth wealth and the Peacock of Prosperity preening her tail feathers, there appeared in Pinal City an aged man who exhibited great interest in the Silver King mill where 20 stamps, working around the clock, were pounding out the stuff of which millionaires are made.

"The man was evidently in need of help and soon went to the office of the company and announced himself as Sullivan, the old soldier, the original discoverer of the vein, and humbly asked for work," reported Harry Brook in the *Quijotoa Prospector,* February 16, 1884. "Although long before he had been given up as dead, and very few of his old acquaintances survived, he was identified beyond a doubt . . ."

On leaving Mason's ranch, the man said, he had crossed the desert to California where he had supported himself by working as a farmhand. Ever hoping to obtain sufficient means to return to Arizona and secure the benefit of his discovery he had labored on, year after year, until at last he learned of the rich discovery of silver by Mason's party and knew he had lost his chance of making the location for himself.

Although, under mining law, Sullivan had no right of ownership in the mine, the Silver King Company provided light work for him and took care of him the remainder of his days.

By 1888 mining circles were hearing rumors that the mighty King was pinching out. Beginning at the top with a blowout more than 150 feet in diameter the ore body had tapered like a cornucopia to the 1200-foot level and there had disappeared. During the 15 years of its working the King had yielded ore to a value of several million dollars—published estimates vary from $6,500,000 to $17,000,000—but taps eventually sounded for the great mine.

"The Silver King . . . one of Arizona's oldest and greatest bullion producers, closed down the first of February and now only five men are employed on the property," stated *Arizona Weekly Journal-Miner*, March 4, 1891. "The ore, apparently, is exhausted, and all effort to discover new bodies has resulted in failure. The 40-stamp mill is now running tailings and only has a sufficient supply to keep operating two months."

In 1914, and again in 1921, attempts were made to reopen the King mine, but results were not phenomenal.

Time passed . . . and in February 1959, when I inquired at a gasoline station in Superior for information concerning the old mining camp of Silver King, I was told that most of the property there was owned by one Richard T. Lobb, born at Silver King in 1885 and for 56 years a resident of that place. No, said my informant, it would not be possible for me to talk with Mr. Lobb because he was off at Phoenix, or somewhere, on business. There was nothing left at Silver King, anyhow. No one lived there; nor would I be able to drive there in my old Mercury sedan. The road, warned the station attendant, was so nearly impassable that nothing but a four-wheel drive vehicle could get over it.

The report was quite typical of the misinformation a ghost town prowler accumulates along the way.

Spurred by curiosity, and with instinct and a U.S. topographical map of the Florence Quadrangle to guide me, I turned north off U.S. 89 at the Superior airport. After skirting the city dump and a large power plant, I began following a rather good bladed road that rambled up a ravine in the desert mountains back of Superior. After four or five miles a few old mine dumps and prospect holes came into view. When these were substantiated by a few crumbling adobe ruins, my pulse quickened! Approaching a "new" ghost town is, for me, always a thrilling experience, and I felt certain the ruins ahead could be none other than Silver King.

What would I find there? Would there be ruins of the Grand Hotel, perhaps? Or the Williams' Hotel, built at an asserted cost of $12,000 and "elegantly furnished" with Brussels carpets? Possibly I might even locate the remains of the church-schoolhouse built by public subscription, or of Ochoa and Buckalew's Hall . . .

Continuing up the winding grade, still pleasantly lost in my tizzy of anticipation, I soon began seeing a few buildings that were better preserved. Beside one small wooden cabin stood a car. Across the road, nearby, sprawled a barrier formed of splintered boards, rocks, and what-have-you. On one of the boards was crudely lettered: "Guided Tour and True Story of Silver King, $1 per person. No Trespassing."

Standing self-consciously back of the barrier were half a dozen city-garbed men, women, and children; also another woman, obviously the guide. Men's trousers, bobbed at the knees and picturesquely tattered, covered her lower extremities. Her upper section was clad in a green sweater and a man's hat.

"We're starting to the ruins now," said the lady in the tattered trousers. "If you want to go along it'll cost you a buck."

"Okay!" I said, as brightly as I could manage, considering my amazement at these developments. "Wait till I get the money and my camera . . ."

"Camera?" The eyes narrowed.

"Why, yes!" I said. "I'll want to get some pictures . . ."

"We don't allow picture-taking here," said the woman. "Not for any purpose." Obviously against her wishes she finally agreed that I might take a maximum of two pictures of the ruins. These, she specified, must be for my own collection, only—not for publication in any magazine or book.

Thereupon, our guide introduced herself as Grace Middleton, part owner of Silver King and a resident there for 19 years, and the group of us started up a rocky, water-gashed road. This, we were told, had been the main street of Silver King. After showing us the former jail—a rather large building of stone construction—our guide began pointing out various shafts and telling us how much silver each had produced, and how deep it was, and when it had been sunk, and by whom.

Almost centering the townsite stood a huge house, almost a mansion. It was easily the most impressive and conspicuous building remaining in the town. Mrs. Middleton called it "The Big House" and said it had belonged to the manager of the Silver King.

"He was an Englishman and had only the finest of everything," explained our guide. "He used it for a guest house, company offices, and pay office."

Grace Middleton, it seemed, had lived in the Big House a number of years and still carried a key to one of its doors. Letting us inside, she conducted us over the lower floor, including the dining room with its 14-foot ceilings, big French plateglass mirror over the mantel, and built-in hardwood wine cabinets. Ascending a hand-carved solid walnut staircase, we found the second floor given over chiefly to bedrooms and bathrooms, each of the latter still equipped with its cast iron tub, lavatory, and old-fasioned flush toilet complete with brown wooden wall-tank and pendant pull-chain.

From the second floor a steep narrow stairway led through the third floor attic to a parapet on the roof.

"That was where they kept a guard stationed to shoot Indians," said Grace Middleton.

Winter's early darkness was creeping over the land as we made our way back down the rutted road to Mrs. Middleton's cabin where she gave each of us a small piece of rock, assertedly ore from the Silver King, and invited us to come back some time. After the other spectators had trundled off down the grade the woman handed my dollar admission back to me. She said she wouldn't charge me for the tour because she always enjoyed reading my stories and then she went on to say that one of her four cats had been ailing that morning and had crept off into the brush.

"Now," she grumbled, "I've got to hunt him in the dark, and my flashlight batteries are plumb dead . . ."

Since she had returned my dollar and paid me a compliment or two besides, I figured I could be magnanimous to the extent of 30-odd-cents, so gave her the two good batteries from my own light. Slipping them into the nickel cylinder of her flash, Grace Middleton started off in search of her missing feline.

"By the way," she called back to me. "If you should write anything about this place, *don't call it a ghost town.* I detest that expression. Besides, Silver King has never been a ghost. There's always been somebody here . . ."

A year later I was in Superior again and thought I'd take another run up to Silver King—hoping, I must confess, that Sister Middleton, meanwhile, had retired to Yuma or Tucson or some other pleasant place, or had mellowed a bit in her attitude toward photographers.

But Luck wasn't in my corner.

As I came in sight of the old ruins I found the same barricade across the road, and my vehicle had barely stopped moving before Grace Middleton emerged from her cabin.

"Hi!" I called cheerfully. "You won't remember me—"

"Yes, I do!" interrupted Grace quickly. "You're that writer woman who was here about a year ago . . ."

After we had shaken hands and inquired politely concerning each other's health and fortune, we moved into the lee of the old cabin where the frigid February wind wasn't quite so strong and the weak winter sunshine felt a trifle warmer, and there we talked of practically everything except "cabbages and kings."

Soon as I thought we had reached a cozy enough footing, I asked her permission to shoot a couple of pictures of the ruins for inclusion in this book.

"Indeed you won't take any pictures!" bristled Grace Middleton. "I told you once and I'm telling you again—I will not have Silver King called a ghost town!"

With that we sort of dropped the subject. Grace said she still enjoyed my stories and, eventually, I said I must be on my way. We shook hands and I started back to my truck.

"Hope you don't feel too bad about not getting any pictures here," the woman called after me.

I stopped. "I don't feel too happy about it," I said. "But I scarcely think it'll keep me awake tonight."

"No," said Grace Middleton. "I hope not . . ."

As I dropped downgrade, out of sight of the old town and its last resident, the woman was still standing in the yard of her cabin, watching me leave. She was holding in her arms one of her 13 cats—a big yellow tomcat. The cold February wind was beating into her pinched face and tugging at her faded hat and clothing, her graying hair. She looked terribly alone. Being the last resident of a once prosperous and lively mining camp probably isn't exactly the happiest business in the world.

Copper Smelter at Helvetia, 1900

—Courtesy James B. Cooper.

"Nothing can be more fascinating than ruins. It is the ruins of the Old World countries which attract the great influx of travelers every year; for in these ruins lie the charm and romance of past greatness and glory, and the haunting presence of the tragedies of long ago."

—MARY HUGHSTON, *in Arizona Highways, April 1948.*

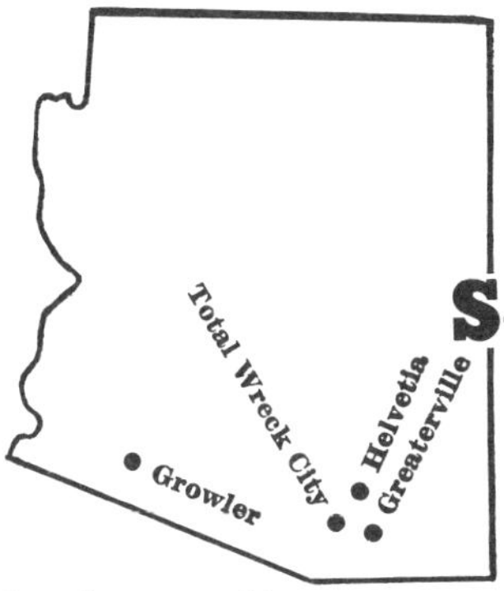

Greaterville to Growler

STRETCHING across southern Arizona for 175 miles, approximately one-half the width of the state, and for more than 120 of those miles bordering on Old Mexico, Pima County is a place of high, pine-forested mountains and dry, rocky desert; of burgeoning Tucson, and dude ranches, and rocketing real estate values. For more than a century this county also has been the scene of recurrent mining booms that have waxed and waned, and in waning have left behind an assortment of ghost towns bearing such fascinating names as Gunsight, Growler, and Greaterville, Silverbell, Twin Buttes, and Total Wreck.

First of these several mining booms, and the one that aroused the most popular interest, was that at Greaterville, in the Santa Rita Mountains southeast of Tucson, where the starring attraction was placer gold.

Since placer mines seldom produce gold over a long period of time, their relationship to lode mines in not unlike that of a desert waterhole compared to a deep well. Once it has been put in production the well may yield good water for years to come; but it takes time and money to drill it—too much time and too much money for a man dying of thirst. But the desert waterhole, ephemeral as it may be, offers immediate water, merely for the dipping . . .

Though Burro Bill might find the prince of all lode mines, he had little choice but to sell his discovery for a fraction of its actual worth. Generally speaking, it was only the sharp promoter who could raise the capital necessary to develop such a claim, erect a mill to treat its ore, establish a water source, possibly build a branch railroad to the property. Long before all this had been accomplished and the first dividend check was in the mail, Burro Bill was out scouting the hills and desert in search of another mine—to be sold, like the first, for a bare fraction of its worth.

But if Burro Bill found a good placer field—golddust and nuggets—the picture was altogether different. Such gold could be recovered with no greater investment than a homemade rocker or sluice box, a gold pan costing less than a buck, or even an ordinary tin skillet. One moment, a nugget might be lying on bedrock where it had lain for nobody-knows-how-many-thousand years. Next moment, in a manner of speaking, it was smiling up from the dark, wet riffles of a two-dollar sluice box—heavy, buttery-yellow, virgin gold, legal tender for all debts foreign and domestic!

So it was in placering that Burro Bill came into his own. He might begin the day with an empty poke and emptier belly, and sunset might find that condition unchanged. On the other hand, one lucky turn of the shovel, one rich pocket beneath a boulder, one yellow-plastered crack in bedrock, might produce the wherewithal to buy bacon and beans for a month, a pair of brogans, a new jackass, even a bottle of bug juice and a painted woman to share it . . .

With placer gold holding such terrific appeal, it is not to be wondered that men's imaginations were fired by reports emanating from the Santa Rita Mountains in 1875.

The new placering field had been discovered sometime during the previous year by one Alvah Smith, described as "a prospector of several years experience in Arizona."

By June 1875, between 350 and 400 men were working the gulches on the southeast side of the range. Greater Creek, Smith's Creek, Sucker Run, Boston Gulch, and Mac's Creek, all had their prospectors feverishly digging the gravel and packing it to water—the poorer ones relying upon their own muscles and sinew, the more affluent utilizing the sturdy backs of burros.

July and August 1875, found the excitement mounting to fever pitch, with daily news reports in the *Arizona Citizen,* at nearby Tucson, acting as added fuel to the flames.

David Burroughs had brought in nearly 10 ounces of nice placer gold, the product of three days work of himself, alone . . . Horace Arden was regularly making an ounce a day in the Smith district . . . Jack Ralston had come in with $150 in nuggets, the largest worth $90.50, the balance in bits from $1 to $10 . . . David T. Harshaw had brought in a sack of dust and nuggets weighing $843—the result of four days labor for three men . . .

"The Santa Rita placers," declared the *Citizen,* "are entitled to rank equal with the best placers ever discovered . . ."

By midsummer, that year, miners of the district were beginning to think about the site for a permanent town.

"Of aspiring localities for the townsite there are three or four," it was stated in the *Citizen* of August 7. "Thus far, Greater Creek appears to have the start, and by far the largest number of settlers. Musick's Camp, on Smith's Creek, and Harshaw's Camp, on Mac's Creek, have each quite a settlement . . . In over 20 years of experience on the frontier," wrote the Tucson reporter, "I have never been in a more orderly camp. I have yet to hear of the first difficulty between Americans in the district . . ."

Herman Welisch, pioneer merchant of the placer field and the area's leading buyer-of-gold, in February 1876, estimated that in the one year these placers had been worked they had produced between $60,000 and $80,000 in gold. One man, in only three months had taken out $8000.

During that summer of 1876 the news from Greaterville continued good. Mr. Welisch had brought in 60 ounces of gold dust valued at more than $1000 . . . Mr. Candelario had found a nugget weighing more than 12 ounces, and only half an hour later a nugget worth $104 had been picked up by Mr. B. Bedolla . . . and in September, Joe Herring had taken $162 out of his claim in a single pan of dirt—one piece, alone, weighing $150.

Unlike many placer fields, that in the Santa Ritas was not only a 90-day wonder, but even after four years of intensive working, the yield was still good.

"About 250 men are at work in the (Santa Rita) diggings," reported *Arizona Star,* August 7, 1879. "The amount of work which has been done, and the amount of gold taken out, is more than our people are aware of. Mr. Hefty, alone, during the time he has been operating, has taken out over $40,000 . . . It is estimated that not less than $250,000 has been the product during the past two years and a half . . ."

Six years later, with the district then in its 11th year, men were still turning in the "bonnie yellow dust." In the *Report of the Director of the Mint,* 1885, it was stated that the Greaterville placers had produced more than half a million dollars in the time they had been worked, and that the largest nugget taken out here had weighed 37½ ounces.

"There are 30 or 40 men working here all year, who take out on an average of $1200 per man, all with the rocker. They frequently pack the water over a mile in canvas sacks on burros," stated the government report.

"One hundred and fifty men are working at the Greaterville placers," reported *Arizola Oasis,* January 18, 1894. "They appear to all be doing well . . ."

As late as 1928 50 persons, including 19 school children, still resided in the old placer camp in the Santa Rita foothills. A typescript, written by Patrick J. Coyne in that year, and now preserved in the files of Arizona Pioneers' Historical Society, supplies almost our only first-hand picture of Greaterville, the town.

According to Mr. Coyne it was a lively camp with several dancehalls, a number of stores, and many saloons.

"Our first public school was started in 1882 with Mr. Harper, a Southern man, as teacher. It was the third school started in Pima County, and I was its first trustee," wrote the old timer. "At one time there were about 500 Mexicans in Greaterville, and it was a great place for smuggling. Sometimes they would bring in ten or twelve burro loads of mescal a day. I was justice of the peace for a number of years. Our deputy sheriff was Bob Kerker. We had a round hole dug in the ground for a jail, and would lower the prisoner down on a rope . . ."

With both its water and gold growing scarcer, year by year, Greaterville's boom gradually withered and died. The town's post office, opened

Ready for a Sunday ride, Helvetia, 1900

—Courtesy James B. C

Ruins, Santa Rita Mountains, 1959

in 1879, was discontinued in 1946; and a few years later I found in the former community only half a dozen Mexican families.

The few adobe-and-stone ruins remaining as tokens of Pima County's greatest placer gold camp are situated a couple of miles north of the Santa Cruz-Pima County line, about five miles west of Highway 83. But even if a visitor does not succeed in finding the ruins or any overlooked nuggets, his trip will not have been made in vain. The Santa Rita foothills are still a delightful place to pass a few hours dreaming of days gone by.

Over the summit of the Santa Ritas, half a dozen miles northwest of Greaterville, the adobe skeleton of Helvetia slumbers in the sun.

Although the earliest mining discoveries in that district are believed to have been made prior to the Civil War, the first mine to be worked seems to have been the Frijole, owned by John Weigle and William Hart, where development got under way in 1880. At that time, and for several years after, ore from the district was treated at the Columbia smelter, near the old Tip Top camp, on the west side of the Santa Ritas.

In the early 1890s most of the important claims in the region were acquired by the Helvetia Copper Company, of New Jersey, and the town of Helvetia began taking form. Within a couple of years its business district embraced several stores and shops, a post office (1899-1921), and two lines of stages giving connection with Tucson. After investing some $800,000 in development work (and recovering approximately half that much in copper), Helvetia of New Jersey was succeeded, in 1903 by Helvetia of Arizona. This company built and operated a smelter at Helvetia, between 1905-07, after which ore was shipped to the Old Dominion smelter at Globe.

Prior to 1950 the Helvetia mines had produced around $4,000,000 in ore, including some 17,000,000 pounds of copper.

Strangely enough, the camp was not as enduring as the mines that fathered it. At the time of my visit, in 1959, it had wasted away to a sprinkling of adobe ruins. The old smelter was completely gone, but its 50-year-old slag pile still spilled into the ravine below like a frozen waterfall of shining jet. Most of the old townsite then was owned, or leased, by Lewisohn Copper Corporation, trucks were still hauling ore from the mines, and the entire property bristled with Keep Out signs.

Every Western state has its mining camps of unusual nomenclature. Rough and Ready, You Bet, Whiskeytown, Hangtown, Tin Cup, Miner's Delight—the list of odd-ball names is all but endless and every ghost town buff, I suppose, has his favorite. My own nomination for the strangest name ever borne by an actual mining town and post office is that of Total Wreck City, in the Empire Mountains, County of Pima, Territory of Arizona. The town, naturally, was named for its leading mine discovered in 1879 by one Jerry Dillon. Observing that the hill in which his new claims were located looked like "a total wreck," Jerry simply hung this name on his mine—and that was that.

It wasn't long before a town of considerable promise was collecting around the mine and mill. In time, Walter Vail acquired the property and when he made application for a post office he assertedly told the Postal Wizards in Washington that unless the new office could be named Total Wreck City, after his mine, they could just plain keep their gol-durned post office! So, Total Wreck it was.

Arizona Daily Star, in its issue of Feb. 13, 1883, reported that the Total Wreck mine was then producing from 65 to 70 tons daily of free-milling silver ore, averaging $65 per ton.

"Total Wreck is a thriving village and is daily becoming more important . . ." it was stated. "There are five saloons, three stores, one butcher shop, four restaurants, one blacksmith shop, one shoemaker's shop, and eight or ten Chinese washhouses . . ."

Four months later *The Star* noted the addition of three hotels, a lumber yard, and a brewery—but the end was already in view.

After producing near to half a million dollars from rich surface ore, the Total Wreck mine began failing sharply in 1884. Though sporadic production in later years has boosted the mine's total output to around $750,000, Total Wreck has not functioned as a town since its post office went out of existence in 1890.

Half way across the state from Total Wreck—but still in Pima County—roam the evicted ghosts of Growler.

Growler is a well-known name in the part of the world lying to the southwest of Ajo, on the road between Darby and Pagago Wells. There are the Growler Mountains, reaching their apex in Growler Peak; immediately beyond the mountains lies Growler Valley and Growler Wash, and down near Bate's Well is Growler Pass and the Growler mine—a nice, generous collection of place names all honoring one John Growler, desert prospector, who made his way into this region before the turn of the century. Until very recently, in the same area, there was also the ghost town of Growler . . . but Growler isn't there, any more.

The Growler mine—gold and copper—never distinguished itself by achievement. It was first worked about 1907 and probably reached its peak of activity soon after 1916 when heavy equipment was freighted from Yuma, by mule teams, over the Camino del Diablo. In the years during and immediatedly following World War I, the town of Growler is said to have numbered several hundred inhabitants.

About 1928 the little ghost town and its mine were purchased by Bert Long and his stepson, Charles Greer, both of Ajo. After nearly 30 years of ownership, and several non-profitable attempts to work the mine, the property was sold to the U.S. Department of the Interior and, subsequently, became a part of Organ Pipe National Monument.

Ghost town fans, over the years, always enjoyed visiting the remote old camp, with its crumbling store and office and cabins, its mine buildings and workings. But one day in May 1962, Dave Rees drove out to Growler, only to find that Growler was gone. It had been bulldozed out of existence. When he queried Monument officials he was told that the old structures had been obliterated as "dangerous and unsightly."

"A park attendant, questioned about Growler, said he didn't know the ruins were the remains of a ghost town," wrote Rees in *Arizona Daily Republic*, May 31, 1962.

So future visitors to the little old ghost of Growler won't find much there but possible bulldozer tracks . . . and I have yet to see a bulldozer track that seemed to me especially interesting.

"Mining stimulated the construction of roads and communication facilities that have made more accessible the agricultural and horticultural resources of the Southwest. It contributed to the building of permanent cities and the advancement of territories to statehood. It brought people, who found other sources of livelihood when the mines had been worked out."

—RANDALL HENDERSON, *On Desert Trails Today and Yesterday.*

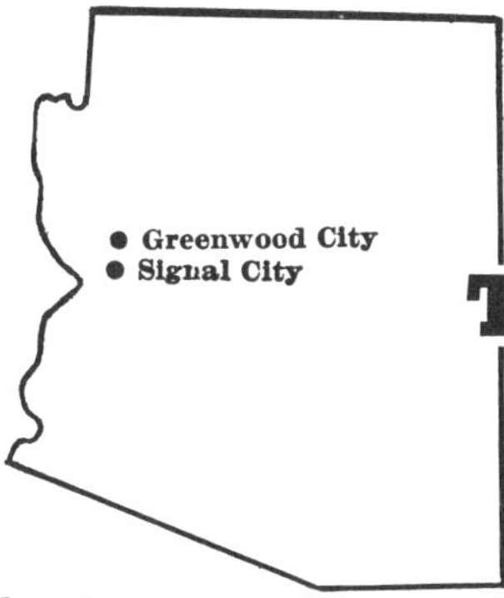

Silver on the Big Sandy

THOUGH CROSSED and bordered some 300 miles by the Colorado River, and bounded another 50 miles by Bill Williams Fork of the Colorado, Arizona's Mohave County is chiefly arid desert laced by spectacular canyons and volcanic ranges. Cleaving through the southern part of this highland expanse lies the long, narrow valley of the Big Sandy River where the early 1870s found a few courageous settlers striving to establish ranches. With cattle and plowed furrows constituting their major interest, these men saw no significance in the fact that a ridge of black rock topped one of the larger desert hills near the valley's lower end. Even prospectors, assuming the cropping to be only a volcanic dike, routed themselves around the base of the hill rather than over its rough summit.

So it was that men had known of the dike several years before anyone guessed that the dark cockscomb was not worthless black basalt but a rich mass of silver chloride!

Discovery of the bonanza was made by two partners, Jackson McCracken and "Chloride Jack" Owen, on Aug. 17, 1874. The hill and lode were named for McCracken; the newly-formed mining district took the name of Owen—and within a year those names were claiming the attention of mining men from Wall Street to the Golden Gate!

Less than six months after the strike it was reported that the "grand McCracken mine" had been purchased by Colonel David Buel at a cost of $240,000. Mining experts who examined the property pronounced it "a second Comstock," and one news reporter told of seeing ore from the mine which assayed $2700 to the ton in silver.[1] Within another month a ten-stamp mill had been installed on the Big Sandy, about a dozen miles from the McCracken; the community of Greenwood City was tak-

[1]*Arizona Weekly Miner,* Jan. 8, 1875.

Opposite page: Portal of Old Timer Mine, on McCracken Hill, 1959

ing form around the mill, and every man willing to work was assured employment at good wages. By March 1875, Greenwood had grown to include a commissary operated jointly by two of the mining companies, plus two other general stores, a blacksmith shop, and a saloon, and population of the district was estimated at 600.[2]

During the several years next following almost every issue of the *Arizona Weekly Miner*, of Prescott, carried encouraging news from the McCracken mines and Greenwood City. Two men from Pioche, Nevada, who had located not far from Greenwood, it was said, had worked two days and sold their prospect for $5000 . . . Colonel Buel was erecting a smelting works . . . There was talk of a narrow-gauge railroad to be built from the mines to the Colorado River, and a new 60-stamp mill was being brought from California.

"Men and families are pouring in from all parts and the road down the Sandy is daily lined with wagons, horsemen and men on foot, packing their blankets on their backs. Greenwood City, on the east bank of the Sandy, about 15 miles above the junction with Bill Williams Fork, is fast growing into a thriving business town. Six months ago not a house was to be seen where now a busy street lined with well-built stores and dwellings, and not a few canvas tents, opens to the traveler's view."[3]

"They have on the dump at the mill 350 tons of ore, at present, and on the dump at the McCracken mine, ready to ship to the mill, the enormous amount of 80,000 tons, all of which will average $60 per ton, making an aggregate in value of $4,800,000!" exulted the loyal *Miner* on Feb. 16, 1877.

Even the Greenwood marts of trade won a pat on the back:

"The Davis House is well patronized, offers the traveler daintily clean beds and luxurious table, and gives him an indefinable feeling of being in a refined home," said the *Miner* on June 22, 1877. "Heavy stocks of goods are carried by Solomon and Meyers, Kimble Brothers, and Levy and Koshland. Money is plentiful, business on a cash basis, and no one seems to have use for bookkeepers. A good physician is there, but the health of the place gives him little to do. There are no lawyers there, the people are law-abiding and cobwebs hang on the scales of justice . . ."

About this time, at a point three or four miles down-river from Greenwood City, there was erected the Signal mill, and here Signal City came into being. Before many weeks had elapsed the new town embraced a goodly business district. A schoolhouse was built, a post office was established in October that same year, a physician hung out his shingle, and on Dec. 28, 1877, the *Miner* carried a headlined story:

[2]*Arizona Citizen,* April 17, 1875.

[3]*Arizona Citizen,* April 4, 1875.

PROGRESS AT THE SANDY

"A man by the name of Murphy . . . was severely cut up at Signal City by another named Mitchell. The affray occurred about some trifling dispute, but it shows that our neighbors in southern Mohave have the true spirit of progress and are advancing toward mining civilization."

More concrete evidence of Signal City's progress was given two weeks later in the *Arizona Citizen,* of Florence.[4]

"The principal street (of Signal City) is lined on both sides for a distance of several squares with commodious and substantial buildings," it was said. "Five general merchandise stores are required to supply the rapidly increasing population . . . said stores carrying stocks that would be a credit to any town in the Territory. Three large restaurants are in full and prosperous operation. Thirteen saloons supply the boys . . . The fine saloon of Messrs Fisher and Nash deserves more than passing notice, their bar and barroom being the finest I have yet seen in the Territory. They also have a quiet clubroom, tastily furnished . . . Quite a number of buildings are going up in this vicinity . . . The Signal mill, lately completed and said to be the most complete in all its appurtenances of any in the Territory, is in full blast, night and day, shifts being worked both at the mill and at the mine . . . The new McCracken 20-stamp mill is nearly completed, and the first stamp will probably fall about January 15. This mill is situated in New Virginia, about one-quarter mile from the Signal mill, and immediately across an arroyo from Signal City. The two towns are virtually one but still bear their respective names. About 60 men, carpenters, masons, machinists, and laborers, are now employed about the McCracken mill . . . They have also erected a row of buildings for offices, machine shops, assay rooms, refining quarters, etc., also a large warehouse. There are about 700 men in Signal City and New Virginia . . . These men get $2.50, $3.00, and $4.00 per day . . . The town, throughout, wears a thrifty, prosperous look. New buildings are going up in all parts of town . . . Mr. Ewing is building an elegant hollow-wall brick residence in the northern part of Signal City, the first brick house in Mohave county . . . Butter very scarce at $1.50 per pound; eggs sell at a bit apiece; milk, when on the market sells for $1.00 per gallon. If the present rate of improvement is kept up, Signal City will, in a year, surprise the world."

Despite its other strides toward progress, Signal City still was far removed from all major centers of habitation and residence there had its drawbacks.

"One can hardly know the isolation of such a camp at that period," wrote J. A. Rockfellow, who arrived at Signal in February 1878. "Practically all supplies came from San Francisco . . . by train to the west side of the river at Yuma, and thence by barge up the river to Aubrey's Landing, where it was loaded on wagons and hauled by long mule

[4]Issue of Jan. 11, 1878.

teams 35 miles upgrade to Signal. The merchants considered it necessary to send orders six months before the expected time of delivery."[5]

The isolated position of Signal City and Greenwood, particularly their lack of adequate mail service, was a matter of frequent comment in Territorial newspapers of that day. One of the more vociferous complainants was *Yuma Sentinel.* In June 1877, after devoting several paragraphs to the progress and promise of Greenwood City, the *Sentinel* concluded:

"The most disgraceful matter remains to be told. Here is a community of over 400 people, building up a prosperous settlement, and actually shipping some $50,000 per month in bullion and ores, and entirely destitute of all mail facilities! But for her own enterprise, Greenwood would have no papers or letters at all," reported the *Sentinel,* explaining that Greenwood's citizens, by subscribing from $2.50 to $10.00 each, per month, employed a private carrier to bring their mail once weekly from Prescott.

On other occasions the matter of Signal's out-of-this-worldness was referred to casually, simply as a matter of routine news:

"Peter, a Mineral Park Indian runner, came into Signal City . . . last week, having made the distance from Mineral Park to Signal, over 90 miles, on foot, in less than two days," reported the *Miner,* March 29, 1878. "Peter was the bearer of papers from Sheriff Comstock to his deputy, P. N. Fisher, and started from the Park with the understanding that if he delivered the papers inside of two days he was to receive $10.00; otherwise, no pay at all. He made it and had time to spare."[6]

But isolation and lack of mail facilities, to a gold or silver mining camp, is of secondary importance to the matter of ore—which Signal City seemed to have in plenty. By the spring of 1878 the McCracken Company's 20-stamp mill was turning out $100,000 monthly, and the Signal Company's mill of ten stamps was yielding another $50,000.

"A monthly yield of $150,000 from one little camp with only two mills is not bad," observed the *Miner* with justifiable pride.[7]

In the summer of 1884, Signal's isolation was somewhat ameliorated by the establishing of a thrice-weekly stageline to connect with Yucca, "carrying U.S. mail, passengers, and freight." Passengers, it was advertised, would be carried at $9.00 one way, and freight at three cents a pound.

"Christmas was celebrated here as usual," reported the Signal "items" in *Mohave County Miner,* December 28, 1884. "The city band, composed of Major McLeod, Mose Levy, Gus Clark, Gabe Levy, Col. Hugh Brennan, and Harry Burns, turned out in full force to serenade their friends."

[5]*Log of an Arizona Trail Blazer,* 1933.

[6]According to the *Alta Arizona* of Jan. 28, 1882, mail sent from Mineral Park to Signal, only 70 miles as the crow flies, necessarily traveled more than 1000 miles to reach its point of destination, going by way of Prescott, Maricopa, Yuma, up the Colorado River to Aubrey, and thence to Signal.

[7]Issue Apr. 27, 1878.

On Christmas Day, continued the account, there were both horse and foot races, with a "grand ball" in the evening.

It was in her observance of Mexican Independence Day, however, that Signal reached her greatest heights of urban sophistication. Though celebrated annually, on September 16, the observance of that holiday in 1888, according to the *Miner,* "was one of the finest that the people of Signal have ever witnessed."

On the evening of the 15th, it was related, speeches were made by three Mexican orators "in a very elaborate way," and on the 16th a big parade took place. Starting from the Eureka House, participants marched up and down Main Street carrying both flags, American and Mexican. That evening a group of Mexican young women took their stand on the platform and sang the Mexican national songs "in good style," following which everyone adjourned to Koshland's Hall for "a grand ball" which was still going strong at daybreak.

But Signal was not given an opportunity to celebrate many such events. Infallible mines are few and far between, and even the great McCracken was no exception. After several good years came dark days when Signal City's mines were plagued by bad management, litigation, fire and flood, refractory ores and increasing production costs. During these years the news from Signal vacillated between hopeful reports that the mill had reopened under new management, and doleful admissions that the mill again had shut down.

Eventually, name of the town all but disappeared from columns of the county seat newspaper, and when the Federal census taker visited Signal in 1890 he found there only eighty-odd persons, mostly Mexicans.

"The mines and mills are shut down and everybody seems to be in the attitude of waiting, to have little to do, and plenty of time to do it in," observed the *Miner* in its issue of August 16, 1890.

After researching the files of Mohave, Yavapai and Yuma County newspapers covering the period 1874-91 and from them tape-recording some 25,000 words of material pertinent to Greenwood City, Signal City, and the McCracken mine, I felt quite well informed on the subject— but I still hadn't visited the site of the old bonanza town on the Sandy. With that end in view, on a pleasant February day in 1959, I turned off Highway 93 near the small way-station of Wikieup and headed south on the 15 miles of desert road leading to Signal.

I had no idea what I would find in the old camp. It might be occupied; it might be completely deserted. The mill might be running, or it might be shut down. Perhaps I wouldn't even be able to locate the old townsite—it happens that way, sometimes.

Of all possible eventualities that trip might yield, I least supposed that within the next few minutes my car would be almost swamped while crossing the Big Sandy River; or that I would be held up five days as result of flood conditions. Neither did I anticipate that it would be a full month before I reached Signal, and in that time I would have made

Former Saloon, Signal City, 1959

Graveyard, Signal City, 1959

a round trip of some 1200 miles to Southern California—but that was how things worked out.

Traveling through a dry, highland desert, where Joshua trees and saguaros lifted their crooked arms, and winter's sun lay warm on the brown rocks, I guided my car along the little bladed road that winds through the rugged hills south of the highway. In about ten miles my wanderings carried me to the brink of the Big Sandy and I was a little concerned to see there was no bridge. Fording desert rivers is not one of my favorite pastimes—possibly because my car has been stuck in them a few times too often. Preliminary investigation, however, showed fresh vehicle tracks leading down into the stream and out the other side and there was no indication that anyone had been stuck in the recent past. If I were to go on to Signal that day there was no alternative but to ford the river, so I followed the lead of the other tracks.

Easing the old Mercury down the sandy ramp into the moving water, I began plowing through the stream toward the far bank. The bottom seemed firm enough and I was just starting to breathe a tentative sigh of relief when both front wheels plunged into a hole. As the cold river water struck the hot metal of the engine a cloud of hissing steam shot high. The drowned motor coughed and died, and the turbid gray water of the Big Sandy began flooding into the car!

With sudden concern for my bedding and food, tape recorder, cameras, and camping gear in the tonneau, I scrambled over the back of the front seat and began moving everything to a higher position on the seats and the back window ledge. As gravel and sand worked out from beneath the wheels I could feel the car settling deeper into the flooding river. Every time its body gave a small, creaking lurch, new spasms of terror shot through me.

With my more valuable possessions moved to places of temporary safety, I frantically considered what to do next.

I couldn't remember passing any inhabited dwelling—not even a desert miner's shack—since leaving the highway, ten miles before. Consequently, though Signal was still four or five miles distant, my best chance of finding help apparently lay in that direction. It was a slim chance, I knew. Even if I were lucky enough to find someone living at the old town it seemed terribly doubtful whether I could reach that point and return in time to save my car and equipment. Not only was my car settling, but I knew now that the river was rising rapidly. The water was already encircling the clutch and brake pedals—several inches deeper than when I had halted, only minutes earlier—and wavelets on the far bank now were lapping at the base of a small willow tree that had been at least two feet back from the water's edge . . .

Without taking time to remove either shoes or stockings I opened the car door on the downstream side and lowered myself into the rolling gray waters—my breath freezing in my throat as the muddy river surged coldly and angrily about my thighs. The rising river, by now, had widened until some 50 feet of water separated me from the safety of the south shore and it seemed to me that never in this world had 50 feet of distance looked so impossibly far! Almost doubtful if I could hold my footing in the flood current—and knowing well enough that if I, a poor swimmer, were swept off my feet I would have little chance of reaching shore alive—it took all the courage I possessed to loosen my hold on the car and strike out for the shore.

Stumbling over that unseen rock bottom, once stepping in a hole that plunged me into the water above my hips, I made my way forward, one agonizing step at a time. After a seeming eternity I could feel that the river bed was sloping upward. Now, the water was only knee deep; a few more quick steps and I was on the dry land—and never had a sandy roadbed felt so wonderfully good beneath my feet!

With one last glance at my faithful old car—a companion of ten years and 165,000 miles, now listing downstream and with its radiator half under water—I started down the long road toward Signal.

Running and walking by turns—running until my heart and lungs seemed ready to burst from the strain; walking until my breathing eased a bit, then running again, always with a terrible apprehension in my thoughts and a prayer on my lips—I followed the tortuous course of that desert way.

Luck, for some reason, was with me—or my prayers were answered. Less than a mile from the crossing I was overjoyed to come upon a small ranch house. When I found there several persons and a pickup truck, I almost collapsed with relief!

Breathlessly I spilled out my story—a story these folks had heard all too many times in the years they had lived near this treacherous ford on the Sandy.

Snatching a stout tow cable, the younger of two men threw it in the back of the pickup, motioned me to get in the seat, and we tore off down the road toward the crossing.

The old Mercury was still standing firm, though the entire lower half of its body was now hidden beneath the gray water, and the river was nearly twice as wide as when I had left it, minutes earlier.

Plunging through the deepening water to the stranded vehicle, my rescuer and I eventually got the cable hooked to my bumper; I climbed in the front seat to hold the steering wheel straight, and the young man floundered back through the flood to his pickup.

A puff of blue exhaust told me when his motor started; the cable tightened, black smoke raised from burning rubber as wheels of the pickup spun on the gravel. Then, slowly, painfully, the old Merc was being dragged over the rocks, out of the water, up on the dry shore . . .

I told the young man if he would pull my car back far enough from the river so it would not be endangered by the rising water, I would be extremely grateful. I had plenty of food and water and warm bedding and would be perfectly comfortable, I assured him. Perhaps I could dry the car, I said, and eventually start the motor. With obvious reluctance, he finally agreed to leave me with my dripping vehicle.

Raising the engine hood I let the warm afternoon sun stream in upon the drowned motor. Later, I was inspired to gather brush and build a rousing bonfire close beside the car in the hope that some of its heat would percolate to the sodden interior. Toward evening I pulled some hot coals out of the bonfire and cooked a skillet of bacon and eggs and made a pot of tea. Shortly after dark, I unrolled my sleeping bag on the ground and went to bed.

I still hadn't fallen asleep when I heard the motor of an approaching truck.

It was the owners of the ranch where I had gone for help—the parents-in-law of the young man who had pulled me out of the river. They insisted I accompany them back to the ranch and be their guest until it was possible for me to continue on my way. All my arguments to the contrary fell on deaf ears.

My hosts were Tina and Angel Madril, native Arizonans of Mexican descent, and two of the kindest folk I have ever known. Tina—small and brown and three times a grandmother, born at Signal and a resident of that locality all her life—hovered over me as if I were one of her daughters, feeding and comforting me, and finally tucking me into bed between crisp, cool sheets.

Next morning I went to work on my car. I'm no expert mechanic, by any stretch of the imagination, but I knew enough to remove the spark plugs and dry them, and paint them and the wiring with ignition seal. Then I drained the sediment bulb on the carburetor, blew out the gas line, and removed the distributor cap and dried it as well as I was able. Mr. Madril was most apologetic that he could not help me.

"It is just that I know nothing about motors," he said in his soft, pleasant voice, with its trace of Spanish accent. "I do not even drive. But even if your motor would start, there is no way to leave here except by fording the river—and the water is even higher this morning . . ."

"Couldn't I go on to Signal?" I asked.

My host shook his head. "No. You see, our little ranch lies in a loop of the river. There is no way to get off our property, with a car, except by fording the Sandy. Every spring there are times when the only way we can get out is to take an old saddle trail that leads over the mountains to Wikieup . . ."

"Why doesn't the county build some bridges over the river?" I asked indignantly.

Mr. Madril shrugged. "I don't know. Perhaps they think we are only poor people down here and do not matter. It has always been so," he continued. "Ever since Signal was a boomtown, more than 80 years ago, the Sandy has been taking its toll of life and property. No one knows how many horses and men have been drowned while trying to cross the river. Every year cars become stuck in the crossings. Some of them are carried away and destroyed by the water; others have their motors and gears ruined by the sand. But still, no bridges are built . . ."

By noon of that second day the Sandy had grown from a desultory stream to a rolling tide, 250 to 300 feet in width, its tossing surface bobbing with brush and trash, even uprooted trees. Although no rain had fallen in our vicinity, we could see masses of heavy, gray storm clouds hovering over the mountains to the north, where the Sandy heads, and knew that there lay the cause of its swollen state.

The third afternoon of my enforced sojourn, Tina Madril asked if I would like to visit the old ghost town of Greenwood City? I was delighted! It had not occurred to me that this predecessor of Signal might be situated on the one side of the Sandy accessible to us.

First of the ruins lay about a mile down-river from the Madril home. Examination of the 84-year-old backwall and stone foundation of the Greenwood mill showed the masonry still standing rugged and firm. Searching over the site, we found one of the old wooden bullwheels; also some heavy chain links, pieces of broken castings, other relics of rust-corroded iron, and a heavy timber studded with stout, square-cut iron spikes.

When I asked Tina where the dwellings and stores had been situated, she led the way through a dense mesquite jungle in which we were forced to bend half over in order to pass through the narrow, thorny tunnels made by generations of cattle that had grazed on the abandoned townsite. From time to time, in the midst of the mesquite, we came upon weathered heaps of adobe, crumbling stone chimneys, and other odds and ends that told of man's one-time presence. Tina called my attention to the remains of several cabins built of a rock similar to volcanic tuff. The material, she said, had come from an old quarry across the river. Hunting farther through the ruins, we found bits of sun-purpled glass, fragments of white Ironstone ware, and broken shards of the brown-and-white earthenware jugs in which some sort of liquor had been imported from Scotland.

Tina entered into the search with delight, her warm, brown eyes shining for pure joy whenever I found something that especially pleased me. When I discovered a rusted muleshoe, she clapped her small brown hands delightedly.

"That is good luck!" she cried. "Now your car start, for sure!"

When I asked if there had been people living in Greenwood City when she was a small girl at Signal, Tina shook her head.

"Oh, no!" she said. "No people here. All my life, Greenwood has been just like it is today—only ruins!"

On my fifth day at Madril's, Angel saddled a horse and rode to the ranch of their son-in-law, Norman, to enlist his aid in fixing my car.

Norman, the young man who had pulled my outfit out of the river, soon found the trouble. The distributor still held a dampness I had been unable to eliminate by only removing the cap. Wrapping a piece of cloth tightly around the end of a stick, he touched a match to it and held the flaming torch under the open distributor, allowing the flames to lick up inside it—something I would never have dared to do, even if

I had been smart enough to think of it! In a very few minutes the distributor was dry as a chip. When the cap and fan had been replaced, and the engine started off merrily, it was one of the grandest sounds I have ever heard!

Norman had barely finished with the car when Mr. Madril brought word that the river was rising again, and that the weather report on the radio said more rain was due in the mountains.

"If you feel you must be on your way, and are willing to take a chance, I'll try to drag you across the river with my Case tractor," said Norman. "You can't make it under your own power, because your fan will suck up water and drown your motor. But I think I can pull you across. Once you are on the other side," he explained, "you can take the old Highline Road to Yucca . . ."

With son-in-law Norman leading the way in his Case, I following after in my Mercury, and Tina and Angel bringing up the rear in their pickup, we paraded to the river—Norman choosing another crossing a little way downstream from the ford that had been my Waterloo. Hitching his tow-cable to my front axle and to his powerful tractor, Norman dragged my car across the river—still deep enough that the water ran in the doors on the upstream side, over the floor mats, and out the opposite side. When the dripping car at last was standing on firm, dry ground, and the motor sprang to life with the first touch of the starter, it was what the late cartoonist, J. R. Williams, would have termed "A grand and glorious feeling!"

Following the Highline Road—an old mail trail used only in times of high water—I made my way to Yucca, driving through cold rain and pounding sleet for much of the 50 miles.

Due to my unscheduled five-day layover at Madril's, the time allotted for my trip had run out and it was necessary that I return to California to take care of business matters.

One month later, however, found me back at Yucca and heading down the graded road to Signal. In the brief space of those four weeks, spring had come to the desert. Flowers were beginning to rouge the flats; even a few blossoms were appearing on the Joshua trees—and no ominous gray storm clouds hung over the mountains to the north.

All too soon I reached a dividing of the ways. One road was signposted to Wikieup and Kingman; the right-hand fork led to Signal.

I still had no idea what I would find in the old town. Almost the last thing I might have expected to encounter there was the sight of a dozen modern house trailers with running water, electric lights, and all the comforts of home—and that is exactly what I found!

Centering a clump of four trailers sat a small brown cabin where I was greeted by a man as lean as rawhide, with keen eyes, a shock of iron-gray hair, and more than 80 years to his credit. Introducing himself as Ben F. Griffith, he invited me inside the cabin, where I made the acquaintance of his pleasant, nice-looking wife.

Mr. Griffith, I learned, was then owner of all the once famous mines on McCracken Hill. From his wiry vigor and sun-browned features I supposed he had been a lifelong habitue of the desert and was surprised to learn that prior to his retirement in 1950 he had been a practicing attorney in Los Angeles, specializing in probate matters, real estate, oil and mining law. Mr. Griffith's acquaintance with the McCracken mine had begun nearly 20 years before when he had been called to Kingman in connection with some of the litigation in which the McCracken has figured. Despite the fact that the mines were then inactive he liked their appearance and in 1944 purchased the entire property, consisting of four patented and 18 unpatented claims. After retiring from law practice the former Angeleno moved to Signal, and in 1952-53 built there a gravity concentration mill of 50 tons capacity—a mill which was not successful due largely to the fact that not enough ore could be put through it to make its operation profitable.

Feeling that the proposition was one in which large scale development was the prime requisite, Mr. Griffith, about a year before my arrival on the scene, had given a bonded lease on the property to the Ari-Vada Development Company, a group of wealthy lumbermen from Humboldt County, in northern California. This company, said Mr. Griffith, was then bringing to completion a flotation-type mill of 150 tons daily capacity and it was hoped that results ultimately would justify increasing the capacity to 500 or 1000 tons daily.

Receipt of this report, I must confess, caused me only to remember the many, many news stories I had read in the old papers . . . "The Signal mill is opening under new management with big plans for its development . . ." to be followed, as surely as night follows day, with a notice that the Signal mill "closed down this week . . ."

Later, I was shown through the mill and had its eventual operation explained to me by "Rip" Thompson, an expert metallurgist pulled back out of retirement to help get the new plant off to a good start. I also met Percy Ramsden, general superintendent of the operation, and Mrs. Ramsden, both second-generation Arizona mining folk who occupied one of the house trailers I had noticed upon first entering the old town. Another of the trailers was occupied by G. C. Bush, millman, and Mrs. Bush, an enthusiastic rockhound. Since their arrival at Signal she had collected several hundred pounds of "pretty rocks" and was assembling a cactus garden in the lee of their trailer home. Third trailer in the "executives' camp" was that of the assayist and his wife, Mr. and Mrs. Harris. Half a mile distant, in a second colony of trailers, lived the remainder of the 15 families then residing at Signal—likely the greatest population the old ghost town had known in half a century. Seventeen men, I was told, were then on the company payroll.

Everyone agreed that I must visit the McCracken mine; and since one of the requisites of that trip was a jeep, it was decided to commandeer the services of Fred and Jean Traylor to provide transportation and to show me around.

Soon as I met the Traylors I understood why their camp neighbors held them in such high regard. I found them living in "lower town" in a nice modern trailer, electrically lighted and refrigerated by their own 3000-watt power plant. Completing their family circle was a small brown dog and six baby chicks, all running loose in the yard and apparently having the time of their lives.

After the three of us had eaten lunch on a backyard picnic table shaded by a big palo verde tree, we climbed in the Traylor's jeep and set forth on the old road to the McCracken mine. As the miles piled up and our way grew steeper and rockier, Fred shifted into progressively lower gears and eventually brought the jeep to a halt where the road terminated high on the southeast shoulder of McCracken Hill. Here we took to a dim foot trail that climbed the ridge through a natural rock garden where great clumps of beavertail cacti flamed with magenta blossoms, and crimson plumes danced on the ocotillos.

Gaining the crest of the mountain, some 2000 feet above the surrounding plain, we felt the quick thrust of a keen, cool wind, and looked out upon a great panorama of alternating valleys and desert ranges visible for 50 miles and more on every side. Leading away in a long dark line to the northwest were the blue Hualpais; to the south lay the dry Harcuvars and Harqua Halas; to the east, the Chemehuevis. Standing on that drafty eminence, overlooking these 10,000 square miles of arid wilderness, it gave me a good feeling to know these desert ranges are still almost wholly unspoiled, largely unsettled by white men, and all but roadless.

Perched on a narrow saddle, only a few feet below the summit of McCracken Hill, we inspected the ruins of a stone structure measuring 21 by 70 feet in floor area, with several rooms in which all wooden work is fastened with square-cut nails. Although known locally as "The Fort," several old timers with whom I later talked told me that this was originally the boarding house for employes of the McCracken.

From the rock ruin we inched our way down the mountain to the portal of the once-great mine. Here we found the rusted remains of the long metal chute through which ore had been carried by gravity down the steep hillside to a flat several hundred feet below, from which point it had been hauled to the mill on ore wagons drawn by 16 and 18 mules. (For a short while, in 1883-4, ore had been freighted to the mill by "steam wagon." As the success of this innovation had fallen considerably short of expectations, it wasn't long until the mules were back in business.)

With the stabbing white beam of Fred's carbide lamp throwing about us a small well of light, we made our way into the blackness of the old McCracken tunnel. Occasionally Fred paused to spotlight a section of wall where the bright sparkle of calcite crystals, the metallic gleam of rich lead ore and dull black of silver, gave a fairy-castle effect to those night-dark walls. We had not penetrated many rods into the mountain until we began seeing a faint glow of light in the blackness ahead and soon could see that its point of entry was a jagged hole broken in the tunnel's roof. Peering upward through this hole, which was barely wider than a man's shoulders, we looked into a subterranean room possibly 50 feet across and twice as long. At points where the hanging wall and footwall pinched together until they were only a dozen-or-so feet apart, we could see the old juniper stulls still wedged in place. This, said Fred, was the great Palace Chamber of the McCracken, whence had come many thousands of dollars worth of rich ore. The upper end of the chamber, he explained, breaks through to the surface not far from the hilltop ruin we had visited earlier that afternoon.

Returning to the jeep we continued north around the mountain to explore the tunnel of the Old Timer mine, and on to the mountain's west flank to inspect the open-cut workings of what Fred called the Blacksmith mine, and the ruins of the large mill in which its ore was worked. (From the location of this mine I believe it must have been the one known originally as the San Francisco and, in the later 1870s and 1880s, as the Signal.)

Completely lost in the day's adventure, I didn't realize the sun had set until the sharp chill of evening began to creep into the air and we decided it was time to turn our steps and our jeep homeward.

Next morning, after the Traylors had left for work—Jean was the bookkeeper for Ari-Vada and Fred was employed in a general capacity—I set out, alone and afoot, on an exploring jaunt over the townsite. My way led me past one of Signal's old false-fronted saloons, still quite well preserved, and past the almost obliterated ruins of numerous adobe cabins. Beside the upright frame of an open-type well, or cistern, set an old school bench. I wondered if this might have been the site of the schoolhouse once termed by the Signal news reporter as "finest in the county." Continuing past the stone masonry ruins of a big mill and up a side wash, I soon reached the old cemetery. Out of 50 or more rock-mounded and thistle-guarded graves, I found but three whose markers still carried identifying inscriptions.

Climbing the hill to the rear of the graveyard I turned to look back the way I had come—back past the old burial ground, down the cottonwood fringed wash, past the old mill ruins, and on to the site where the new mill was being erected by Ari-Vada. There were still no buildings of importance at the mill-site—the company having elected to set up the machinery first and fit the building over it, later—but there was activity and life. The morning sun was glinting on the bright tin roof of the new office building. A truck was moving through the yard. Workmen were hurrying about their appointed tasks, and I could hear the powerful drone of a bulldozer, the staccato ring of hammering.

Sitting on the quiet sunny hillside, looking across the old townsite to the new mill then taking form, I wondered if its eventual operation would profit its owners? Would it make new wealth for Arizona, new jobs for workmen; would the old town come to life again? Or, was this mill, too, destined to join the ranks of other noble efforts tried and found wanting?

Before I left Signal, later that week, I was well satisfied with my visit to the old bonanza town. I had taped close to 10,000 words of notes to augment those taped earlier from the old newspapers, and had explored both the townsite and the mines responsible for its boom. One matter, however, still nagged at my conscience. Except for Tina Madril I still hadn't talked with anyone who had known Signal in its livelier days. I still hadn't "shook the hand that shook the hand" of Signal, the boomcamp.

"You might talk with Amy Neal," said Kingman's pleasant librarian. "She was born on the Cornwall Ranch, in Sandy Valley . . ."

Calling at the John Neal home, in Kingman, I learned that both Mrs. Neal and her husband remembered Signal when it was still an operating town, though definitely on the downgrade.

"You've read of 'Chloride Jack' Owen, one of the discoverers of the McCracken mine?" inquired Mrs. Neal. "He was quite a close friend of my father and used to be around our ranch a lot when I was a little girl in Sandy Valley."

With those words, I knew I had struck paydirt! Here, at last, was a living link with Signal, the boomcamp.

"Jack was originally from Georgia," went on Amy Neal. "At some time in his life he had known Abraham Lincoln's family and had been especially fond of the President's son, Todd. After Jack sold his share in the McCracken for $25,000, I believe, he decided to return to Georgia and 'live it up.' Before long word came back to Sandy Valley that Jack was living in style. 'Even has a nigger to dress him!' hooted the old timers who had known Jack when he had scarcely a shirt to his back. To those grizzled old miners the idea of Jack having a valet was the height of folly!

"With wine, women and song—and the alleged valet—it didn't take Jack long to run through his $25,000. When it was gone he came back to Arizona and resumed prospecting, but he never made another good strike. In 1909, when I was living temporarily at Prescott, Jack turned up in the county hospital there," said Amy Neal. "I called on him at the hospital and we had a nice visit about other times. The old man must have realized that his last days were close upon him. Anyhow, he said that because I had always been kind to him he was going to will me everything he owned in the world—'a pair of nice, warm, gray blankets.'

"Chloride Jack died soon after that, but I never claimed my legacy," said Mrs. Neal. "I knew the hospital could use the blankets to good advantage, so I just left them there . . ."

After visiting an hour or so with the Neals, I was standing at the door bidding them goodbye when Amy said, "You should talk with Jennie Berger. Her father was one of the first merchants at Greenwood City and Signal . . . "

A handsome statuesque woman with dark eyes and graying hair, Mrs. Berger readily admitted her early connection with Signal. Her father, Moise Levy, she said, had been born in Alsace Lorraine and, as a twenty-year-old youth, had developed a desire to try his luck in the California gold fields.

"Grandfather thoroughly disapproved of the idea," said Mrs. Berger. "He used every argument at his command, every threat, and entreaty. When father went right ahead making preparations to sail for San Francisco, grandfather gave him one last warning—'Moise, don't you expect fried pigeons to fly into your mouth!' "

But Moise did very well—though not in the gold fields. Soon after his arrival in San Francisco he found work in a shoe factory. By saving most of his wages—as well as the 500 marks a month sent him by his father in Alsace—young Levy, by 1875, had accumulated sufficient money to go into business for himself, and it was he who opened the first store in Greenwood City. He soon took into partnership a fellow worker at the shoe factory, young Henry Koshland, a Bavarian. A year later the Levy-Koshland partnership was joined by Moise's younger brother, Gabriel, and Henry's brother, Wilhelm.

"In 1878 the Levy-Koshland store at Greenwood City burned to the ground, including all its contents," recalled Moise Levy's daughter. "As Henry Koshland stood looking at the blackened ruins that represented all their savings and hopes, he burst into tears. Stepping up beside his partner, father laid his hand on the other's shoulder. 'Henry,' he said, 'If I thought we could get our store back by crying, I'd join you and we'd both cry. But both of us know we won't get it back that way, so we've got to figure another way to do it!' "

Feeling that the new town then taking form around the Signal mill offered a better opportunity and future than the old settlement of Greenwood, the partners built at Signal City a store even larger than the one destroyed by fire. For many years this was the largest mercantile establishment for miles around. Moise, in time, became justice of the peace and a member of the schoolboard at Signal. Gabriel was appointed postmaster; and in 1883, each of the four partners was naturalized as a citizen of the United States.

Came 1890 and Moise Levy and Wilhelm Koshland decided to return to Europe to attend the International Exposition in Paris, and to visit their respective families. At a dance one night, in his home town in Alsace, Moise met a lovely young girl. A whirlwind courtship followed, and when the forty-year-old merchant started back to Arizona he took with him his sixteen-year old bride, Leonie.

"Father and his bride—who eventually became my mother—left the train at Yucca Station and boarded a stagecoach for the last 42 miles to Signal," said Jennie Berger. "Mother, who had never seen any place like the desert, was terribly frightened by it. When a coyote trotted across the road ahead of the coach and father explained to her that it was sort of a wolf, she was scared speechless. Then, when they arrived at Signal and the poor girl saw a rough mining town, populated chiefly by Mexicans wearing knives and guns in holsters and swaggering through the dusty streets, she was positively terrified!

"Mother never got used to Signal, but she was a good sport about it. Every now and then something would happen that would convince her we were all going to be murdered in our sleep; but father, invariably, would reassure her that everything would be all right—and it usually was."

When Wilhelm Koshland had found his parents quite enfeebled he had elected to remain with them in Bavaria and had never returned to the United States. Later, Henry Koshland ran a cactus spine in his finger, developed blood poisoning, and died; and when Moise and Leonie had been married 13 years—and the lovely Leonie was still but 29 years of age—Moise died as result of a brain tumor.

"This left Uncle Gabe the last of the four partners," said Mrs. Berger. "After father's death, mother and I left Signal, but Uncle Gabe remained there, in business, until shortly before his final illness in 1932."

When I was thanking Mrs. Berger for her help, and bidding her goodbye, she mentioned that I should talk with Wilford C. Babcock. "He should be able to tell you quite a lot about Signal," said Mrs. Berger.

As I had been told the same thing by the Traylors, at Signal, and by Mrs. Neal, I made Mr. Babcock's home on Oak Street, in Kingman, my next port of call.

Mr. Babcock, I learned, had come to Mohave County in 1896, at the age of 15. "By the time I began driving stage between Signal City and Yucca Station, Signal was pretty much of a ghost town," recalled the old timer. "Uncle Gabe Levy was the last merchant still doing business there—and you should have seen his store! By that time he was getting old, the novelty of storekeeping had worn off, years before, and he seemed to have lost interest in the place.

"None of his merchandise was ever put away systematically. It was just heaped, here and there, on the floor, on benches, piled on the counters. Yet, Uncle Gabe knew exactly where everything was! He seldom got out of his easy chair to wait on a customer. He just sat still and told folks where they could find whatever it was they wanted. Neither did he bother to keep any books; yet, he knew exactly what every man owed him.

"Uncle Gabe clung to Signal like a burr to a saddle blanket. Not only was he the town's last merchant, he was also the last of its original inhabitants. When he finally had to give up the store and post office, just before his death in 1932, he had been in business at Signal more than 55 years . . ."

Shortly after my visit to the old bonanza town on the Sandy, I received word of the death of Ben Griffith, owner of the McCracken mine. Still more recently, through Jean Traylor, I have learned of Signal City's return to complete ghost town status.

"The Ari-Vada Corporation folded up about three months after Fred and I left Signal," wrote Jean. "Part of the property was attached and sold by the Mohave County sheriff in October 1960. All that is left there now is the concrete slab where the mill stood, and the slab and walls of the little office where I worked; and, of course, the usual rubble one finds in almost any abandoned town . . ."

Reading Jean's letter, my mind went back to the old newspaper files I had scanned, and their eighty-year chronicling of Signal's good years and bad. To that collection of vacillating hope and despair I added, mentally, one last line:

"The mill at Signal has shut down."

Adobe cabin of "Uncle Charlie" Newton, McMillen, 1950

"In this country . . . it is significant and pathetic to see how almost every stamp mill has been abandoned, sold at auction, transported to other districts, tinkered, rebuilt, and sold and removed again, in disastrous repetition. While the sun shines, such an enterprise must make hay in in a hurry; for the rainy day is certainly coming."

—ROSSITER W. RAYMOND, *Mineral Resources Report, 1871.*

Hoyden of the Gila Hills

OLD CHARLIE McMillen was in no frame of mind for jogging along a rough and rocky trail, not even with the early Spring sun of southeastern Arizona lying warm on his aching carcass. Tugging his mount to a halt in a clump of scrub oaks, the grizzled desert rat spilled himself from the saddle.

"Gotta stop, Dory . . ." the words trailed off wearily. "Gotta take five . . ."

"Gosh, no, Charlie! Not at this time of day," protested the old man's partner, young Theodore Harris. "Come on. Let's go on to the spring where we want to camp. It's only a couple of miles farther . . ."

But the old man had slumped in the spattered shade of the desert oaks and was already lost to the world and its multiple vexations.

Irritated by this needless delay so near camping time, Dory Harris dismounted and ambled across to a lichen-grown ledge where he gave vent to his impatience by sinking the point of his prospecting pick in the outcropping. A small piece of rock rattled down from the ledge. Dory picked it up and was drawing his arm back to fling the rock into space when he noticed that it seemed unusually heavy. But Dory Harris was not much of a miner. Old Charlie was always ribbing him about the "iron pirates" he lugged into camp as potential gold and silver, and even after examining the piece he still couldn't have sworn that it, too, wasn't only pyrites.

Shaking his snoozing partner awake he injected the sample into the other's line of vision.

"Look, Charlie—is this some kind of ore?"

With only a cursory glance at the specimen, Charlie McMillen muttered darkly and had begun resettling his weary bones when some extraneous message seemed to penetrate his befogged consciousness.

"Dory!" he croaked, snapping to sitting position. "Let's see that sample again!"

After scraping the piece with the blade of his jackknife and subjecting it to the searching scrutiny of a pocket magnifying glass, the old man lifted rheumy eyes to stare at his partner for a speechless moment.

"Where'd you get this?"

Stumbling across to the indicated outcropping, Charlie grabbed the other's pick and began driving at the ledge, snatching and examining each piece of rock as it was loosened.

"By the Eternal Graces, Dory!" breathed the old miner hoarsely. "We've hit it! *This is native silver . . .*"

The discovery claim was named the Stonewall Jackson; the date of location was March 6, 1876.

In those days of the 1870s Arizona Territory was a hard, demanding land, scourged by hostile Apaches and offering little hope of longevity; yet, no sooner had news of the strike reached the outside world than multitudes began converging upon the new camp. On the heels of those who came seeking riches in the bowels of the desert mountain, followed tradesmen, gunmen and gamblers. A town was born, and out of deference to an old prospector who had spent most of his life in a hitherto fruitless search for precious metal, the new camp was named in his honor.

Within six weeks the rocky trail so lately pioneered by Harris and his partner had been replaced by a wagon road soon to connect with the likewise new camp of Globe City, some 18 miles to the southwest. Dozens of adobe buildings soon were in course of construction—some to serve as gambling halls and bawdy houses, some as saloons whose doors would never be locked. Timber cut in the Pinal Mountains was whipsawed into rough lumber for bars and tables and counters.

Of the many mines ultimately developed in the surrounding hills, none was destined to equal in richness and importance the great Stonewall Jackson.

In a resumé of mining news from the Globe District, *Arizona Citizen* reported July 7, 1877:

"The latest sensation is the new development on the Stonewall Jackson mine, owned by McMillen and Harris. The shaft is down 52 feet and a drift is being run each way along the vein. There is . . . a streak of almost pure silver. I do honestly believe that of every 100 pounds of first class ore, 75 of it is pure silver. In fact, I was told by the foreman that every sack of it was worth from $1000 to $1200 . . ."

From ten tons of average ore mined that same summer of 1877, a lot of 6800 pounds, hand-sorted for exhibition at San Francisco, was found to assay $19,000 to the ton—an average of $9.50 for each pound of material! "The ore . . ." wrote a mining reporter of that day, "is literally full of metallic silver, the metal forming 60 to 70 per cent of the entire weight."[1]

In its issue of August 25, 1877, the *Herald*, of Silver City, N. M., reported that the paystreak in the Stonewall had widened to 20 inches, and that a lot of 2300 pounds of selected ore, sent to San Francisco, "netted the owners $18,000." In the same issue it was stated that McMillen and Harris had sold their great mine to "a San Francisco gentleman" for $120,000.

[1]*Mining and Scientific Press*, San Francisco, November 3, 1877.

By the close of that year of 1877, McMillen Camp, McMillenville, or just plain McMillen, as the place was impartially called, numbered some 300 inhabitants served by a more-than-adequate number of grog shops, two stores, three blacksmith shops, one carpenter shop, a bakery, barber shop, and a post office established November 12, that year. An unidentified correspondent, writing from McMillen Camp under date of November 29, 1877, stated: 'Standing in the street you can count ten buildings in active construction . . . We also notice a new blacksmith shop being erected by Messrs. Hyatt & Company, near Peter Tompkins' boarding house . . . Peter has rather more boarders than he can well supply, having about 40 from the McMillen Company's mine, ten from the General Lee, and outsiders too numerous to mention . . . The McMillen Company is taking out and locking up in their tunnel the usual amount of silver ore."[2]

With bullwhips cracking, drivers shouting, the iron-tired wheels of massive freight wagons grinding on the rocky streets and the rumble of blasting echoing from mine tunnels and shafts, McMillen's days were filled with the music of honest endeavor—even if the music that filled her nights was of different calibre. From half a dozen saloons blared the jingle-jangle of honky-tonk pianos, the shrill laughter from Damsels of the Dark, clinking of glasses, slip-slap of cards, heavy boots thudding on rough plank floors, and always a roaring crescendo of rough language not infrequently punctuated by gunfire.

McMillen, for the most part, was a man's camp. History records that one Christmas celebration there featured a Yule tree hung with nothing but "cigars and tobacco, dynamite and fuses, grub and bottles of whiskey."[3] Another Christmas whingding in McMillen was reported as a vigorous affair highlighted by eggnog in the morning, followed by broad-and-high-jumping, foot racing, stone throwing and leapfrog, in the afternoon.[4] Other McMillen merriments, as chronicled in various issues of the *Silver Belt* included such diversions as itinerant organ grinders, occasional Mexican circuses, and dances.

Throughout 1878 and '79 McMillen's star continued in its ascendency, every week seeming to bring more encouraging reports from the new camp in the Gila hills.

"There is little about the camp that partakes of a temporary or transient nature," declared *Arizona Citizen* in its first issue of the new year, 1878. "Most of the buildings are substantial and look as if they were intended for years of service . . ."

With the Vernal equinox, the editor of the *Citizen* was still ringing the praises of McMillen Camp and its matchless Stonewall Jackson mine. "I dare say that, without doubt, there has been more silver taken from the mine in the last four days than any other mine in the Territory will produce in a month," exulted the Florence news hawk. "I saw silver

[2]*Arizona Citizen*, December 7, 1877
[3]Wolle, *The Bonanza Trail*, 1953
[4]*Arizona Silver Belt*, January 3, 1880

to amount of $10,000 or more taken from the mine last Saturday—the result of one blast . . ."

"A trip to McMillenville would not only convince a casual observer that its citizens are generous and hospitable to strangers, but that the place is thrifty, enterprising and flourishing," reported the *Silver Belt* of May 9, 1878. "Improvements are steadily going on, buildings springing up on every side, and business is generally prosperous. With its energetic inhabitants, and being surrounded as it is by the best mines in the district, a bright future for it is inevitable."

Almost on the heels of this cheerful prognostication, the *Silver Belt*, on June 13, copied from the *San Francisco Stock Report* a page article entitled *An Arizona Bonanza.* Its praise for the Stonewall was unconstrained. "There is now on exhibition in the basement in the rear of the San Francisco Stock Exchange, Pine Street, seven tons of ore from the Stonewall Jackson ledge, of the McMillen Mining Company, Arizona," it was stated in part. 'The ore runs from $10,000 to $20,000 to the ton, and is the richest ever exhibited in San Francisco.'"

In January 1879, the McMillen Company began grading the site for a mill, and on April 27, that year, the first run was made. This original five-stamp mill soon was replaced by one of 20 stamps; and with completion of the Southern Pacific Railroad through Arizona Territory in 1880, freight wagons laden with bars of silver bullion began traveling the 100 miles of rough road between McMillen and Casa Grande, nearest shipping point on the new rails. Other freighters were busy hauling wood for the boilers and salt for amalgamation—the latter purchased for two-bits a pound from the Cox brothers who evaporated it from the waters of saline springs on a tributary of Salt River.[5]

The Stonewall's fabulously rich ledge was traceable for nearly a dozen miles—nor were owners of the mine unduly troubled by the fact that much of its paystreak lay within the forbidden domain of San Carlos Apache Indian Reservation. Relying on the usually dependable procrastination and pigeonholes of Congress, the owners assertedly made haste to strip this portion of the vein so that by the time Uncle Sam got around to protecting the interests of his red-skinned nephews, the cream of the Reservation's silver already had been cast into bullion.

But even the great Stonewall was not invulnerable. In 1885, after assertedly running up a production total of more than $3,000,000, her paystreak petered out. Close of that year found but seven of McMillen's smaller mines still in operation, and by 1887 the streets of this once-exuberant mining camp knew only the hollow tread of a Faithful Few who remained behind, firm in their conviction that the camp would come back bigger and better than ever.

Unfortunately, it has not done so.

[5]J. H. McClintock, *Arizona,* Vol. II, 1916.

⇦ *At Left: Headframe, Stonewall Jackson Mine, McMillen, 1950*

As for Dory Harris and Charlie McMillen, their fates and fortunes ran a close parallel to that of the town.

After taking out ore to a reported value of $60,000, the partners, as reported earlier in this chapter, sold the Stonewall for $120,000 cash. With half that booty burning holes in his pockets, Old Charlie, according to popular report, "hastened to drink himself to death."

This would imply that Charlie McMillen was little better than a bibaceous lush with an uncontrollable appetite for the distilled kernel and fermented grape. Newspaper accounts of that period, however, do not bear out this derogatory portrayal of the old miner; yet, it does seem evident that something soon happened to his share of the proceeds, as it was not long before Charlie McMillen was again back at work.

On November 30, 1877, about three months after McMillen and Harris sold the Stonewall, the *Arizona Citizen*, in a collection of "News Items from Yuma", reported: "McMillen, discoverer and seller of the famous Stonewall Jackson mine, is here with a party. They are fitting up a team of their own to go out and operate in the Globe District."

In its next issue following (December 7, 1877) the *Citizen* stated: "Mr. Charles McMillen, of Stonewall Jackson fame, has been engaged as the superintendent of the Hannibal. Charlie intends to open up the mine at once and, under his management, rapid developments may be expected . . . We congratulate the Hannibal Company on securing his services."

It is difficult to imagine a position of such responsibility being given to an inveterate tippler; yet, the very fact that Charlie McMillen had accepted a wage job barely four months after the sale of his great mine, would seem to indicate that the proceeds of the sale were already gone with the wind.

Dory Harris, the educated Easterner, according to report, purchased a seat on the San Francisco Mining Exchange and in 90 days had lost his entire stake. Returning to Globe he is said to have taken a job "pearl diving" in a cheap hashhouse, got religion, joined the Salvation Army, and married an Army lassie. In the *Silver Belt* of April 24, 1880, it was reported: "T. H. Harris, one of the discoverers of the Stonewall Jackson, left Globe for Southern Arizona on Thursday."

With that departure, Dory Harris seems to have disappeared like a muleshoe dropped in a well.

In the ranks of McMillenville's Loyal Legion of The Devoted, none was more steadfast than "Uncle" Charlie Newton. Long years after the old camp was dead and deserted each pleasant evening would find Uncle Charlie sitting in a homemade rocking chair on the rickety porch of his little adobe cabin, puffing his pipe and gazing out over the silent ruins of this near-forgotten town . . . rocking, smoking, drowsing, dreaming, waiting. But, at last, even Uncle Charlie went down the long trail to the Great Camp where the paystreak never pinches out.

When I visited the little ghost town of McMillen, in 1950, not much of consequence remained—a few old foundations, part of a corral, a few old mines, the headframe of the Stonewall Jackson, tons of discarded milling machinery, rusty horseshoes, scraps of iron and leather, bits of trace chains, square-cut nails, bottles . . . that was about all. As I wandered over the deserted flat my feet seemed to turn automatically toward the little adobe cabin in the gulch below the Stonewall Jackson, where Uncle Charlie had lived so many years. From across the wash the old place appeared much as I had remembered it—squat and plain, bare and weathered. More intimate inspection disclosed the footprints of neglect—the half-open door, the floor buried in clutter. Packrats had preempted every cranny; a pair of slim brown birds darted from their nest in the narrow wedge between roof and wall, and a cottontail scurried for cover as I stepped across the worn threshold.

Prowling about these musty little rooms where Uncle Charlie had gone his quiet way, I found the interior of one cupboard still lined with yellowed sheets of the *Arizona Silver Belt* published in Globe City, Territory of Arizona. Dates of issue were in 1878, and the news was of mines and mining, and Apache raids and stage holdups and life in a raw, new land. Another cupboard was lined with yellowed pages from the San Francisco *Chronicle* and the *Arizona Republic,* of Phoenix.

A slim gray lizard pattered across the splintered floor, halted in a wedge of sun and tilted his head to rivet beady eyes upon me. It was very still.

I made camp that night in the yard of another old adobe cabin across the wash. About sundown a little bunch of red-and-white cattle meandered down the silent street. A coyote, in the hills back of town, gave a few yips in the early darkness. Later in the night a flurry of breeze came whisking around the adobe where I was camped. Like a fussy woman shopper it paused to finger the dry needles of the crooked old piñon over my bed, examined a piece of loose tin on the cabin roof, then whisked down the ravine and on to wherever night breezes go.

Soon after breakfast next morning I left the old town of McMillen to its gray adobe ruins, its small gray lizards . . . and the gray ghosts of its departed dreams.

I have never gone back.

Road to Dripping Springs, in the land of the Papago

"It is . . . a matter of routine in a prospector's life to go to every new place that is discovered; and if he gains nothing by it, he consoles himself by the reflection that at least his education has been advanced to the extent of knowing that there is nothing there."

—Rossiter W. Raymond, *Resources Report, 1870*

Stampede to Papagoland

Spreading like a saddle blanket across the Arizona-Sonora border is the realm of the Papago Indian, a strange and enduring land.

The Papago farmer of centuries agone—a planting song on his lips and a planting prayer in his heart—dropped his treasured seeds and with a bare, brown foot pressed over each the sun-warmed desert earth. If the Papago gods saw fit to smile on those efforts, the cotton seeds he had planted, plus the weaving skill of his women, yielded cloth to cover his nakedness. The corn seeds gave meal for tortillas and pinole; the squashes and beans of his harvest supplemented the edible fruits and herbs so sparingly provided by the desert.

Life, in the Papago's land of little rain, was hard and demanding. Yet, it was a good life—as lives go. Elsewhere, throughout the world, men were striving and fighting and dying in a never-ending struggle to retain their meager shares of the Good Earth. But not in Papagoland. Except for those times when the predatory Apaches came swooping down to pillage his fields and villages, the Papago had known peace through all the years of his tenure—and for good reason. None other could live where he lived; none coveted his wild, rocky, arid acres between the Gila River and the northerly tip of the Gulf of California.

After untold centuries during which the Papago was accountable only to the gods of his fathers, his tribal chiefs, and his own conscience, the picture began changing.

Mexico's conquest and the exploration of *Pimeria Alta* in the early 16th century placed the Papago, theoretically, under the rule of Spain and King Charles I—a nation and ruler of which he had never heard.

Three centuries later, a revolution waged hundreds of miles distant between men he had never seen, fighting for ideals and principles of which he knew nothing, had the effect of replacing Spain's flag with the new flag of Mexico—and the Papago, again theoretically, became a Mexican subject.

The Treaty of Guadalupe Hidalgo, under terms of which the United States acquired ownership of all our Southwestern States situated south of Oregon and west of Colorado, had little if any effect on Papagoland; but with the Gadsden Purchase of 1853, and surveying of a new International boundary, the Papago's realm was bisected. Members of the tribe whose natural home was in the south portion of their claimed territory continued as citizens of Mexico; but approximately 5000 Papagos, residing north of the new line, became wards of the United States government.

Uncle Sam's newly-acquired nephews made few demands upon him, neither did they give him any trouble. Except for helping him fight their mutual enemy, the Apache, they went on living as they always had lived—planting, reaping, hunting, gathering and preserving the fruits of the giant cactus, observing their religious ceremonials, husbanding their meager water supply. No treaties were necessary, for nothing had been taken from them. Papagoland continued to be theirs and no one bothered to post its nebulous boundaries with Keep Out signs, because no one wanted in.

No one, that is, except occasional wandering prospectors—and the Papagos had watched prospectors come and go throughout more years than any man can say.

If it were possible to know the full history of the first mining activities in the Papago country, it would be an epic more fascinating than fiction. Who were the miners? Whence did they come? How did they work? How much did they profit or lose? When and why did they depart; where did they go? But the answers to these questions are buried in the yellowed archives of Spain; or, more likely, lost in the dust of antiquity.

Work is believed to have been begun on the Baboquivari gold mines, in the eastern part of Papagoland, about 1748. The Picacho mine, in the Cobabi Mountains a little to the northwest, likely was being worked about this same time—possibly earlier—its ore being ground in arrastres at the town of Fresnal.

In his *Adventures in the Apache Country*, J. Ross Browne wrote in 1863: "A day's journey through the . . . foothills of the Baboquivari brought us to the first of the inhabited rancherias . . . the small . . . town of Fresnal, a collection of adobe hovels built at this place . . . for the reduction of ores stolen from the Cahuabia mines." Wallace Elliott's *History of Arizona Territory*, 1884, also states that ore from these mines was carried "to Fresnal . . . where can be seen today the remains of more than 45 arrastras . . ."

According to a Spanish manuscript or report, assertedly discovered in an old mission church in Mexico some eighty-odd years ago by one A. H. Hanson, ore mined in the present vicinity of Covered Wells trading post, in the heart of Papagoland, was being shipped to Muleje, Baja California, for reduction as early as 1774.

In this same general area, about six miles north of Quijotoa Mountain, Elliott's *History* describes a placer field three miles square which had been extensively worked as early as 1774 by Pedro Lopez, a Castilian priest.

"The workings of the placers are remarkable," states Elliott. "The most of the ground is a perfect honeycomb of working shafts, from five to twenty feet deep, covering the goldfield so close together that it is almost impossible to ride over the ground without danger. These shafts, or pits, are connected by underground workings from which the gold was evidently taken. The deepest shafts are those farthest removed from the base of the mountain. Some of the dumps of the deep workings are very large and have been found to be rich enough to work with profit, as the methods used by the ancient gold miners of that region were so rude and primitive that none but the coarse gold was taken . . ."

Spaniards also are believed to have worked mines in the vicinity of Ajo as early as 1750. One of these early Spanish mines—later known to Americans as "the Old Bat Hole"—had an incline shaft cut on a 60-degree angle, with platforms of mesquite logs four feet long and three feet wide placed at ten-foot intervals as rest stations for the slave miners who carried the heavy loads of ore out in rawhide bags strapped to their backs.

As a natural consequence of these early mining endeavors throughout the Papago country, Anglo-American miners had been, for some time, eyeing that region with both curiosity and cupidity. The only thing that held them back was the uncomfortable circumstance that all this area still belonged to Mexico. With negotiation of the Gadsden Purchase, however, Captain Peter R. Brady, surveying for the proposed Texas & Pacific Railroad, immediately collected a number of silver-copper ore samples from several of the old mines and had them assayed in San Francisco. The American Mining and Trading Company was organized and in 1854 Captain Brady headed a party of some 20 men who relocated the old Spanish-Mexican mines at Ajo. The first ore taken out under the new management was carried overland by L. J. F. Jaeger's packtrains to the Colorado River (at $105 per ton), thence being shipped down the Gulf and on to the nearest smelters in Swansea, Wales.

Having witnessed the sporadic prospecting and mining activities of Spaniards, Mexicans and Anglos over a period of possibly 130 years, the Papagos likely attached little importance to the peregrinations of Alexander McKay who wandered into their desert domain in the winter of 1878-79, chipping off a sample of rock here, scrutinizing a ledge over yonder. In February 1879, McKay found what he later described as "some pretty good looking mineral indications;" but when he failed to find any actual mineral except a little iron, he wandered on—still chipping, still scrutinizing.

About three years after McKay's discovery of the iron capping, a friend of his, Arthur Weldon, found beneath that capping a good showing of copper which carried some silver, and McKay put his name on three locations—although he still didn't think too highly of the prospect.

One day, not long afterward, while prospecting in the vicinity with a small group of companions, McKay climbed to the top of Quijotoa Mountain—not in search of the proverbial "lost burro," strangely enough, but in search of a silver mine. When he returned to camp his companions asked the usual question, "What luck?"

"The Comstock of Arizona," answered McKay, "is on top of that mountain!"

On May 11, 1883, locations were made by McKay, Albert Weldon, James A. Roark, and George Teitsworth, and the group also located for W. C. Davis and Martin Medley, not members of the party.

The first ore samples from the new Peer mine, taken to Tucson, assayed $1500 to $4500 per ton.

That same month a small force of men, numbering at times as many as 20, began work on the mountaintop mine. Through an arrangement of cables, ore was let down over the bluffs on the east side of the mountain and supplies were hoisted. Before the end of that first summer considerable ore had been taken out and shipped profitably to several different smelters. (One lot of five tons, shipped to the Benson smelter, returned $2500 to the partners.)

To the Peer mine, in August, was added the Peerless—its discovery having been made in a manner possibly unique among Arizona mines.

One scorching day, according to the story, Alexander McKay's hat blew off, rolled over the edge of the precipice, and lodged in the brambles about 70 feet below the rim. Having no other hat there available, and not relishing the thought of working bareheaded under the fervent rays of southern Arizona's August sun, Mr. McKay insisted upon recovering the headpiece. When his companions, with good-natured badinage, let him down the cliff on a rope the hat was easily retrieved—but the veteran prospector, still suspended in mid-air at the rope's end, then insisted upon doing a little prospecting. Breaking a few pieces of rock from the face of the precipice, McKay told the boys to "pull away."

When he was safely back on the rim, he and his companions examined the ore samples and found them to be rich in silver! The Peerless mine was staked on the spot, according to the story, and the samples were later found to assay $3600 to the ton.[1]

Soon after discovery of the mines, two brothers, J. T. and W. R. Logan, began digging a well on the east side of Quijotoa Mountain—which was soon to lose its ancient Papago name and be rechristened Ben Nevis Mountain; and George J. Roskruge, of Tucson, surveyed a new townsite to be known as Logan City. Due to its situation on the promontory's sharply-tilted east side, the new townsite possessed what one writer of that day facetiously described as "the finest drainage of any town on the Pacific Coast."

The rather spectacular discoveries on Quijotoa-Ben Nevis Mountain naturally occasioned quite a stir in Western mining circles—a stir which soon gained the attention of San Francisco capitalists and opportunists. Prominent among the latter was W. S. Lyle.

Without any advance warning of a projected visit, Lyle suddenly turned up, one October morning, at the Peer mine. He examined the

[1]*Arizona Daily Citizen,* October 1883.

still-shallow workings, sampled the claims, and before starting back to Tucson, a few hours later, had bonded the group for 14 days at $450,000!

At Tucson, Lyle telegraphed the Comstock Kings, James C. Flood and John Mackey, who hurriedly dispatched J. B. Low, of Virginia City, Nevada, to examine the properties. Low's favorable report resulted in the deal being closed by telegraph on December 7, 1883.

One week after transferral of title, there were organized in San Francisco the Peer, Peerless, Weldon, and Combination Mining Companies, each for the avowed purpose "of mining precious metals and reducing ores in Pima County, Arizona." Each of the four companies was capitalized at $10,000,000, divided into 100,000 shares having a par value of $100 each.

An extension called The Crocker, located in the names of M. M. Bruce, Frank Litchfield, Herbert B. Tenny, Fred A. Adams and Dan H. Crocker, none of whom, it is said, ever saw the claims, was purchased the following January 16 by Flood and incorporated as the Crocker Mining Company, also capitalized at $10,000,000. As the shareholders in each of the five companies were the same and the mines were located in a body and all under the same supervision, they constituted, in effect, a single company representing $50,000,000 capital stock!

"This may be said to be one of . . . the heaviest mining corporations on the Pacific Coast, or the whole United States; or, for that matter, in the world," wrote Bascom A. Stephens, author of the *Quijotoa Mining District Guide Book,* published in 1884.

Almost immediately after the sale to Flood and Mackey, three other townsites—Virginia City, New Virginia[2] and Brooklyn—were laid out contiguous to Logan City; and on the opposite side of the mountain, Allen City was founded by General J. B. "Pie" Allen, perennial follower of boom camps. Confident that Comstock and California money assured long life and prosperity to any one of this quartet of camps, boomers began flocking in from every direction and sale of business and residential lots transcended mining as a top money-maker. Lots, sold originally for $100 to $250, within six weeks were bringing $750 to $1000 each.

On a morning in January 1884, a representative of the *Arizona Citizen* was among the hardy Tucsonians who rose sufficiently early in the chill winter dawn to make his toilette, partake of breakfast, and be ready to board the Quijotoa stage before it left town at seven o'clock.

The lengthy communique of that peripatetic reporter, published in the *Citizen* of January 28, 1884, describes well the route and stage trip from Tucson to the Quijotoas, and also conveys some idea of the excitement and ferment that gripped men borne aloft on the high tide of Big-Things-About-to-Happen.

"On Sunday morning the *Citizen* representative climbed on top of one of Pedro Aguirre's Concord coaches and had the good fortune to

[2]Both of these towns were named for Virginia City, Nevada, whose growth they were supposed to emulate.

have Bill Wilkins for a driver and a jolly crew of good natured passengers for companions," wrote the frontier scribe, whose contribution was signed with the pseudonym "Stephano."

"The road leads directly past . . . Silver Lake and the old Mexican smelter at the crest of the hill . . . Beyond the smelter the road heads for a double-knobbed black hill and passes to its left into the great Papago Valley . . . There are two waterholes to the right as you first enter the valley. The road across . . . is smooth and hard, and a bee-line. At Bolles' Well we found James Bolles, the proprietor, cleaning his well. It is 165 feet deep. Good dinner is served here. Here a change of horses was made, and a quick run of 11 miles was made through Coyote Pass to Dobbs' Well. Here another change of animals occurred. The road is by the northern skirt of the Baboquivari mountains, and passing the Papago Indian village called Coyote. We halted at a new station called Quinlan's in honor of our new councilman. A stone cabin is at the crossing of a sandy wash. An iron pipe brings the water down the canyon half a mile. Here a third change of horses took place. The road from Quinlan's leads directly west across a grass plain to the Comobabi mountains. There it enters a rounding pass of easy ascent to a Papago Indian village, and then there is a quick descent of four miles to Wood's Well, formerly known as Haynes. The transfer was made last Saturday. Mr. Wood is a son of Judge Wood of the Probate Court. At this place a royal supper was served. Tender venison and cow's milk supplied by the Papagos was on the table. Horses were changed for the fourth and last time and the race began for Logan, distant by the supervisor's odometer 18 miles and a fraction. It was dark, but 'shining in the distance the lights of Logan lay.' The stage took us for a mile through a line of tents and houses in all sorts of conditions but all progressing toward completion. At 9:45 p.m. the stage stopped in front of R. Gibson's store and was immediately surrounded by a crowd. Greetings from Tucsoners were numerous and hearty. Mail is distributed at Gibson's, for the present, and as faithfully as though he were sworn in as postmaster.

"The sound of the blast on the dark hill above the town was heard all last night, showing that the Bonanza Company are working with their old-time activity. This blasting sounds to the people of Logan and New Virginia like the music of the spheres. 'Do you hear that? Ain't that like Old Comstock, eh, my boy?' said an old miner to me last night.

"I have not had time to look around as I am writing this in time for Mr. C. B. Sessions to take with him on the stage this morning, which leaves for Tucson at 7 o'clock. C. G. Lewis keeps a first-class lodging house, which I know by last night's experience, and I will test his board this morning. . . . J. C. Perry is the pioneer lawyer of the camp and is doing a fine business. Eugene W. Aram has come with his law books and is pitching in to work. J. S. Mansfeld has a beautiful building which will be his news stand . . . One place has a sign, 'Tucson Beer.' Seepage water has been struck in the Logan Well. Everything is life. Bustle and activity is on every side. McAuliffe, the curiosity man, is here and has purchased several lots and is making money. He is at a standstill to know whether he will return to Tucson and sell cactus canes at five cents apiece to burst the competition, or remain in Logan and start a bank.

"Last night the bark of the festive coyote was mingled with the dreams of wealth. J. N. Mason has 100,000 feet of lumber en route. There are 40 houses on the main street, and 50 tents. All the lots on Main Street are bought and few for sale by the purchasers. Only one fight has happened in town and that was when a foolish fellow endeavored to resist a deputy sheriff.[3] For want of a jail the foolish fellow was tied to a tree until sent to Tucson . . . There is a Boston doctor here. Milk is 60 cents a gallon, whiskey 70 cents a gallon, and water 10 cents. Take your choice.

"There is a tide in the affairs of men which taken at the (J.C.) Flood leads on to future fortune, and the Loganites have taken it . . ."

On February 16, 1884, about three weeks after "Stephano" made his afore-described visit to Logan City and reported his findings to the *Citizen*, residents of the Quijotoa district hailed with enthusiasm the appearance of Vol. 1, No. 1 of their own newspaper, *The Quijotoa Prospector*, a six-column, four-page weekly. Editor and publisher of the new sheet was none other than Harry Brook, formerly with *The Wasp*, at San Francisco and, more recently, with the *Pinal Drill*, at Pinal City, A. T.

As there was invariably a rush to establish the first newspaper in any new bonanza camp, the earlier issues of the *Prospector* were set up and printed in the office of a small weekly in Tucson while Brook awaited the arrival of a first-class printing outfit ordered from San Francisco. Since the office in question had barely enough equipment to do its own work, these earlier numbers of the *Prospector*, Brook later recalled, presented an astonishing array of type of all shades and sizes, some of which had been knocking around on the frontier for a quarter of a century.

"In a canvassing trip for subscribers through the embryo camp before the first number had been issued," wrote Brook in the *Los Angeles Times*, "I took in $315 within a couple of hours, at the rate of $5 a year, of which $30 went for 'hospitality' as three-fourths of the business houses were saloons . . ."[4]

"Where Quijotoa now stands there were, on the first of January, a couple of tents," declared Editor Brook in a lengthy editorial centering the first page of the first issue of this new boomcamp's first newspaper. "All around was a dry and deserted land. No water within ten miles.

"There are now a mile of houses and a population in and around the camp of at least 1200 . . . The opening of 1885 will see a city here compared with which Tombstone, Virginia City and Leadville are mere villages . . . 'Toppy' Johnson, A. J. Frazer, E. W. Dobbs and others, are

[3]Thomas D. Casanega.

[4]Quoted in Joseph Miller, *Arizona, the Last Frontier*, p. 89.

the owners of the New Virginia Townsite, which is rapidly settling up. . . . Third street has been cleared off and they will soon have a scraper at work on other streets. The Logan Townsite Company is composed of Wm. R. Logan, president; James M. Quiggle, vice-president; James H. Shotwell, secretary; and George A. Roskruge, general manager. They are about to build an office at the corner of Roskruge avenue and Adams streets where parties wishing to purchase lots should apply.

"J. Melstedt, who owns a chop-house here, has followed nearly every mining rush on the coast. Amongst other places he has been at Washington Camp, Total Wreck, Bodie, Harshaw, La Noria, Panamint, Darwin, and Pioche. S. Marks, of Contention, who is in the general merchandise business here, has a building 75 feet by 25 which would do credit to any town. Vicente Flores, of Tucson, has a butcher shop in New Virginia. Dr. J. B. Grady, from San Diego, has his office in the Lewis House. Mrs. Christie has a neat, homelike lodging house on Logan avenue. Pedro Bustamente opened the first barber shop at Quijotoa. Mrs. L. A. Henry, of Tucson, keeps the Star Hotel on Logan avenue. The Pioneer Saloon, on Logan, was the first saloon in camp. The Tucson Beer Depot is kept by John Lawrence, from Tucson. H. H. B. Summers, of Florence, is expected here shortly to open a law office. F. H. Mason is the polite and obliging manager of the Quijotoa lumber yard. Pierre Sautriaux has a grocery store in New Virginia . . . J. B. Skidmore, from Temple, has a feed and provision store, also a corral, at New Virginia. The New Virginia Corral, the largest in the camp, is kept by Messrs. Knox and Sea. Mr. Knox is from Pyramid City, N. M., and Mr. Sea is from Tucson. The Headquarters Saloon, at New Virginia, is kept by A. J. Frazer who has been mining around Harshaw and is interested in the townsite. He has a fine building. Joseph Bitschnau . . . has a bakery in New Virginia . . . A. Nevins & Co. have one of the largest general stores in town . . . Hermann Loptien, the well-known 'Jolly Hermann' of Tombstone, sells his excellent beer at the Arizona Brewery, where he will soon commence brewing . . . Charley Lewis, the well-known pioneer of Florence . . . is keeping the Lewis House on Logan avenue . . . G. Uripe, one of the earliest settlers here, has a corral on New Virginia where Robles' Stage stops. He always has men on hand for adobe making, building, etc., and keeps wood for sale. C. S. Shotwell, who has a general merchandise store here, was a partner of J. M. Wilkins, the leading grocer of Tucson. Job M. Seamans, our watchmaker and jeweler, has also an establishment at Tombstone . . .

"Messrs Hutchings and Nash . . . are keeping a saloon on Logan avenue. Glassman & Co., from Tucson, will soon start brewing at the Logan City Brewery. They, at present, sell imported beer. James Carruthers . . . has a butcher shop here . . . Messrs. Hibbard and Casanega, of Tucson, are running the Headquarters Saloon on Logan avenue. S. Platt, who has a large store here, is from Los Angeles, where he also has an establishment. Murphy & Co. keep a fine assortment of liquors at their saloon. Mr. Hassett, from San Antonio, Texas, keeps a bakery and provision store on Logan avenue. F. Y. Hauck, of Washington Territory, has a blacksmith shop in New Virginia. Messrs. Tryon and Marsh, who keep a saloon here, formerly had one at Harshaw. H. Myers, of Tombstone, runs a fine store on Logan avenue. R. Gibson was appointed postmaster but has not qualified . . . The mails are, at present, very carefully handled by Mr. Ruffner, of Teitsworth & Ruffner . . .

"Five lots have been purchased on Roskruge avenue by parties who intend immediately to put up a three-story hotel and furniture manufactory. M. L. Wurts, of New York City and recently from Contention, has opened a little lodging house on Logan avenue. W. B. Walker, the assayer, from Tombstone, has built himself a neat lumber office and is rushed with work . . . Mr. Proctor, from Calabasas . . . has a butcher shop on Logan avenue. W. Douglass, from Johnson, Dragoon Summit, is running the Pioneer Hotel. N. Bannon, from Amador county, California, has a saloon in town. John Stahle, a Californian, has a shoemaker's shop on Logan avenue. Matt Johnson, from Benson, runs a saloon on Logan avenue. S. Mund, Logan avenue merchant, has adobes made to build a store of that material. Mr. Mund is an old Californian, having been at San Francisco as early as 1854. Frank A. Milich, proprietor of the Big Bonanza Saloon, formerly had a wholesale produce and commission business at Tombstone. Messrs Moss and Driscoll are putting up a fine building on Logan avenue for a wholesale liquor business. Mr. Driscoll formerly had the Oriental at Tombstone. Mr. Moss ran the Cliff House at San Francisco. B. Cook, from Globe, has a barber shop on Logan avenue. Mrs. J. S. Mitchell, who has a hardware and liquor business in Mansfeld's building, was formerly of Lordsburg . . . Mrs. Mitchell is well known in the Southwest, having for years followed the van of civilization. Douglass Snyder, one of our leading merchants, has a large establishment in Tucson . . . Mr. Snyder represented Pima county in the last legislature and distinguished himself very honorably. Mr. Jones, from Tombstone, has a large corral on Logan avenue. The St. Nicholas Chop House is run by Travers and Sheehan from Clifton, A. T. . . ."

That all this merchandising activity, these widely assorted business concerns operated by persons drawn from points as far distant as New York City and Los Angeles, were available for chronicling only six weeks after the embryo camp had embraced only "a couple of tents," is one of the most concrete examples anyone could offer of the magnetism exerted by every new boom camp of promise, and the widespread enthusiasm and optimism that attended its birth.

During the first several months after founding of the four camps generally lumped together under the single name, Quijotoa, shortage of water constituted a problem of sorts.

Reporting on conditions at Quijotoa, *Phoenix Gazette* remarked on January 6, 1884: "Drinking water is quoted at from 50 to 75 cents a bucket. It is surprising that saloons can afford to furnish liquor at 12½ cents, with water commanding such a high figure . . ."

"We would advise parties who contemplate going to the Quijotoas to postpone their trip for two or three months, as there is neither water nor work there and it is no place for a poor man to go at present," stated *Arizona Silver Belt,* of Globe, on February 2, that year. "About 800 people are in the camp, and more going in every day . . . Chinamen are not allowed in the camp, at present. They are putting up some good buildings of lumber and canvas. No adobes are made, as water is too scarce. It has to be hauled five to ten miles and sells at 10 cents a gallon."

Again, on March 3, the *Silver Belt* remarked: "Water is scarce at the Quijotoas, but there is a liberal supply of whiskey as there are 14 saloons in Logan City and New Virginia . . ."

But the problem of water's lack was slated to vanish like the disturbing details of a bad dream, when an abundant flow was struck in the Logan well, on the morning of April 8, 1884, at a depth of 452 feet. With that discovery the burro-back water packers went into retirement and the four adjacent camps on the east side of Quijotoa-Ben Nevis Mountain—by then merging into the single metropolis of Quijotoa City—went into orbit.

That same month of April 1884, saw the inauguration of daily mail service between Tucson and the Quijotoas, with Benabe Robles, owner of a line of stages and carrier of a weekly mail to Gunsight, winning the contract. In April, too, was completed a telegraph line giving direct connection between the Quijotoas and Tucson.

In anticipation of the day when the rising new bonanza camp would conceivably have need for world-wide communication even more rapid than that provided by the crack teams and drivers of Benabe Robles, the Arizona Telegraph Company had been formed by a group of Tucson men. Sevently-two miles of line was surveyed, poles were set, and wire imported from England was strung, at an over-all cost of $16,000. From the time the first message went out over the new wires on April 2, 1884, business prospered so vigorously that within a few months the line had returned to its builders all costs of construction, operation, and maintenance, and was paying handsome monthly dividends.

In addition to Robles' daily stage line, two other lines, both offering daily service, were in operation by midsummer, 1884.

"Ho! for Quijotoa!" read the advertisement of the E. A. Saxe line. "Casa Grande & Quijotoa Stage Line leaves Casa Grande daily for the Quijotoas, via Christmas Gift, Vekol, Reward, and Copperosity mining camps. Through in 12 hours . . . Good roads, good horses and fine coaches."

The third line, operated by Pedro Aguirre & Co., advertised daily service between Tucson and Quijotoa, and thrice-weekly service between Tucson and Calabasas, Arivaca, Oro Blanco, Esperanza, and Sombreretillo. Thirty pounds of baggage was allowed each passenger, and the fare between Tucson and the Quijotoa district was nine dollars.

During one six-month period in 1884 more than 1500 mining locations were made in the district, and the *Report of the Director of the*

Mint said of Quijotoa: "This now famous bonanza camp is attracting as much attention among mining men as did the first discovery of gold at Sutter's Mill in California in 1849."

Ore was shipped to outside smelters until August 23, 1885, when the first carload of "paydirt" came down Ben Nevis Mountain over the company's brand new tramway to the company's new mill, and Territorial newspapers jubilantly reported "everything working nicely."

But "everything" didn't work nicely for very long.

After the original mountaintop scabbing of rich ore had been harvested, the Bonanza firm, anxiously seeking a new ore body, drove a tunnel all the way through Ben Nevis Mountain. But the pretentious passage, to quote Charles Dunning: "disclosed little more than the daylight at the other end . . ."

Millions of dollars had been poured into the mountain. Possibly half a million dollars had been returned by it . . .

Mining activity in Quijotoa was at a still-faintly-hopeful standstill when fire broke out in the Nevin Building on June 19, 1889. Fanned by a hot, dry desert wind, with little fire fighting equipment available, the entire town, in scarcely more than an hour, was reduced to a blackened ruin.

In connection with this fire, an interesting story is told. First published, I believe, in the *Southwest Observer*, the story was later retold by the Western historical writer, Tom Bailey, under the title, *The Ghost in the Mine*, published in *True*, "The Men's Magazine," in November 1946.

According to this copyrighted story, three travel-worn and taciturn strangers had arrived in Quijotoa on the Robles' stage from Tucson and were waiting in Mrs. Stewart's restaurant for their meal to be served, when the fire was spotted and the entire town was thrown into an uproar.

After the conflagration had burned itself out and a tally was taken to see who, if anyone, might have perished in the flames, Mrs. Stewart recalled the three strangers who had been waiting for service in her restaurant. When these men were nowhere to be found it was feared that they must have been victims of the fire.

Since the mines already were closed, there seemed little point in rebuilding the town. Most of its inhabitants soon pulled out for new locations and Quijotoa became a ghost town—nor did it seem that this was entirely a figure of speech.

"Persons who chanced that way in the following years reported strange goings-on," wrote Tom Bailey. "They heard weird cries and bells ringing. One party of four which camped on the townsite in the fall of 1891 left hurriedly at midnight after listening for hours to groans which sounded like those of a dying man . . ."

The mystery of the weird noises was cleared up in 1896, according to Bailey's story, when a disheveled, unkempt hermit was found living in the old mine workings with a human skeleton he called "Silver" kept as sort of a mantel piece.

On complaint of a man named Ed Whipple who, assertedly, had effected his escape after being captured and confined by the aged eccentric, the hermit was taken to jail at Tucson, where he subsequently attacked one of the trusties assigned to feed the prisoners.

By slow degrees and extended questioning, according to the story in *True*, it was learned that the hermit's name was Fred McComb; the trusty's name was John Cronkhite, and the skeleton preserved by the hermit in the Quijotoa mine tunnel had been that of a man named Matt Silver. These three, assertedly, had robbed and killed a farmer, Will Hodges, near Lincoln, Nebraska, on February 9, 1889, and had made their way to the decadent boomcamp of Quijotoa just in time to witness the fire that destroyed the town.

In the confusion occasioned by the fire, it seemed, Cronkhite had taken the stolen money and skipped, whereupon Silver and McComb had hidden out in the abandoned mine tunnel until Silver sickened and died. McComb, the hermit, had preserved the body of his former partner, (carefully wrapped in chicken netting "so the rats couldn't get him"), and to keep unwelcome visitors away from his hideout had devised the ghostly repertoire of bells ringing and midnight screams.

As seven years had passed since the murder in Nebraska, during which time the chief witness had died, the case was dropped. Examined by a board of doctors, McComb was adjudged "sane" and released, and both he and Cronkhite disappeared.

"Some said that McComb went back to the Quijotoa mines and continued to haunt them, for as late as the 1920s visitors to the old ghost town reported weird cries that made their scalps tingle," wrote Bailey; and when the *Arizona Guide Book* made its appearance in 1940, it was stated of Quijotoa: "Though its buildings are now reduced to crumbling walls and it has no human inhabitants, bells have been heard ringing and doors creaking on their hinges . . ."

"Today, Logan City is just a spot about eight miles south of Covered Wells on the Ajo highway in the Papago Indian Reservation," wrote Bernice Cosulich in *Arizona Daily Star*, April 10, 1935.

"In the midst of the tortured contours of the Quijotoa Mountains may be seen the old boom-camp site. There is just one structure standing, this an adobe house with roof tumbled in and sidewalls tapering at the top with wind and water washing. The foundation markings of wooden houses are dimly seen, but the old streets are lost among the desert vegetation. Three steam boilers, in their brick housings, with three cyanide tanks, are gentle jogs to recollections of what was once a bustling, ambitious camp."

But Bernice Cosulich wrote that description more than a quarter of a century ago.

Today it is difficult to locate even the site of Quijotoa City, successor to the one-time ambitious bonanza camps of Logan City, Virginia City, New Virginia, and Brooklyn; and if any bells have rung there in the last ten years, I have not heard them.

"It is no uncommon thing . . . to meet an old, gray-haired man who has been prospecting and mining all his life and has not yet made a dollar, but is buoyantly sure that he will strike it soon."

—Atlantic Magazine, December, 1883.

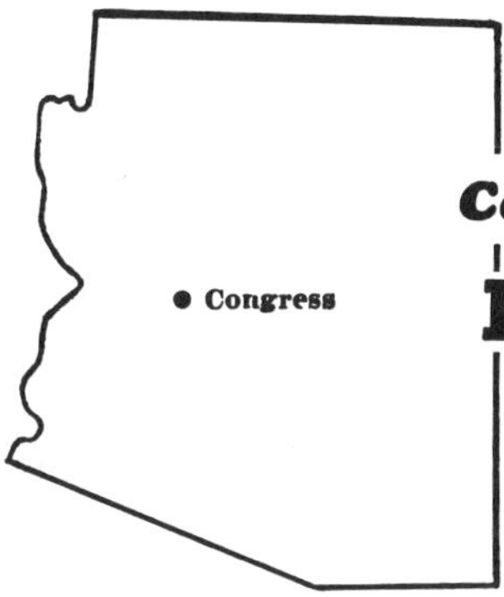

Congress had a Good Session

IN SOUTHERN Yavapai County, in the foothills of the Date Creek Mountains—where there are no dates and very few creeks and the mountains aren't especially mountainous—a scattering of rock and adobe ruins mark the former site of Congress City and its once celebrated gold mine.

Original settlement in this vicinity was not at Congress but at Martinez, three miles to the south, a stage station on the La Paz-Fort McDowell road. From a one-building way-stop, Martinez eventually developed into a small town. Here, in the early 1880s, lived a prospector, Dennis May, who is credited with discovery of the Congress Mine.

Inasmuch as gold mining had been the chief occupation in this region through all the 20 years since Abraham Peeples' party had discovered Rich Hill, eight miles to the southeast, it seems likely that the outcrop staked by May had been noted and sampled by earlier prospectors but had been passed by as worthless, possibly due to the fact that its gold content was not free milling but was so coated with iron that it would not amalgamate.

In any case, it was May who located the Congress claims and May who subsequently sold them to "Diamond Joe" Reynolds, a colorful and dynamic personality then known as "the boss" of transportation on the Mississippi River.

Reynolds' interests were many and diverse. Prior to buying the Congress Mine he had risen to ownership of the *Diamond Joe* line of steamships on the Mississippi; had founded the city and health spa of Hot Springs, Arkansas, and had built a railroad to that resort. With the same limitless enthusiasm given to these earlier ventures, Reynolds now threw himself into development of the Congress Mine and the construction there of a model mining camp. As superintendent of operations he wisely engaged Frank Murphy, one of the all-time "greats" of Arizona mining circles.

Due to the isolated location of the Congress at the time of its purchase by Reynolds, it was necessary for its ore to be wagon-freighted 60 miles to Prescott at a per-ton cost of $17. From that point the ore was shipped by rail to the main line of the AT&SF at Ashfork, thence to Socorro, New Mexico, for reduction. Since these pyramiding costs of transportation virtually eliminated any chance of profitable operation, Reynolds immediately began pushing for a branch railroad to be built south from Prescott, via Skull Valley, Hillside, and Date Creek, thereby giving the mine direct connection with the AT&SF.

It was during this formative period that the Congress Mine and its "model" camp were visited by Charles H. Akers, Yavapai County recorder, whose enthusiastic report was published in the *Arizona Weekly Journal-Miner* of October 1, 1890.

"Congress City is a thriving town with a population of 350 people," related the county official-turned-scribe. "Its regular, well-kept premises and systematic arrangement show that the development and rapid growth it is now enjoying was anticipated by the far-seeing men who planned its construction. Each side of the main street is beautifully adorned by rows of umbrella trees, famed for their beauty and perfect shade. Beds of blooming roses grace the yards in front of the residences, and everything bespeaks the thrift and prosperity of a contented and prosperous people who are building for themselves permanent homes in a locality that is certainly well chosen.

"A distinct and appreciable addition to the accommodations of the town is the restaurant and lodging house kept by Gus Shamp," enthused Mr. Akers. "Guests at his house have nothing but words of praise for the excellence of the fare and the scrupulous order and cleanliness which prevail throughout . . . Kastner and Co. conduct the only saloon in the town and . . . contribute to, rather than detract from, the high standard of of morality which prevails . . . as they keep a quiet and orderly establishment, holding all attempts at excess strictly in check . . .

"Congress stands as a tribute written and riven deep in the rugged tablets of the Bradshaws to the sagacity of Jo Reynolds and the energy, skill, and good judgment of his superintendent, F. M. Murphy," it was declared.

Work of building the branch railroad—eventually to be known as the Santa Fe, Prescott and Phoenix—was getting along nicely when Diamond Joe Reynolds suddenly sickened and died while on one of his periodic visits to the Congress Mine in March 1891. How ironic that the body of this dynamic "boss of transportation" and devotee of progress, should have to be packed out to Prescott on mule back—the fastest method of travel then available between Congress City and the outside world.

Following Reynolds' death the Congress remained closed pending the settlement of his estate and completion of the branch rail line building south from Prescott. Not until the closing week of 1894 was it announced that the mine had been purchased from the Reynolds' heirs at a rumored price of $1,000,000. The mining syndicate acquiring the property was headed by E. B. Gage, Phoenix banker prominent in Tombstone mining circles; N. K. and Wallace Fairbank, Chicago capitalists and meat packers; and Frank Murphy, the mine's erstwhile superintendent.

With completion of the railroad (and a personal visit by President William McKinley) the mining camp of Congress climbed back on the band wagon and in scarcely any time at all the *Journal-Miner* was reporting all sorts of interesting activities at Congress. A new lodge of Red Men had been instituted with 60 charter members . . . The Congress lodge of Woodmen of the World was preparing for a Grand Ball on New Years Eve. The A.O. U.W. was doing this, and the Odd Fellows were doing that, and a community Christmas tree was being planned by ladies of the Congress Sunday School.

There were also other activities such as street fights in which "women of the town" shot it out on the street in Frankie-and-Johnny fashion. In one such fight, in May 1897, a bullet allegedly dispatched by one "Mamie," toward the person of one "Sadie," went wide of its mark and struck an innocent and respected bystander, Thomas Concannon, who died in only a few minutes.

"Concannon," remarked the *Journal-Miner,* "was quite seriously stabbed about a year ago by a man who mistook him for another . . ."

But these were only normal growing pains and by the autumn of 1897, Congress, with 1500 inhabitants, was considered one of the liveliest, best equipped, and most successfully managed mining camps in all Arizona Territory.

In October 1897, a Prescott newspaper reporter left the train at Congress Junction, as the former stage station of Martinez was coming to be known, and there boarded a horse-drawn stage for the last three-mile lap to Congress City. Why the county seat reporter elected to travel this last lap by stage isn't quite clear, as the company had built its own three-mile spur from the mine to the Junction and over this road was operating a 56-ton oil-burning Brooks' locomotive. In its one round-trip daily, the little train hauled to the Junction concentrates, highgrade ore, and passengers, and took back to Congress passengers, supplies for mine and mill, and stock for the various merchants.

"We landed in the city at 9 a.m., Sunday, and found the town alive with the hum of business," wrote the peripatetic newshawk from Prescott. "The roar of machinery and the mill, the labored breathing of the hoist . . . the wheeling of ore to the bins and refuse to the dump, coupled with business activity . . . told us plainly that Sunday, as a day of rest, is practically unknown in the mining camp . . ."

A 40-stamp mill, it was set forth, was in operation night and day, seven days a week. Ore was being taken from three shafts; the deepest then being No. 1 at 1750 feet. New hoists, soon to be installed in Shafts 2 and 3, however, each would be capable of hoisting a 20,000-pound load at a rate of 1000 feet per minute from 3000 feet depth. Four hundred and twenty five men were then employed in the Congress mine and

[1]*Arizona Weekly Journal Miner,* October 13, 1897.

Lower Town, Congress, 1959

President William McKinley visits Congress, 1900

—*Courtesy Arizona Pioneers' Historical Sc*

mill, and the company payroll of $30,000 monthly was being met with agreeable regularity.[1]

Thus far, except for Joe Reynolds' death, Congress had experienced no great tragedies or even major setbacks—but her turn was coming. Almost every desert mining camp of any importance eventually had its devastating fire, and Congress was to be no exception.

It came in April 1898.

Believed to have started about 3 o'clock in the morning in a Lower Town restaurant adjoining the Silver Dollar Saloon, the blaze spread rapidly. In this section of camp, where were located the privately-owned business concerns, the saloons, restaurants, and most of the miners' dwellings, water for fire fighting was non-existent and, in a matter of minutes, the flames had consumed nearly every commercial building in town. (In Upper Town, separated from Lower Town by an uninhabited buffer zone, were situated the mill, company offices, hospital, boarding and lodging houses, the company store, and homes of the superintendent and other V.I.P.'s.)

Lost to the early-morning blaze were the large general mercantile owned by B. Herschkowitz, C. J. Sullivan's drygoods store, the C. C. Inglehard grocery, Tim Loy's lodging house, four Chinese restaurants, and seven saloons. The fire also cost two lives. Gertie Graham, a so-called "girl of the town," evidently was overcome while asleep and was burned beyond recognition; and James Vidano, proprietor of the Silver Dollar, who occupied living quarters in the rear of his saloon, was burned so badly that he died the next day.

But as long as ore continued to pour forth and payrolls were met with regularity, mining camps were remarkably resilient. Though Congress reeled momentarily under the body blow dealt by the fire, it was not long until she had rebuilt and beginning of the new century found business proceeding as usual and the goose (or geese) hanging high.

Early in that year of 1900 the camp in the Date Creek foothills was visited by the editor of *Arizona Graphic*, an illustrated journal devoted to portraying and promoting the advantages of life in Arizona Territory, and in its February issue the *Graphic* carried a lengthy story concerning the Congress Mine and its camp.

In the course of identifying the four chief owners of the Congress, the magazine editor referred to N. K. and Wallace Fairbank, of Chicago, as "capitalists and renderers of lard."

"There is no covert satire in this description," the Phoenix scribe hastened to add. "I wish merely to identify the gentlemen. I would as soon get rich rendering lard as publishing an illustrated paper . . ."

The mine, it was stated, had been operating without interruption since 1894, during which time, according to Superintendent W. F. Staunton, it had yielded some $5,000,000 in ore, "with enough ore in sight today to keep the stamps pounding for 50 years . . ." In the main workings there were then 30 levels, and ore was being taken from six shafts,

Like most pencil-pushing visitors to Congress, the editor of the *Graphic* waxed enthusiastic concerning the town's remarkable water system.

Pumped at company expense over a 500-foot divide from Martinez creek a mile from the camp, 80,000 gallons of water daily was delivered to huge tanks from which it was dispersed to the mill and to the public. Actually, the mill and Upper Town required some 200,000 gallons of water daily, but since that used by the mill and concentrators was cleared in a series of tanks and pumped back up the hill for re-use as long as any of it remained, it was possible to make 80,000 gallons cover the camp's daily needs.

This bit of good water husbandry, however, does not seem to have impressed the magazine man as much as did the unique system employed for distribution of domestic water.

"The camp is supplied with a single inch faucet in front of the company's store, where a keen-eyed clerk keeps tab on everyone who comes," wrote the editor. "Each family has a 50-gallon whiskey barrel. Around the barrel's waist is a piece of rubber belting, usually a piece of worn-out concentrator carrier. The local blacksmith cleverly hammers out an iron rod into a sort of short wagon pole that divides to curve around to the center of each barrel head. And there you are with a water cart that is its own wheel! Jack and Jill roll the empty barrel up the hill to the company store and load up with 50 gallons. On the homeward trip, down the rocky streets, gravity does all the work save that of guiding the revolving tank. Those who haven't attained the dignity of owning their own tank wagon make use of children's toy wagons, setting within each two five-gallon oil cans.

"The Lower Town hasn't any water system but it is supplied by a sturdy German who owns a spring somewhere around the mountain. But Lower Town," commented the editor wryly, "doesn't need any system in its water supply . . . because three-fifths of the business houses are saloons . . ."

This same year there was established at the neighboring town of Congress Junction, *nee* Martinez, a weekly newspaper, *The Martinez News-Herald*, published by Eli S. Perkins, and entered in the post office at Martinez, Arizona, as second class mail matter.

Among advertisements carried by the Martinez news sheet were those of numerous shops and services in Wickenburg, Phoenix, and Prescott, all interested in bidding for the Congress miner's paycheck. In addition, the businessmen of Congress and Martinez showed by their patronage that they, too, were aware of the advantages of advertising.

The *News-Herald* of March 5, 1904, for example, carried paid messages from a wide assortment of local shops and services, including Adolfo Chretin's O.K. Barber Shop; R. L. Brady, wholesale and retail dealer in general merchandise; and the Red Front Saloon, operated by Sam McLees and Katie Oliver. J. C. Tovrea, of the Congress Meat Market, characterized himself as a dealer in all kinds of meats, fresh and salt, live poultry, etc., with "free delivery to all parts of the city." J. D. Vietti, it was quite evident, thought well of his Red Front Saloon. Pointing out that his was the oldest saloon in town, Mr. Vietti also drew the readers' attention to his "club rooms" equipped with card tables, roulette, pool, billiards, faro, "and sports of all kinds."

Congress Consolidated Mines Co., Ltd., advertised itself as "The Big Store—Prices Always Right," and noted its excellent stock of drygoods, notions, clothing, furnishing goods, shoes, hat and millinery departments, etc.

Martinez concerns advertising in the same issue included George Burnell's Golden Gate—"choice wines, liquors, cigars, miner supplies, hay, grain, and livery;" also Dick Bullard's Congress Saloon, and Martinez Mercantile Company. There was the professional "card" of H. H. Holbert, mining engineer, at Martinez; and it was noted that the Harrisburg Stage, Ed Zeiger, proprietor, left Congress at 8 a.m. on Monday, Wednesday and Friday, returning at 6 p.m. on alternate days.

During the years when the Congress Mine operated under the direction of W. F. Staunton, seven incline shafts were sunk, the deepest to 4000 feet, with extensive development down to the 3900-foot level. In the course of this work the mine produced nearly 700,000 tons of ore yielding a total of 388,477 ounces of gold and 345,598 ounces of silver; or approximately $8,000,000 in all, an average of roughly $12 per ton.

Underground work at the Congress ceased to be profitable around 1910, and with the closing of its mines the town in the desert foothills soon withered away.

It was almost 50 years after the ghosts took over before Mom and I visited Congress for the first time. Many previous occasions had found us in Congress Junction, *nee* Martinez; but, for some reason, we had never turned on the unpaved side road that leads off through the creosote brush to the old mining camp, three miles to the north.

About two miles from the junction we began seeing the stone and adobe ruins of former dwellings and business houses, and a few wooden cabins, long empty and deserted. Reaching the buffer strip that had divided Upper and Lower Congress, we found our way barred by a padlocked gate and a No Trespassing sign. In the forbidden territory, perhaps a mile beyond the gate, were visible the mine and mill buildings and other scattered ruins. Promising ourselves that we would obtain the necessary permission and return later to visit this one-time realm of the V.I.P.'s we turned back into Lower Town where the "common" people had lived, and where three out of every five business houses assertedly had been saloons.

It was fun prowling around the old townsite. Some of the dwellings had been nicely plastered or wall-papered within, and had included as many as four or five rooms. In one house we found an old-fashion leather-covered trunk with brass fittings, its domed lid lined with a lithograph of two English gentlemen riding to the hounds. The leaves of one calendar pad had been torn off as far as October 1903. One cupboard held a lot of mouse litter and several empty bottles, one of which had contained Jamaica ginger; another, Mustang Liniment.

Mom got interested in picking up fragments of old China dishes and bits of colored glass. In some parts of the townsite these glass fragments covered the ground surface almost like mosaic. In an outbuilding behind one house I found some old horseshoes and muleshoes corroded with rust, and a pair of blinders from a horse's bridle. The leather of the blinders was dry and twisted and so brittle it could be broken between the fingers, almost like a potato chip. Evidence of the Big Fire, or of some later blaze, was everywhere. Miscellaneous pieces of metal were burned and blackened; glass, melted by the intense heat, had rehardened in misshapen masses, some of which still held embedded bits of charcoal.

The thin, gray whips of ocotillos, and one palo verde tree, gnarled with age, whispered dryly in the late summer breeze and cast upon the packed and sterile earth their feeble shades, as thin and pale as ghostly spiderwebs. But of the rows of umbrella trees that had lined the streets, 70 years before, and the beds of roses that had graced the yards, we saw no remaining scrap of evidence.

I was examining an old safe, fire charred and rusted, when Mom called from the front yard of a half-fallen adobe cabin.

"Wait till you see what I've found!"

It was a tiny China doll, no more than an inch in length, but with movable arms and legs. Although the little object likely had known more than half a century of exposure to the cruel wind and weather of the desert, it gave no indication of those long and lonely years. Even the black hair painted on its head, its rosebud mouth and blue eyes, were still clearly defined and unfaded.

It was a strange souvenir to find in a ghost mining camp in the Arizona desert.

"No man feels safe half a mile out of the settlements. Twenty-three of my own personal acquaintances have been killed, stripped, and mangled by the Indians during the past year."

—Vulture correspondent to the Alta California, May 15, 1865.

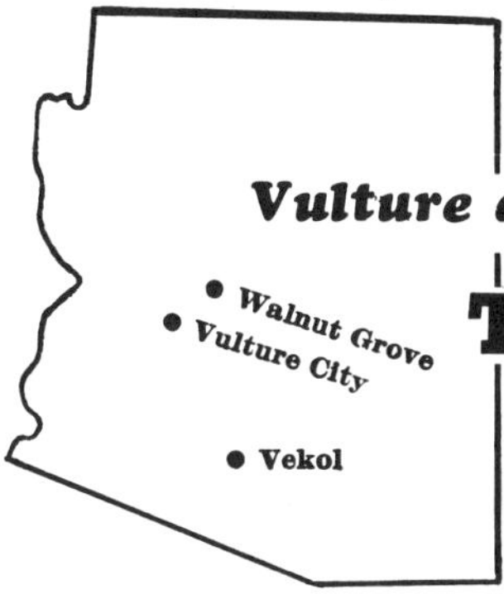

Vulture and Vekol, Tragedy Camps

THE mining camps of Vulture and Vekol have little in common except their alliterative names, the fact that their mines were more than commonly rich, and that the original locator and developer of each met a tragic end.

In the desert hills, 15 miles southwest of the present Dude Ranch Capital of the World, the Vulture mine was discovered in 1863 by Henry Wickenburg, then about 43 years of age. Born Heinrich Heintzel, in Austria, the future Arizonan assertedly had found it advisable to flee his native land after selling a load of coal from his father's property instead of turning it over to the state as required by law.

There are several versions of the circumstances attending the discovery of the Vulture—one, of course, being the overworked bromide of the Strayed Burro. Likely the true version of the discovery was much as described in a personal interview with Wickenburg, published in the *Florida Press,* of Wickenburg, and reprinted in the *Arizona Miner,* June 6, 1868.

According to this account, Wickenburg and an unnamed companion, prospecting in central Arizona, had spent several months in a fruitless search for gold. When the companion at last fell ill and was forced to remain in camp, Wickenburg continued the search. One day he sat down to rest on a bleak, dry hillside. He was weary and discouraged, spirit broken, his money virtually gone. He was also thirsty—and the nearest water was in the Hassayampa River, nearly 15 miles away.

At this moment of slack tide, almost miracle-like, the Austrian emigrant saw that the quartz outcrop on which he was resting was richly laced with free gold!

Hurrying back to camp, Wickenburg poured out the story of his wonderful discovery, only to find that his partner refused to put any faith in it. Scoffing at the report, he would have no part of investigating the outcrop or posting location notices, and soon as he was able to travel he took his departure. Wickenburg remained behind to mine his treasure.

Even in the Indian-scourged lands of central Arizona, in 1863, it was not long before word "got around," and other hopeful miners began converging upon this desert region between the Big Horn and Vulture mountains. Realizing that he could not develop his mine without help, and could not hire workmen to do so, Wickenburg entered into an agreement whereby these later arrivals would work his mine, paying him $15 for each ton of ore removed.

Within the next year or two, according to Farish, the Vulture was being busily worked on this basis. Carried on burro back to the Hassayampa, the ore was pulverized in individually-owned arrastres, as many as 40 of which were in operation at a single time. The ore was rich. It is said that a wandering prospector, in 1864 or '65, found a deserted arrastre along the Hassayampa, its owner presumably having been killed by the Apaches. Beside the arrastre lay a small pile of Vulture ore, which the stranger preempted and ran through the mill. From it he realized more than $100 in gold for only a few hours work—most of which was done by his burros.

In 1866 Wickenburg sold four-fifths interest in the Vulture mine to Benjamin Phelps, of New York City, taking a down-payment of $20,000 on the agreed price of $85,000. Phelps immediately built a 20-stamp mill—soon increased to 40 stamps—on the Hassayampa about a mile from the new town of Wickenburg. At this mill site, according to Raymond's *Resources* report of 1871, the company maintained its office, storehouse, assay office, boarding house, "sleeping apartments," and outbuildings, all of adobe with shingle roofs. At the mine site—Vulture City—the buildings were chiefly of stone. (In later years, after the Vulture's ore had become less abundant, even these buildings were torn down and milled for a yield of $20 per ton.)

"The company," noted Raymond, "also cultivate a garden of twelve acres in the Hassayampa bottom a short distance below the mill, in which they raise all the vegetables required in their large boarding house. The cooks are Chinese . . ."

The Vulture's ore was pockety, some of it extremely rich, and highgrading flourished. Persons around Wickenburg, it was declared, were notoriously bow-legged—not as result of riding horseback, but from packing off in lunch-buckets such vast quantities of highgrade ore!

As he grew older and more disillusioned, both with mining and with people, Henry Wickenburg turned his attention to agriculture and soon developed a nice farm in the Hassayampa bottoms. Because of this farm however, he carried to his grave a bad facial disfigurement where part of his right cheek had been shot away about 1870.

Wickenburg, strangely enough, was not sensitive about the scar and told amusingly of the near-fatal incident.

According to the story (as given to the Arizona Pioneers' Historical Society by Dan B. Genung, in 1939), a renegade living in the Wickenburg area was out hunting ducks one day and went tramping across Henry's field, mashing down and ruining his fine stand of wheat.

"Ven I tell him to stop and get oudt," said Henry, in his heavy German accent, "he say, 'Shut up, you Dutch sonuvvabitch, or I blow your head off!'

"I say, 'Shoost you blow avay! Und, py Gott! he *did* blow!"

Fate and the years were not kind to Henry Wickenburg, either in mining or farming.

Although his Vulture mine produced millions of dollars in gold, even in Henry's lifetime—estimates run as high as $15,000,000—virtually none of that bounty ever enriched the pockets of the Austrian immigrant. Most of the $20,000 paid down by Phelps on the Vulture, subsequently was expended by Wickenburg in a fruitless effort to collect the remaining $65,000 due. Also, he seems to have derived no benefit from the one-fifth share in the mine retained by him at the time of sale.

Soured on mining, he retired to a shabby adobe cabin in the Hassayampa bottoms and there developed the aforementioned farm and orchard.

After 20 years of backbreaking toil and effort, Henry Wickenburg's riverbank fields and pastures were picture-book pretty on that February day in 1890 when the Walnut Grove dam gave way only a few miles upstream.

As a result of those catastrophic flood waters, Wickenburg's beautifully-tilled fields were buried hopelessly beneath thousands of tons of sterile rock and sand.

After this disheartening blow, Wickenburg seemed to lose all interest in life. Near penniless, discouraged and despondent, he often remarked to friends that he was old and tired of living.

At last, on a summer evening in 1905, when the old man was about 85 years of age, he was found lying dead in the trees near his cabin.

A bullet wound was in his right temple, and an old style Colt revolver lay at his side . . .

Not too many years later, Vulture City became a ghost town—and a ghost it remains, today.

It was April 1958, when I decided to visit its site.

After following an unpaved county road 15 miles southwesterly from Wickenburg—all that way through pleasantly wild highland desert grown to ocotillos, palo verde, saguaros, and several varieties of smaller cacti then bursting into flower—I came to the sideroad leading off to the remains of Vulture City.

Near the summit of a low hill, possibly half a mile to the north, huddled several old buildings—great, dark hulks, whose original function I could only guess. One of them appeared to have been a mill; another looked like a hoist house, and a two-story stone structure had the general appearance of a boarding house. In addition, there were numerous smaller buildings—enough of them to quicken the pulse of any ghost town buff.

Unfortunately, there was also a gate across the road leading to those buildings. The gate was solidly padlocked, and a curt sign warned that no visitors would be tolerated at the buildings unless accompanied by the caretaker.

Two-burro-power Ore Mill, known as an Arrastre

—Courtesy Arizona Pioneers' Historical Society

Nameless Grave, Vulture City, 1958

Although the caretaker's cabin set just beyond the gate, all my shouting and horn-tooting failed to raise anyone; and since rather unpleasant things had happened to a couple of my friends who had treated that warning sign a bit too casually, I hesitated to snap my fingers in the face of Fate, as it were.

Continuing southwesterly along the main road I soon spotted the old Vulture City cemetery lying about 100 yards to the right, inside fenced cattle range.

Climbing through the barbed wire barrier I spent close to an hour prowling over the old burying ground. Paling fences that had enclosed grave plots were fallen to ruin. Weathered headboards bore faint traces of inscriptions, all too indistinct to be read. Other graves were heaped with rocks but completely lacking in markers. A few faded crosses lay prone in the desert grass and thistles.

In all the cemetery I found only one headstone bearing a legible inscription. It marked the grave of a little girl, Francisa Valencia, born September 19, 1875; died December 12, 1881.

As I drove back past the caretaker's cabin I stopped again to look longingly up the side road toward that clutch of old buildings on the hillside to the north. But no caretaker answered my shouted greetings—and I still lacked sufficient nerve to disregard the snarls and bared fangs of those Keep Out signs.

Some day, when circumstances and the stars are more propitious, I still hope to visit the wasted cadaver of Vulture City.

Breaking of the aforementioned Walnut Grove Dam brought tragedy not only to Henry Wickenburg but to many others, as well.

Although situated in the hills south of Prescott and completely surrounded by mines, Walnut Grove was not primarily a mining camp. The settlement that developed into the town began accumulating as early as 1863 or '64, and on October 24, of the latter year, the First Territorial Legislature seriously considered the feasibility of moving Arizona's new capital to this incipient city on Hassayampa Creek. The movement failed and, six years later, in 1870, the town still claimed but 70 inhabitants. When Wheeler's Survey party visited here, October 15, 1871, Lieutenant Lyle, of that body, wrote: "At Walnut Grove is a settlement where we found well-cultivated lands."

So the ranching community went its quiet, pastoral way until the middle 1880s when the Walnut Grove Water Storage Company was formed and Wells H. Bates promoted a large dam which was brought to completion in 1887 in a narrow canyon of the Hassayampa, just below its confluence with five large tributaries. Though primary purpose of the reservoir was to impound water for downstream mining activities, such as washing gravel and driving machinery, the lake formed behind the dam quickly became one of the most popular recreation spots in Central Arizona. Covering 1100 acres to an average depth of 60 feet, this water body measured two-and-one-half miles in length and three-fourths of a mile wide. It was soon stocked with fish, bordered by cottonwood trees, and dotted with sailboats and canoes. Several hotels and camps along the lakeshore catered to summer visitors from Prescott, Phoenix, and surrounding mining camps.

In 1889 the company owning the lake began construction of a new dam about 15 miles below the main storage reservoir and in January, 1890, brought this latter project to completion at a cost of approximately $1,000,000.

Thus was the stage set for the night of February 21, 1890, which saw enacted one of the most terrible tragedies in the history of Arizona Territory.

Water in the lake was high, and each passing moment was bringing more of it out of the canyons on the west slope of the Bradshaws. When the upper dam—like a woman going into labor—first began to hint of its distressed condition, a messenger, Dan Burke, mounted on a swift horse, was dispatched to warn men at the lower dam, as well as all ranchers and others living on the downstream side of the big reservoir. Burke, unfortunately, proved unworthy of the trust placed in him. Reaching Cameron's House, a saloon, the faithless courier paused for "a quick one," followed by several others. Meanwhile, another courier, William Akard, dispatched from headquarters, arrived at Cameron's to find Burke "so drunk he could scarcely walk across the floor."

By this time it was too late to save anybody or anything below the dams.

Yavapai County Sheriff Bucky O'Neill, in a dispatch to the *Arizona Miner*, February 24, wrote: "The scene of desolation along the Hassayampa below the dams beggars description. A tornado could not have made such a complete wreck. For miles the waters, turned free by the breaking of the dams, have filled the bed of the creek with dead bodies, enormous trees, cacti, and every other kind of debris . . . Those who saw it said that it came down in an almost perpendicular wall, 90 to 100 feet high . . . Immense boulders, weighing tons, were thrown around as a child might toss a ball. Enormous trees were broken into sections and torn in shreds. Iron bars were broken and twisted out of shape, and an ordinary flatiron was picked up, carried five miles, and then embedded in the walls of the canyon 80 feet above the present level of the stream . . ."

Immediate cost of the flood was between $3,000,000 and $4,000,000 in property destroyed, and more than 80 lives lost, in only a few minutes time.

In reminiscences of the Genung family, given to Arizona Pioneers' Historical Society and now preserved in the files of that organization, it is stated that most of the bodies recovered were buried at the Rorock Ranch, and at Wickenburg. Eye-witness Genung also recalled that a saloon belonging to Bob Brow was washed away, along with its safe, which contained about $4000 in cash and checks and several thousand dollars worth of jewelry.

"Though there was lots of searching," the old timer reported, "they never found this safe or its contents."[1]

Third camp in this tragic trio is Vekol. It lies down in the Papago country, in the southwestern corner of Pinal County, where the only living settlements bear names that sound more like Saudi Arabia than like Arizona—names such as Chiapuk, Totopitk, Sif Vaya, and Kokatk.

Back in the early 1860s, when Arizona was still plenty wild and a man's scalp was in constant danger of being lifted, the Territory was enriched by the arrival of a new settler, John D. Walker, of Illinois. Enterprising, honest, and public spirited, Walker would have been a material asset to any community, and Arizona was soon reaping the benefits of his character and energy.

In addition to large-scale ranching operations, Walker quickly became interested in mining and had a hand in developing several important properties. From time to time, during the next dozen years, Walker heard rumors of a rich silver deposit in the Papago country south of Casa Grande; but when he mentioned the matter to Papagos employed on his ranch they would either disclaim all knowledge of the location, or would refuse to disclose the long-held tribal secret.

It was Juan José Gradello, a Papago, who eventually volunteered to take Walker to the site of the mine. Surface indications showed the deposit to be as rich as rumor had said, and on Feb. 5, 1880, the property was located in the names of John D. Walker, his partner, Colonel Peter R. Brady, and Juan José, the Indian. That same year the Indian's interest was purchased by Brady and Walker, and in the fall of 1881 Walker's brother, Lucien, acquired an undivided one-third interest in the mine, which had been named the Vekol, a Papago word meaning "grandmother."

The Vekol paid handsomely from the grassroots down. Before the close of 1881 the main tunnel had been sunk to a depth of 118 feet, through rich ore all the way, and the partners were taking out and shipping to San Francisco and Denver about ten tons monthly, averaging $300 per ton. From this they were netting some $2000 each month, and were stockpiling around 150 tons of second-grade ore running from $40 to $90 per ton.

On April 29, 1882, the *Weekly Arizona Enterprise* reported, "The last shipment of ore from the Vekol . . . was the richest cargo yet sent out from that wonderful mine. It consisted of 22 tons, and every ton of it milled over $400! The owners of this rich mine . . . refused an offer of $100,000 for the property last week. They have a splendid well at the mine, now, which furnishes water for about 400 Indians and a large amount of stock, besides the water used at the mine . . ."

[1]Sheriff O'Neill, recounting this same incident only three days after the flood, identified the saloonkeeper as Robert Drew, and stated that the amount contained in the safe was "in the neighborhood of $7000."

Less than two months later it was reported that the Vekol then embraced some 1000 feet of tunnels and drifts from which more than $36,000 worth of ore had been shipped, plus 800 tons of second-grade ore on the dump, this averaging $100 to the ton.[2] About this same time, 30 sacks of Vekol ore, weighing 24,994 pounds, was sold on the open market in San Francisco to Helmrich and Hanson, for $3774.09, thereby netting the owners almost $3500.[3]

Although it has no direct bearing on the history of the Vekol mine, an interesting sidelight on labor costs of that era is contained in the *Weekly Enterprise* of February 3, 1883, in which it is stated that Messrs. Walker, Murray, and Slavan had been awarded the contract to build a county road, 70 miles in length, from Casa Grande south to Gunsight. A force of 14 men, it was stated, were already at work on the road and would have it completed within a few weeks. The contractors, noted the *Enterprise,* "estimate the probable cost of building the road at $300."

In March, 1884, John D. and Lucien Walker paid P. R. Brady $65,000 for his third interest in the mine—and the news continued good.

"The Walker brothers have shipped eight carloads of Vekol ore to Kansas City," stated the *Enterprise,* May 10, 1884. "It will average $400 per ton . . ."

"J. D. Walker and brother are making steady shipments of their highgrade ore from the Vekol mine to Kansas City," was the good word on February 28, 1885.

"The Vekol 10-stamp mill is yielding good results . . . It turns out a bar of bullion every 24 hours . . . None of the bars weigh less than 100 pounds, and a majority weigh 115 pounds." (August 8, 1885).

"The Walker brothers shipped from their Vekol mine, Saturday, eight bars of bullion, valued at $10,200; making in all, since their mill started, about 35 days, $31,809.61 . . . The mill is now turning out $1000 per day, and the expenses, per month, at mine and mill, are only $5000," exulted the *Enterprise* September 12, 1885.

On the month following (October 17, 1885) the same loyal news medium noted: "The Walker brothers, with the open-handed generosity for which they have become known, purchased a fine residence in Los Angeles for the widow of Mr. Huston, the man who was killed in their mill recently. They also gave her money to live upon. This is a graceful act, and one that but few men would have performed."

While this beneficence was, indeed, a "graceful act" it was not one that threatened to bankrupt the Walkers—not with the Vekol's little 10-stamp mill in only six months of that same year (July 1-December 31) producing silver bullion to a value of $133,973!

Current activities of the Vekol company were interestingly described in the *Enterprise* of May 29, 1886:

"There is, perhaps, no other mining camp in the United States so quiet and orderly," it was stated. "There is not a saloon in the camp

[2]*Arizona Weekly Enterprise,* June 17, 1882.
[3]*Enterprise,* July 8, 1882.

and, therefore, no drunkenness.[4] Indians and Mexicans are encouraged to save their wages and accumulate money. The company own the only stores at the camp, but they advise their employes to purchase only such articles as they actually need . . . they do not compel their employes to purchase of them, and neither do they charge monopoly prices. Their goods are sold as cheaply as in the town of Casa Grande . . .

"They never turn a stranger away, weary or hungry. No matter who he may be, Indian, Chinaman, Mexican, or American, he is welcome to food and lodgings . . . until he is strong and able to resume his journey. . . . Even the Indians who reside in the surrounding country regard Judge John D. Walker not only as a trusted friend and adviser, but as their oracle in all things. He speaks their native language with a fluency equal to the most erudite among the natives, and he often hears their complaints and arbitrates their differences with exact and impartial justice."

By the close of 1886 the town of Vekol had a fine new store and boarding house, both in large, well-built adobes. A young woman from Michigan, Miss Ida Warfield, had been employed to teach the fifty-or-so students enrolled in the school, and the Walkers even had established and stocked a good library for use of their workmen. Approximately 100 men were employed in the Vekol mine and mill, about a third of these being married men with families, who lived in clean, comfortable dwellings provided by the company. Single men patronized the model sleeping quarters and boarding house.

For 31 straight months the Vekol mill had an average output of $16,000 per month—and then came tragedy.

Bad relations between partners is always bad business—and all the more bitter if those partners happen to be brothers.

Ever since that February day in 1880 when John D. Walker and P. R. Brady located the Vekol, Walker had been the recognized head of the company, its heart, soul, and guiding hand. Beside this titanic personality, Brother Lucien seemed to have only come along for the ride. Chances are, this situation galled Lucien, and when John D. Walker, in 1890, suffered a light stroke which left his health somewhat impaired, it was only natural that Brother Lucien should begin casting covetous eyes at the driver's seat.

What was not so natural was that Lucien and a third brother, William, should have seized upon John's temporary infirmity to have him committed to a California hospital for the insane! At the same time, Lucien and William had themselves appointed guardians of John's considerable fortune.

Since there was, apparently, no basis for holding John Walker in the insane hospital he was released in only five weeks; but his brothers, even then, refused to relinquish their guardianship.

[4]A paragon of sobriety, John Walker let it be known he would discharge any employe found patronizing a saloon, and would not tolerate the presence of intoxicating liquor on the townsite.

The bitter lawsuit that followed was one of the most celebrated civil cases ever tried in the courts of Arizona. Territorial newspapers carried pages upon pages of testimony, as well as full coverage of new developments in the case, and editorial comment upon it. The last, almost without exception, was fiercely partisan toward John D. Walker, the persecuted brother.

Adding fuel to the flames was a surprise marriage. At their boyhood home in the East, the three Walker brothers had been friends of a Miss Eleanor Rice. The friendship between Eleanor and John eventually ripened into love, and when John left the East for Arizona Territory, in the early 1860s, there was an "understanding" of sorts between them.

For reasons best known to himself, John let nearly 30 years pass without sending for his affianced—though he corresponded with her regularly and she remained true to him in her letters and her prayers. Now, in 1891, feeling possibly that he had not long to live and wishing to secure to Eleanor a comfortable fortune, John Walker sent for the girl he had left behind in Quincy, Illinois.

On April 18, 1891, Eleanor Rice and John Walker were wed at Tucson and left immediately for an Arizona health spa for their honeymoon and John's recuperation from the ill health which was plaguing him.

With this unexpected move, Brothers Lucien and William stepped up their efforts to have John recommitted to a hospital for the insane.

In the language of the *Arizona Daily Star,* of April 23:

"The two (Eleanor and John) had not been long at the springs . . . when Lucien and William appeared on the stage, prepared for a scheme of the most diabolical infamy, and had John put under arrest, or pretended arrest, and put on a train for the purpose of kidnapping him and taking him into a foreign state. At Tucson they found the hand of justice ready to be laid upon them and John was released by *habeas corpus* from the bloodthirsty grasp of the two brothers. Not willing to give it up so, the said Lucien and William had John again put under arrest as a criminal felon. Again the case was tried and . . . John was released.

"In the name of God and suffering humanity, will this community permit such highhanded outrages to go on endlessly?"

The case of Walker vs. Walker has entirely too many ramifications to enter into here, and there is no point in trying it again, with all the principals and every witness long in their graves. It must be sufficient to say that this, unfortunately, was not a TV drama in which the bad guys get their just dues and the good guy comes out on top with banners flying.

After several more months of litigation and persecution, Judge John D. Walker, the old pioneer, died July 2, 1891, at the Napa (California) Hospital for the Insane.

At the time of his death his estate was valued at $1,500,000, including $80,000 in cash in a Los Angeles bank.

This much money, quite naturally, brought out a swarm of claimants—including a 21-year-old girl who represented herself as the daughter of John Walker and a Pima Indian woman whom he had assertedly married according to the ritual of her tribe.

The bitterly contested case dragged on and on, for 17 bitter, costly years. All the way from the Probate Court of Pinal County to Washington, D. C., and the Supreme Court of the United States of America.

There, in 1907, a verdict was rendered in favor of Lucien and William Walker. Although there seemed to be little if any doubt that John D. Walker had been the natural father of the girl, Juana Walker, her claim was rejected by the Supreme Court on the grounds that marriage between persons of white and Indian blood was then illegal and void in Arizona Territory. Therefore, ruled the Court, a child of such union had no right of heirship from the father.[5]

The show was over, the curtain fallen on the final act. There was nothing more to be said or done.

Reviewing the long bitter imbroglio, *Arizona Daily Star* called the tragedy of John D. Walker's incarceration and death "one of the saddest of the many sad experiences of the pioneers in this Territory."

Fate of the Vekol mine and mill and the model community grown up around them, strangely paralleled that of their developer.

Under the watchful eye and guiding hand of Judge Walker, the Vekol mine had been one of the great silver producers of Arizona Territory, and for ten good years had poured forth treasure with remarkable generosity. Estimates of its proluction range from $1,500,000 to $5,000,000—though the lower figure is probably nearer correct.

After the death of Judge Walker, strangely enough, the mine never again distinguished itself. It was almost as if his passing had cast on it a curse.

Vekol, the model community with school and library and free bathhouse—but without saloons—followed the mine to oblivion. Its post office, opened in 1888, was discontinued in 1909, and by 1912 the well-planned settlement was abandoned.

Today finds Vekol a lonely and tragic spot. On a dry, desert flat, at the foot of an equally-dry desert mountain, huddle a few old adobe buildings, a few foundations, and the usual rust and rubble of deserted mining camps. One of the remaining ruins—a large, plastered adobe—was the store. Another was the boarding house; a third, the assay office.

And, of course, there is the stone backwall and other ruins of the old Vekol mill where so many bars of silver treasure were produced . . . treasure that contributed to the enrichment of Arizona and aggrandizement of the Walker fortune, but brought to the founder of that fortune only heartache and disillusionment.

[5]*Arizona Daily Star,* September 15, 1907.

Last Building in Metcalf, 1951

"Arizona . . . is a land of contrasts in scenery . . . Nowhere is this statement more strikingly true than in the Sierra Blanca . . . well-watered glades covered with a luxuriant growth of grass and flowers . . . a wealth of autumn coloring . . . deep and sombre forests of evergreens . . . combined to complete this perfect landscape. The impress upon the mind of such a view is final and can never be forgotten . . . The Sierra Blanca . . . is one of the most inviting portions of our country yet remaining for civilization to occupy."

—Written of Arizona's White Mountains by Lieut. George M. Wheeler, Annual Report of the Chief of Engineers, 1875.

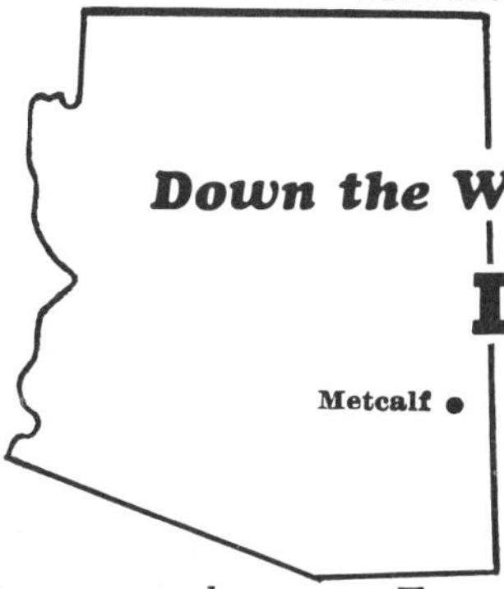

Down the White Mountains to Metcalf

IT IS SEVERAL YEARS since I have been over the Coronado Trail through the White Mountains from Springerville to Clifton. I am told that this is now a splendid road, largely paved, beautifully engineered, and adequately wide. It was not always so. Even as recently as a dozen years ago, when Mom and I traveled this road to the old copper mining ghost town of Metcalf, driving it was not an undertaking for the timorous and faint hearted.

We could have reached Metcalf more expeditiously, of course, from the south—except that I wanted to travel the Coronado Trail. My interest in the road had been whetted by Lieut. Wheeler's Survey Report of 1875, and by a few descriptive lines in the American Guide Series book, *Arizona,* published in 1940. I had read part of this last mentioned description to Mom—the part about wide vistas of green valleys, and distant mountains "shading from lapis lazuli to purple." For obvious reasons I hadn't felt it necessary, or advisable, to read that part of the description where the Guide Book warned: "For 135 miles, as it winds and twists over mountains abounding with game, the road is wide enough for only one car (*car going down backs to a turnout if another car approaches from below*)."

A passenger's equanimity is less disturbed if some things are left unmentioned.

Dawn was barely breaking as we left Alpine on the clear, frosty morning of October 3, 1951. Autumn's rich colors were painting the aspens and oaks, and banks of purple asters and yellow sunflowers edged our road—which would not see another town and only two or three occupied habitations in the next 100-or-so miles.

We crossed canyons where small streams chuckled around the rocks. Once, in a bright green valley, we surprised a sleek doe and her twin fawns, and only a little way beyond five big Merriam turkeys crossed the road a few steps ahead of our car! This was the first sight of wild turkeys Mom and I had experienced, and no creature we had seen in our rambles had thrilled us so much as these four turkey hens and their strutting gobbler. Just seeing them on this frosty morning made us feel as if we were living back in the days of the Pilgrim Fathers, getting ready for the First Thanksgiving, and things like that. A little farther along the road we came upon another bunch of four turkeys drinking at a stream. The water was sparkling clear; and when we stopped to sample it and fill our tank, we discovered that the quieter eddies of the brook were wearing a thin skin of ice. The road was bordered by timber almost unbelievably dense. Pines and firs stood so close together it would have been nearly impossible to ride a horse through the forest, and the quakers stood like teeth in a comb—tall and slender, straight as flagstaffs.

It was still quite early in the morning when we reached Hannagan's Meadow. At the edge of a pretty green flat, about 20 acres large and almost completely covered with wild blue iris, sat a small, rustic hunting lodge. In addition to selling gasoline and motor oil, a few groceries, beer, and fishing tackle, the lodge served meals, offered pack horses for rent and guides for hire—all from a single building.

When we stopped to inquire about the road ahead, an army of multicolored lion dogs set up a great clamor of baying and barking. They were still going strong when a pleasant-faced woman appeared at the door of the lodge and sent them packing.

When we asked about the road to Clifton, the woman shook her head. It was, she declared, "the worst 100 miles of road in the entire United States and Canada." It was always bad enough, she explained, but now it was being worked by a road crew and was all plowed up and full of big boulders.

"My husband and I came over it yesterday and ruined two good tires," she said. "I wouldn't go back over it today for all the tea in China!"

Beyond Hannagan's we climbed to the crest of the range and began following its rugged spine upward and toward the south. Wherever there was a break in the dense forest growth we found ourselves looking out over a greater expanse of country that most city dwellers are privileged to view in their entire lifetime. Every range of mountains—tier upon tier of them from our feet to infinity—was black with timber. Canyons plummeted away on either side of the road, some of them so deep it seemed as if the bottom must have dropped out of the world!

With our road growing steadily more primitive, we passed a marker noting the elevation as 10,500 feet, and continued on, toward the south.

About noon that day we made camp in Crook National Forest. Our campsite was surrounded by tall pines and firs. There was a spring of pure water, almost as cold as ice, and two little log snow shelters, each with three sides and a tight roof. A big stone fireplace, facing the open front of each shelter, was so constructed that its heat flooded the interior and made it toasty and comfortable. With our folding camp table and canvas stools set up inside one of these shelters, a tarp hung to close off part of the open front, and a pile of dry pine wood stacked beside the fireplace, we were ready for whatever cold the night might bring.

Leading away from camp, down the mountainside to the south, was a narrow wagon road signposted to Crabtree Flat, two miles. The name seemed to fascinate Mom. Right from the beginning she insisted on calling it Crabapple Flat and, shortly after lunch, suggested we might walk down to "the crabapple orchard."

"If we can find some nice crabapples," she said, "I'll make some jelly when we get back to camp, and you can bake some hot biscuits for supper . . ."

The soft gray dust of the trail was closely imprinted with deer tracks, and twice we glimpsed white-tails as they bounded away into the roadside brush. Mingled with the sharp little hoofprints of deer were the round pad marks of rabbits and skunks. Once we saw the long fingered pads of a big old boar 'coon; and we studied another track that looked as if it might be that of a wildcat. All along the little canyon road grew tall ferns and wild flowers. In only a short way we counted 15 species of blossoms—including a tiny fairy garden of delicate white violets!

But there were no crabapples at Crabtree Flat. All we found there was a rather primitive one-man sawmill that made us think of the Wright brothers' first plane. A few pine slabs lay in the yard, and a thin wisp of smoke was curling from the screened top of a sawdust incinerator. There didn't seem to be a living soul anywhere around the place. As a matter of fact, the only person we had seen all day was the woman at Hannagan's; nor had we seen another car—not even a Jeep, or truck.

After a wonderful night in the little log shelter—a night that was frosty cold and starlighted and deathly still—we continued on our way. The curves became sharper and narrower, the grades steeper. Switchbacks and hairpin turns followed one after another.

When the road had been growing steadily worse for close to 50 miles, we reached the vanguard of the road workers. They had just finished tearing loose a lot of boulders and heavy roots which still covered the roadbed, along with hub-deep dust. Grinding along in low-gear, lurching, roaring, and floundering, our gallant old Mercury made her way over this rock-and-dust chaos until reaching a stove-sized boulder that even she could not surmount. As the road was altogether too narrow for us to pass around the boulder on either side, we had no choice but to wait patiently, half an hour or so, until a bulldozer came along and pushed the rock off the edge of the grade. For almost a full minute we could hear it crashing down the canyonside through the trees and underbrush, bumpity, bumpity, bounce.

Not far beyond this point we began dropping down rapidly from the high mountains towards the desert and the San Francisco River. The spruce and firs and tall yellow pines gave away to piñons and junipers and whole hillsides covered with chaparral. Eventually we entered a deep gorge, where our road clung to the right sidewall like a pencil mark.

There were high, red cliffs on either side, and a few trees in the bottom. At one point we could count seven switchbacks, one below the other.

While we were still passing through this canyon, but seemed to be nearing its mouth, Mom called my attention to a steep, rocky hillside on the left of our road. It was scalloped with retaining walls where now-vanished dwellings had stood, and on the flat at the foot of the hill were a few old foundations and one last remaining cement building.

Bad as the road had been, and eager as we were to begin exploring a new ghost town, it gave us a feeling of regret to know that our trip through the White Mountains—the Sierra Blanca—was ended.

These retaining walls and the ruins below marked our destination, the site of Metcalf, once one of the largest copper mining camps in the Southwest . . .

While scouting Apaches under Captain Chase in the early 1870s, Robert B. Metcalf had taken time out to discover the rich deposit of copper that would develop into the Longfellow and Metcalf mines. Several other claims in the area also were staked by soldier-prospectors—who quickly learned that an embryo copper mine is no proper plaything for a man drawing soldier's pay. In consequence, mines of the region saw little development until after Leszinsky Brothers, of Silver City, N. M., acquired an interest in the leading properties. Leszinskys had everything necessary to the successful production of copper, namely: almost unlimited capital, and know-how.

In its issue of September 20, 1873, *Arizona Citizen* reported that Leszinsky and Company were successfully smelting copper on the Francisco River, and only a few days since had loaded 21 teams with some 150 tons of the metal.

The original Mexican blast furnaces, using charcoal for heat and hand bellows to provide a forced draft, soon were replaced by a more modern reverberatory furnace with a 120-foot brick stack. With these improvements—and with copper selling at 20 cents a pound—the town of Metcalf began to boom.

Early in their period of ownership the Leszinskys built the first railroad in Arizona—a 20-inch gauge from the mine tipple at Metcalf to the smelter at Clifton, five miles down the canyon. Along with the ore, transported by gravity, rode mules in sufficient number to pull the emptied cars back up the hill. About 1879 the mules lost their job to a four-ton locomotive built in Pittsburgh. Knocked down and shipped to the AT&SF terminal at La Junta, Colorado, the little loco was freighted overland from that point, some 700 miles by ox-team, to Clifton, where the component parts were reassembled by Henry Arbuckle, a former railroad engineer. When "Dad" Arbuckle completed his assigned task he had created a strange looking contraption and Metcalf's gambling element is said to have laid odds that it would not run.

But the diminutive engine, christened *"Little Emma,"* ran plenty good and Arbuckle pulled the throttle on it for many years afterward.

"His train crew was wholly Mexicans," noted McClintock. "A queer lot of railroaders, usually wearing the large, high crowned straw hat of their native land . . ."

With endless streams of copper pouring from her several mines, Metcalf prospered. Stores, saloons, and dancehalls were established. Also a bank, a hospital, hotels and boarding houses, Chinese laundries, a post office, a Catholic church, railroad station, a newspaper, and a schoolhouse. By 1916, with the Shannon, Arizona, Standard, and New England copper companies all operating full blast. Metcalf had a claimed population of 5000, including some 700 youngsters of school age. As her original school facilities were no longer adequate to care for this many students, the district erected a fine, new $50,000 school comprising two stories and a basement, with numerous classrooms. It was the finest building Metcalf would ever have—for the end was not far distant.

With copper prices nose-diving in the early years of the Depression and mines closing throughout the West, Metcalf began to feel the general pinch. As each mine closed down and its laborers departed for other camps where there might still be a chance of finding work, business concerns were left with fewer patrons and more red ink. Valuable homes were sold for a few hundred dollars each—or were left standing intact by owners who expected to return to them, but never did.

In 1940 the fine big school building was purchased by Phelps Dodge Corporation for $1000—one-fiftieth part of its original cost—and was razed so that its campus might be used as a dumping ground for the Morenci pit[1] where, by the middle of World War II, some 45,000 tons of copper ore were being mined daily. When PD needed still more space for waste disposal, the Metcalf cemetery was sacrificed. With approval of the Greenlee County board of supervisors, and the Catholic bishop of the district, more than 1100 graves dating from the 1870s to 1938, were moved to a new site south of Clifton.

On that October day in 1951 when Mom and I made our pilgrimage to Metcalf, the mountain back of the old townsite was still embroidered with retaining walls, each of which had held in place a tiny bit of earth so that a house might be built upon it—but, now, all the houses were gone and only the terraces remained. Tunnels hewn from solid rock to enable passage through the narrow canyon of *Little Emma* and her laden ore cars, were still visible, and a few ruins rose from the bleak flat like the snaggle teeth of an old crone.

Across the canyon from these last reminders of Metcalf, the great Morenci waste dump was growing steadily higher, longer, mightier. Trains moving slowly along its rim, high above us, looked no larger than small black caterpillars; and every time another trainload of rock was spilled down the face of the dump, the canyon roared and reverberated and the ground beneath our feet seemed to tremble.

[1]*Arizona Citizen,* June 28, 1940.

If the ghosts of Bob Metcalf and the Leszinsky brothers ever return to this scene, they must be completely dismayed by this great openpit operation of Phelps Dodge, where more copper is produced in a single shift than the men of their era could have mined in several lifetimes of toil with double-jack and steel.

Former Stage Depot, Dos Cabezos, 1959

"The rest of Arizona is booming. Multi-million-dollar buildings. Superhighways. New subdivisions. Growing industries. Population expanding. And old Pearce rots and melts away, year by year, and fewer old dogs snap at flies in the sun, and the weeds take over the streets, and the dry desert winds find more hiding places for the red sand."

—DON DEDERA, *Arizona Republic, September 22, 1959*

Old Pearce Drowses in the Sun

DOWN IN THE BROAD, flat Sulphur Springs Valley, in almost the exact center of Cochise County, the old mining camp of Pearce and her adobe ruins drowse quietly in the sun.

But, stranger, beware! Tread softly, and be careful what you say. Old Pearce is a light sleeper, and even though her heart beats but feebly and her pulse is weak she's still as touchy as a she-bear with a toothache!

What makes the old town angrier than almost anything is to be called a ghost. When she finds herself included in this rollcall of phantoms, I have every reason to believe she'll cast off her winding sheets, spit on her work-calloused hands, and come out of her corner fighting. But that's an occupational hazard that has to be faced. With her Commonwealth mine one of the greatest gold and silver producers ever discovered in Arizona, it would be an unpardonable thing to omit Pearce from this roundup of towns with more past than foreseeable future . . .

It was "Uncle" Jimmie Pearce, a Cornishman, who started the ball rolling.

Uncle Jimmie had labored in the mines at Tombstone almost since the boom started at that place, 14 years earlier. Through most of that time "Ma" Pearce had operated a miners' boarding house. Frugal, hard-working folk, Ma and Jimmie had never been known to throw away a brogan as long as it might be repaired, or to discard a leaky teakettle if there was a chance of soldering it one more time. They had done well, generally speaking, and saved their money; and now they were glad they had done so.

It was 1893 and the whole country seemed to be going to hell in a handbasket. Men were speaking worriedly of a "panic." Silver already had been dealt the body blow from which it would not recover in two lifetimes. Mines all over the West—even the great Tombstone mines—were closing. Banks and businesses were failing; towns dependent upon mine and mill payrolls were becoming ghost towns.

To say that the future didn't look especially rosy to an unemployed, middle-aged miner, and his graying wife, might be classified as a masterpiece of understatement.

The Pearces' two sons were sure they had the answer. For years, while growing to manhood in Tombstone, the boys had talked and dreamed of almost nothing but ranching—of being cowboys, and riding horses, and raising cattle. That wasn't Jimmie's sort of life. Jimmie was a miner, straight to the core; but with everything in the mining world falling to pieces, it almost seemed as if the boys might be right. Like they said, maybe he and Ma *could* get a little ranch somewhere near Tombstone, have a garden and some biddies, build up a small herd of cattle, and sort of take things easy the rest of their lives . . .

Before the end of that year Jimmie and Ma had let the boys talk them into making the change, and spring of 1894 found the Pearces settled on a ranch in Sulphur Springs Valley, 25 miles northeast of Tombstone.

In the spring of the year, when new grass is greening, baby calves are being dropped, and meadowlarks are nesting, Sulphur Springs Valley is a pretty place, even today. Likely it was still more beautiful back 70 years ago when Jimmie Pearce began exploring it on horseback. Almost every pleasant morning he would saddle his cayuse, stick a bite of lunch in his pocket, hang a canteen of water over the saddle horn, and ride off. He always told Ma he had to look about the range, or the cattle, or hunt horses, or something. But Ma knew Jimmie, and she knew the old adage, "Once a miner, always a miner;" so she wasn't at all surprised when Jimmie came home one night and said he had found a mine.

Uncle Jimmie said he had ridden to the top of a low hill, about noon, and after gazing out over the countryside for awhile he had sat down on a rock ledge to eat the lunch Ma had fixed for him that morning. As he sat eating he had idly picked up a piece of quartz and knocked it against another piece of rock—as miners and ex-miners have been doing since before the days of King Solomon. When both pieces of the rock fell apart, Jimmie had seen with amazement that both were dark with chloride of silver and rich in free gold!

On the day following, Jimmie and his sons returned to the scene of the strike and staked five claims—one for each of them—another for a married daughter, and one for Ma. Then Uncle Jimmie Pearce hit the road to Tombstone and the nearest assay office.

There wasn't a stingy bone in Jimmie's body. Now that each of the Pearce's had one claim, the old Cornish miner began spreading the glad tidings among his friends, at the same time urging every one of them to come on out and locate some ground—"Cut yourself in on a good thing!"

When the assayist reported that Jimmie's samples ran $22,000 to the ton in silver, and $5000 in gold, friends needed no further urging. They came—ranchers, storekeepers, unemployed miners—afoot, astride, by buckboard and wagon.

First-comers staked claims adjoining those of Uncle Jimmie Pearce and his family. Later comers staked next to the first comers. Almost before a man could say "Gold strike!" claims had been located for miles in every direction, and a raw, rough camp was taking form to the immediate northwest of the discovery hill.

First place of business in this new town of Pearce was a saloon, a simple affair consisting of two barrels of whiskey set beneath a rag lean-to, with a plank laid across the barrels for a bar. This makeshift Palace of Pleasure, known as Gentry's, soon was followed by more elaborate places, and by stores, a hotel, boarding house, other shops and services.

The Pearces named their mine the Commonwealth, and Uncle Jimmie, his sons, and son-in-law, went to work with crowbars and picks, breaking off rock along the ledge. 'Tis said that the first carload of ore taken out in this manner yielded them $10,000.

With word of the fabulous discovery bouncing from peak to peak, like a heliograph signal, it wasn't long before the Pearces were besieged by promoters seeking to lease, option, or buy, the property. Jimmie and Ma were satisfied to let things run along as they were doing—in all their days they had never had it so good!—but the boys could smell big money in the offing and wanted to expand operations, or to sell.

In the end, an option agreement was entered into between the Pearces and John Brockman, Silver City banker. Under its terms, Brockman was to make a small down payment which would give him a specified time in which to prospect and develop the property, priced to him at $250,000. In only 60 days, according to the story, the New Mexico man had stockpiled more than this amount of ore and the sale was consummated.

Before signing away her rights to one-fifth interest in the Commonwealth Mine, it is said that the canny Ma Pearce insisted upon a special clause being written into the agreement.

In addition to the $50,000 that was her rightful share of the proceeds, Ma demanded a lifetime franchise to operate a boarding house at the camp!

Taking possession of the property in the spring of 1896, Brockman immediately began construction of a 200-stamp mill. Wagons drawn by 24-mule teams were soon hauling fuel oil from the nearest railroad point, at Cochise Station, 16 miles distant—and bars of bullion were being hauled back.

Three shipments of bullion, having a total value of $120,000, had been shipped by the Commonwealth Mine during the previous month, reported *Arizona Republic*, May 3, 1900.

"The driver and two guards, who accompanied the bullion from the mine to this station on the *Southern Pacific*, were a little nervous when they arrived here on the last trip," stated the news release from Cochise. "The story was current that Alvord and Stiles, and Bravo Juan, were to hold up the bullion train, but they did not appear. The bars were run in such chunks that the weight was too much for a person to pack off, and there was really no danger from the attack except from the anger of the desperados when they found that they would have to leave empty-handed . . ."

One month after the appearance of that news story, the Commonwealth mill burned to the ground. Another was quickly erected, and from 1900 to 1903 the owners are said to have netted more than $3,000,000 annually.

In 1903 the mine was purchased by Count de Pourtales, for $1,000,000, and was operated under the new ownership for two years until water began flooding the lower levels and ore values declined with depth. In 1905, Pourtales leased the Commonwealth to A. Y. Smith and D. T. Swatling, mill superintendent, who built a cyanide plant and proceeded to treat nearly 200,000 tons of mine ore and more than that amount of mill tailings, at fair profit.

On March 24, 1910, the big Commonwealth mill was again destroyed by fire. The mill had just been completely remodeled and the first test run had been successful "far in excess of the most sanguine hopes," when fire broke out in the engine room, presumably as the result of crossed wires. Within five minutes the mill was transformed into a spectacular furnace, the red glow of the flames being visible for miles.

"Twenty minutes from the discovery of the fire, the skeleton of the big, 300-foot building stood sentinel over a mass of twisted, ruined wreckage," reported *Courtland Arizonan,* two days after the fire. "The mill proper, including all of the recent improvement, engines, generators, shafting, new machinery . . . was lost. The estimated loss is conservatively placed at $160,000, with insurance protection of about $60,000 . . ."

One month later occurred the death of D. T. Swatling, co-leaser of the mine, at the age of 53 years. After Swatling's death the property was purchased by the Montana Tonopah Mining and Milling Company, which built a new 350-ton cyanide mill, and resumed production on a larger-than-ever scale.

Population of Pearce, at its peak, was likely around 2000. In its issue of December 15, 1919, the *Tombstone Epitaph* carried a long feature story on its sister city and its prosperity. In addition to various stores and shops, it was noted that Pearce had then "an excellent school . . . good hotels and restaurants, garages, soft drink parlors, and a moving picture show . . ."

By 1917 the value of the ore had declined to barely five dollars a ton and the Commonwealth eventually closed down—though much of its silicious ore, valuable for flux, would still be shipped to the Phelps-Dodge smelter at Douglas.

Total production of the Commonwealth is variously estimated between $10,500,000 and $30,000,000. No one will ever know how much the great mine actually gave forth, since highgrading flourished there throughout its palmier years. Those "in the know" have said that dishonest miners at the Commonwealth could—and often did—carry out in their lunch buckets each night stolen ore equivalent to a week's pay.

During the big Depression of the 1930s, Pearce's population began drifting away until only 20-or-so old timers continued to make their home in the town. At last there was only one operating store and the post office left in the business district, the store operated by Mr. and Mrs. Albert Rothe, the post office by Mrs. Gladys McLeod, a native daughter of the town. Then, Mr. Rothe died and the store closed, and only Gladys McLeod and the post office, and a few old pensioners, remained. In 1961 the Old Store, as it is known, was purchased by Mr. and Mrs. Ike Cornish and their son, Chuck, from Chicago, who have reopened it as a combination country store and museum.

My most recent visit to Pearce was in the late summer of 1959. After visiting with Mrs. McLeod for awhile, Mom and I explored some of the vacant buildings that dot the townsite and finally parked in front of the old adobe office of the Commonwealth Company, situated to the left of the Courtland road and about a quarter mile from the ruins of mine and mill. Though the adobe walls of the office were still standing reasonably true and stout, its roof and floor had vanished, along with windows and doors and everything else of intrinsic value. At the building's rear stood a tall date palm, the only one visible anywhere around. Its trunk was draped with dead fronds of years past; and even the fronds still living rustled dryly and sibilantly in the hot wind of that summer day.

Within the building, the floor of one room was almost knee-deep in old correspondence, canceled checks, invoices, duplicate purchase orders, income tax reports, payroll records, et cetera, most of the papers bearing dates of more than half a century earlier. Knowing that all this material would soon be ruined by rain, nesting mice, and other forms of destruction, Mom and I spent about an hour combing through this historical largess, salvaging some of the more interesting letters and other memorabilia.

From the office building we walked around the foot of the hill to the old millsite and spent another hour prowling over and through the remains of massive foundations, tangled lengths of cable, old railroad rails, and other assorted rubble—all heavy and corroded by rust.

When there seemed to be nothing more of interest to discover at the millsite, we trudged back around the mountain to our waiting truck, and headed for Dos Cabezos.

Thirty-three miles northeast of Pearce—past Three Sisters Buttes, the Kansas Settlement, and lettuce fields almost without end—is this little village of the Two Heads, almost a ghost town, itself.

Back about 80 years ago, quite awhile before Uncle Jimmie Pearce made his fabulous discovery on the hilltop to the south, Dos Cabezos seemed to hold forth a lot of promise. Gold and silver mines in the surrounding ranges looked good, and the mining camp that took form at their feet soon boasted a nice assortment of hotels, saloons, and other places of business—even a newspaper. But the mines didn't prove out so well, and with litigation, over-rated prospects, and man's cupidity, the little town of the Two Heads ceased to prosper.

The day of Mom's and my visit to Dos Cabezos found its peaceful old streets swept by sunshine, its empty, crumbling adobes shaded by leafy mounds of China-berry trees. The only operative business we succeeded in finding was a small fourth-class post office. Even it was closed until later in the day. Farther along main street stood a long, white-stuccoed adobe ruin, which had evidently embraced several business units. It now stood roofless beneath the sky, and its likewise floorless interior was densely grown to bird-of-paradise shrubs.[1] With neither loving hands to tend them, or water to slake their thirst, these pretty plants had flourished so exuberantly that their feathery green boughs protruded from every empty window frame and doorless doorway in the building.

Following a dusty road that led across town, past more crumbling adobes, and more trees and flowers, we came eventually to an old adobe stage depot partly fallen to ruin. Castoffs and junk littered walls and floors. In one corner of the building stood an old steel safe, its door secured with a huge brass padlock. Behind the building were piles of ore samples and deer antlers, and a cactus bed edged with Indian mortars and metates. A little Mexican boy, who sauntered over to chat with us, said he thought the old barn was owned by a man named Judson.

Before leaving Dos Cabezos, Mom and I visited its cemetery on a grassy hillslope not far from town. We found many of the headstones carrying dates in the 1880s and 1890s, but inscriptions on most of the wooden markers, undoubtedly the oldest burials in the plot, have been almost entirely destroyed by wind and weather and the passing years.

From the cemetery we gazed out over the little adobe town and on across the dry flats to the high, Dos Cabezos Peaks. In all that wide sweep of land we were unable to glimpse one moving creature, one car. No smoke curled from any chimney within our view, and not one whisper of sound came to our ears.

If any spectral shades of the past were still haunting these scenes of their earthly endeavors, at least they were doing it quietly.

[1] *Caesalpinia gilliesii,* a native of Argentina and Chile, which has escaped from cultivation in Southern Arizona and now grows wild with such enthusiasm it is often mistaken for an endemic species.

"It is a curious and little appreciated fact that the miner is the scout of civilization. He braves the savage, the desert's heat, the arctic's cold. Alone he fearlessly penetrates regions wherein his foot is the first to tread . . . Following the prospector and the miner come the trader, the cattle rancher, the farmer, the homeseeker . . .

—*McClintock's ARIZONA, Vol. II.*

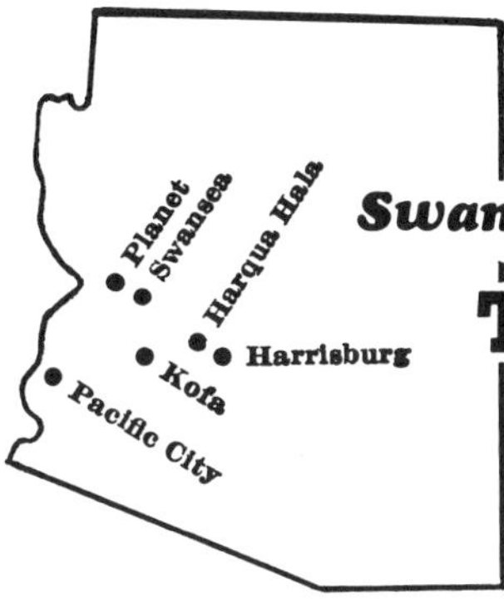

Swansea, Kofa, and Harqua Hala

TUCKED AWAY in the dry, desert hills of Yuma County are scores of once-prosperous mining camps where shift whistles no longer blow, and the welcome clangor of the boarding house dinner gong is stilled forever. Planet, Kofa, Pacific City, Ehrenberg, La Paz, Gila City, Castle Dome, Swansea, Fortuna, Harqua Hala, Harrisburg—the roster is long, the company distinguished.

Except for Gila City, Ehrenberg, and La Paz, treated in an earlier chapter, one of the oldest of Yuma County's ghost towns is Planet, situated a mile south of Bill Williams River at a point 12 miles east of its confluence with the Colorado.

History's pages have little to offer concerning one Richard Ryland, asserted discoverer of the Planet mine, in 1863. In those days of a century ago Yuma County was a dangerous and little-known region. But for a few small river ports, white settlements were virtually non-existent and most prospectors who dared the twin hazards of hostile savages and desert would not willingly settle for anything less than gold or silver. But Mr. Ryland was not so choosy.

Mr. Ryland—whoever he was—went forth and found a copper mine, the second mine of such nature to be discovered and worked in Arizona Territory.[1]

In 1864 the Planet company, embracing five mining claims—the Planet, 2700 feet; the Ashley, Wash and Sentinel claims, of 2100 feet each; and the Mountain Chief, 1800 feet—was incorporated in San Francisco, and actual production was soon under way.

Reporting in 1867 on the operation at Planet, J. Ross Browne wrote: "The Ashley claim has been so far developed as to show evidences of a copper ledge about 600 feet in length . . . an average thickness in the ledge of about 10 feet. The character of the ore is malachite, assay 30 pct. copper . . . The Planet claim has been worked since the spring of 1865, and about 800 tons of ore have been taken out . . . average assay,

[1]The great copper mine at Ajo, owned and operated by Americans since 1854, had been worked even earlier by Mexicans.

40 pct. It has been sold in San Francisco at an average of $100 per ton. . . . The cost of transporting ore to San Francisco was, at first, $60 per ton, It is now $28 per ton, and will probably soon be but $18 or $20. If the company could erect a warehouse at the mouth of the Colorado and store the ore there until a cargo for a large vessel accumulated, it could be shipped from the mines to Boston or Swansea (Wales) at a total cost of not more than $25 per ton. Mr. Thompson . . . has erected furnaces for smelting the ore taken from the Eliza mine . . . 1000 feet distant from the Planet. According to late report the company have two small furnaces running, turning out copper from 91 to 96 pct. fine. . . . A lot of this copper has been sold for 15 cents a pound . . ."[2]

Before too many months had passed, a small town, bearing the company's name, had grown up around the mines and furnaces, and news stories of the doings and developments at Planet began finding their way into Territorial journals.

The Planet furnace was closed down for a few days pending a change in the management . . . Owners of the Planet mine were drilling a well . . . There had been a cutting scrape in John Stiles' saloon at Planet. Jack Earle, a teamster, had cut a man twice in the abdomen and the victim was not expected to live . . . A water-jacket furnace, while being freighted in for use on Planet ore, had been carried away by a flash flood on Bill Williams River and no trace of it had been seen since . . . It was rumored that a large number of Chinese were soon to be hired at the Planet mine . . . Richard Gird, an early smelter man at Planet, had made some valuable locations in the Tombstone Mountains east of the San Pedro River . . . and Johnny Carr had taken a contract to haul bullion and supplies to and from the Planet Mine and Franconia Station, on the Atlantic & Pacific Railroad, a distance of 70 miles, and already had four 16-mule teams at work . . .

Operating sporadically over the course of 52 good years and bad, the Planet claims produced close to a million dollars worth of copper before tossing in the towel, in 1917. With that development there was no longer any need to maintain a desert town 115 miles northeast of Yuma. After hanging on grimly until 1921, the Planet post office bowed out of the picture—and the ghosts took over.

Not many years before the citizens of Planet began packing their skillets and folding their tents, a new mining camp was born in the Buckskin Mountains, a dozen-or-so miles east of the older camp. Its name, Swansea, honored the ancient smelting city in Wales.

The mother of Swansea was Clara Consolidated Gold and Copper Company, an ambitious parent financed largely with French and Belgian money. Before developing its ore body the Clara built a costly and magnificent smelter, and a branch railroad—the Arizona & Swansea—giving it connection with the AT&SFRR at Bouse. After this elaborate build-up it must have been slightly disconcerting when unexpected dif-

[2] *Resources of States and Territories West of the Rocky Mountains, 1868,* pp. 456-7

ficulties forced the company to cease operations almost immediately, and go into bankruptcy.

Much of my information on Swansea has come from Robert Bunker, of El Monte, California, who took a job as bookkeeper for the company in 1913, when the property was under lease to a Judge Thomas. Soon after Bunker began work at Swansea, Judge Thomas died suddenly and the whole deal was plunged into trouble again, and into Probate Court. O. M. Souden, of the United States Bank, in Los Angeles, trustee in bankruptcy for the estate, refused to advance any funds to carry on work.

"Fortunately," recalled Bunker, "we had $10,000 due from the smelter at Humboldt. When Ernest Lane, our foreman, tried to get an advance from the smelter and failed, the miners—about 150 at that time—agreed to stay on until funds became available. We operated for the Thomas estate until termination of the lease. As the price of copper was increasing and we had good contracts with the smelter, we were very successful. After termination of the Thomas lease we continued to operate for the Bankruptcy court. Copper went to 29 cents a pound, and by 1916 or '17 we were able to pay off every creditor. (Some of the heavy stockholders had come over from Europe and bought up claims at 10 cents on the dollar, so got rich quick . . .)

"After every claim had been paid, the company was reorganized. Lane, who had been so successful, was promised a lease—but it never materialized. Instead, the property was leased to Charles Clark, of the United Verde, at Jerome, and the Swansea Lease Company was formed with Clark president; J. Ross Clark, vice president; C. O. Whittemore, secretary; F. E. Show, treasurer, and W. H. Comstock, assistant treasurer."

Lane and Bunker stayed with the new organization until 1918, then quit to go into mining for themselves . . .

Swansea, the town, flourished for a period of years; even had a newspaper, the *Swansea Times* (1908-10), owned and edited by Mrs. Angela Hammer. Between 1908 and 1930 the Swansea is said to have produced some 27,000,000 pounds of copper. Eventually, however, the mine was running too spasmodically to support a permanent population and business houses, and with the price of copper falling in the early years of the Depression, operations ceased.

Years after Swansea became a ghost town its site was visited by John W. Hilton, noted desert artist and writer.

"From a distance," wrote Hilton in *Desert Magazine*, January, 1941, "Swansea had the appearance of being a prosperous operating mining camp. The first building on the left was an adobe structure that had been the railroad station. Behind it was the wreckage of an old passenger coach and the cab of a locomotive that had been robbed for scrap iron. An interesting object was a gasoline-driven car with a canopy top that looked for all the world like an old surrey on railroad wheels. What a picture it must have made, chugging along through the cactus-studded hills, loaded down with passengers bound for the bright lights and excitement of Bouse!

"Crossing the wash we drove up a street lined with rather well preserved cottages and stopped in front of the largest one. Here we met the Lanes who are making another attempt to bring Swansea back to life. They have leased the property and in a few weeks expect to begin work in one of the shafts and start ore rolling to the mill. The machinery, about a million dollars worth, appears to be in good condition and needs only men like Ernest G. Lane, and Ernest, Jr., with vision and ambition to start the wheels turning again . . ."

But that vision and ambition availed Ernest Lane but little. After sinking his life's savings in the mine he finally died there . . . and Swansea is still a ghost.

Though located only 25-or-so miles north of Yuma, the ghost town of Pacific City, in the Trigo Mountains, is not a place to be visited casually some Sunday evening after supper. Not only is it a whingdilly of a spot to locate, but the sandy, rocky, all-but-obliterated trail leading hence from Fisher's Landing, on Martinez Lake, is much better suited to hooves and wings than to rubber tires.

Pacific City began life in 1879 or '80 under the name of Silent, after Judge Charles Silent, prominent mining man of that area, and then an associate justice of the Territorial Supreme Court. Situated in Black Rock Wash, about five miles east of the Colorado River, the new town soon became the trading center for several rich mines of the immediate vicinity. Included among these were the Red Cloud, about a mile to the north; the Black Rock, a mile to the east; and the Silver Clip, five miles northeast. A small smelter built at Pacific City in the 1880s was not especially successful and closed after three years of attempted operation. The town was also a way-station on the Yuma-Ehrenberg stage line, and with miners, mill hands, teamsters and other residents, likely had a peak population of close to 500 men, plus a few women and children.

Due to scarcity of water for mixing adobes, and the terrific expense of freighting in lumber for building, a considerable number of Pacific City's inhabitants excavated caves and dugouts in the relatively soft sandstone banks flanking the wash. While some of these caves were scarcely larger than a coyote's burrow, others were good-sized rooms as much as a dozen feet square, their fronts neatly closed off with stone masonry and a door.

There is little evidence that Pacific City ever had an elaborate business district. One large wooden building, called La Cantina Plata, served as a combination saloon, dancehall, and general store. The post office, established in 1880 under the name Silent, was in charge of Paul Billicke who later starred in a fatal shooting scrape with young Don McLeod.

In a news release published January 29, 1881, *Arizona Sentinel* remarked that the camp of Pacific City was "busy as a hive of bees," and that Stein's and Billicke's saloons added to the spirits of the boys.

"Silver District is, today, one of the liveliest mining camps in the Territory," stated the *Sentinel* on April 23, 1881. "The vast capital which has been laid out in the district already, together with the large amount which will be expended in the course of two or three months more, in the erection of mills and furnaces, is enough to put this district beyond a doubtful success. A wealthy Boston company, who just paid $135,000 down for a group of mines in the district, is about to erect a 40-stamp mill on the grounds which, when in running order, will give employment to at least 400 or 500 men."

The Red Cloud Mining Company, represented as "a strong New York corporation owning extensive claims in the district," was then erecting a smelting furnace "of immense magnitude," and was employing nearly 100 men in their mining operations, it was said.

The early spring of 1882 found everything loping along merrily.

"Postmaster C. T. Norton is engineering the Harvey and Gilchrist store now, in the old Norton Building," stated the *Sentinel* in its issue of February 11. "Mr. Johnson is making arrangements to open a tony caravansary and, if rumor does not error, there will be a demand for biled shirts and sto' clothes in the near future . . . Pacific City has taken on quite a metropolitan air occasioned by the music of busy hammers and rasping of saws . . ."

Again, on May 23, 1882, the *Sentinel* carried a long news story concerning the camp in the Trigos, concluding with the remark: "Judge Crawford and John Stanley will deal out the 'O Be Joyful' as pards in the old Norton store at Pacific City. A stag dance is in order . . ."

The stag dance, presumably, was held—likely several more of them. But Pacific City, despite its ambitious name, was not to know the permanence and lasting notoriety of Tombstone and other contemporaries. Revenue from her mines was diminishing even before the disastrous blow dealt silver in 1893, and the economic chaos of that year brought her story to its conclusion.

Not much remained at the site of Pacific City in 1959. Probably even less is there, today. A few rock foundations, mill tailings, a pile of slag from the smelter, assorted rubble, the big roof of the old store-saloon without any sidewalls to support it . . . and, of course, the caves gouged out of the sandstone banks.

Not even the most predatory person can carry away a cave . . .

Tucked away in the bare desert mountains, 50-odd miles northeast of Yuma, are the meager remains of Kofa.

Kofa was never much of a town, as towns go. It was just a good, honest mining camp that served a good, honest gold mine—the King of Arizona. Unlike some Western mines the King wasn't given to shady promotion deals and false starts and stops. Its owners were decent and honorable; its workmen, for the most part, quiet, law-abiding folk. In consequence, Kofa just wasn't the sort of place that produces many newspaper banner lines.

Freight Team at Kofa, 1902 —Courtesy Ella MacBeat.

Harqua Hala, 1960

When Charles E. Eichelberger, prospecting in the southwest foothills of the S. H. Mountains, in 1896, discovered an outcrop rich in gold, he enlisted the aid of Epes Randolph, Arizona manager of the Southern Pacific Railroad, and H. B. Gleason, and these three organized the King of Arizona Mining Company. Due to complete lack of water at the mine site, a small amalgamation mill was built, in 1897, on the Gila River at Mohawk, 35 miles south of the mine. Later, that same year, deep drilling produced a good well much closer to the mine and here a new and larger mill was erected.

The King of Arizona was phenomenally rich, almost from the grass-roots. Vast quantities of ore, close to the surface, ran as high as $2000 to the ton in gold, plus some silver content, and Eichelberger and his partners had taken out hundreds of thousands of dollars worth of treasure before selling The King, in 1899, to Col. Eugene S. Ives. The purchase price, reportedly, was one-quarter million.

Among Western mining companies of that era it was the custom to brand company property with the initials of the company—in this case, "K. of A."—and when the camp's new post office opened for business in 1900, it was under the coined name, Kofa. Later, this same name was applied to a butte rising directly east of the mine and, eventually, was extended to include the entire range, thereby supplanting the inelegant title formerly borne by those rough desert mountains bordering La Posa plain.

Between 1899 and 1910 The King's mine and mill operated constantly, averaging 200 tons daily. From the spectacularly rich ore near the surface, values gradually declined to an average of $40 a ton and at 750-feet depth the paystreak was yielding but $3 a ton—an amount insufficient to cover the costs of mining and milling.[3]

Before The King closed, in the summer of 1910, it had produced some $3,500,000 in gold and silver and was considered to be one of Arizona's all-time greatest gold mines.[4]

One day in 1906 a Mexican prospector exhibited to Felix Mayhew, at The King, a sample of rich ore which he had found a couple of miles to the north. Working on this lead, Mayhew discovered what developed into the great North Star Mine and a year later sold his claims to the Golden Star Mining Company for $350,000. Like The King, the North Star was almost phenomenally rich—one ore streak on the footwall assertedly running from $6 to $20 per pound in gold! The natural result of this opulence was that highgrading flourished, and the mine's official yield of $1,100,000 between 1907 and 1911, is thought to be far short of the amount actually taken from it.

With both The King and the North Star closed, the equipment and buildings that had served them and their workers gradually deteriorated —though not so rapidly as in many no-longer-active camps, since care-

[3]*U. S. Geological Survey Bulletin* No. 620.

[4]*Arizona Bureau of Mines Bulletin* No. 137.

takers have kept watch over both mines through most of the 50-odd years since their closing.

I had visited the site of The King in the middle 1950s, shortly before the big Monterey-style office building and boarding house was torn down; but it was not until the spring of 1960 that I visited the North Star. Mr. Parker, then caretaker of The Star property, said I was welcome to look around the millsite, but he doubted if I would find there much of interest.

It was true. There was little left but cement platforms, splintered boards, tangled cables, several black tunnels leading into the mountainside, cast iron and tin, and other miscellaneous debris of no present use to man or beast. There were also several tons of "Swedish potatoes"—the small, flint-hard cobblestones formerly imported from Europe for use in the ore-crushing process. Down on the flat below the mill ruins were the usual last reminders of a vanished town. Broken bottles without number! Broken glass and dishes, the rusty skeleton of a baby cab, bits of charcoal where cabins had burned, and an old steel safe lacking its door.

Driving back four or five miles toward the highway I camped that night at the site of the deep well that had provided water for operating both the King and North Star mills.

It wasn't an especially scenic campsite. There was one small adobe house, its walls standing ruggedly firm but roofless. There was a jumbled conglomeration of mammoth timbers, tangled cables, wire, and a length of heavy chain, its links fused solidly by the corrosive action of half a century. There was also a mountainous heap of creamy white ash that once had been ironwood trees on the now starkly-denuded landscape; and all over the flat were millions of newly-sprouted plants, about two inches high.

They looked as if they might develop into desert poppies; but, life being as it is, they probably turned out to be Russian thistles.

On U.S. 60-70, 56 miles east of the Colorado River, is the town of Salome, "Where She Danced," and where the only ghostly thing is the little brown cabin and neighboring grave of Arizona's beloved humorist, Dick Wick Hall. From Salome a desert road leads south, past the town jail, and back into the hills nine miles to the ghost town of Harqua Hala.

Like the King of Arizona and the North Star, Harqua Hala's appeal was based on gold. The first locations in the new district were made November 11, 1888, and by Christmas of that year a stampede was under way.

"The latest reports from the Centennial District in the Harqua Hala Mountains confirm the first accounts of the rich deposits of mineral in that district . . ." enthused the *Mohave County Miner*, January 12, 1889. "The rush for the new camp is unprecedented in the history of Arizona. Prospectors are already in the district from every camp in the Territory, and people are headed for the Centennial from every state in the

Union. The first discoverers have 11 claims which they hold at $300,000, and claim to have taken out $70,000 in gold, already." Mining locations, it was stated, had been staked for 20 miles in every direction.

The resulting boomtown soon embraced the usual stores, saloons, and boarding houses. Even a newspaper, the *Harqua Hala Miner,* blossomed under the stewardship of Capt. John McCasey.

Although Harqua Hala's ore ran as high as $500 to the ton in gold and one week's run of a small amalgamation mill is said to have yielded $36,000,[5] life at the new camp—like that of true love—did not always run smoothly. The original locators soon sold; claims were jumped, litigation followed. By the close of 1890, however, most of the legal difficulties had been ironed out and the Bonanza and Golden Eagle veins had come into possession of A. G. Hubbard and George W. Bowers. After building a 20-stamp amalgamation mill in 1891, piping water from a source six miles distant, and otherwise expending a total of some $275,000,[6] Hubbard and Bowers sold their Harqua Hala properties to an English syndicate at a reported price of $1,250,000. After remodeling the existing mill, sinking a new shaft, and building a 150-ton cyanide plant to treat the tailings, (which ran from $3 to $5 per ton in gold), the British company is said to have produced around $750,000 in bullion. By 1897 both the orebody and the tailings dump were suffering from malnutrition so that the company was no longer showing sufficient profit to satisfy its stockholders, and was forced to close down. Following several unsuccessful attempts to resume operations, the property was sold at public auction in August 1899. Its purchaser was A. G. Hubbard who paid for it barely $5000—only a tiny fraction of the amount he and Bowers had received for the same mine six years earlier.

For the next 30-odd years the Bonanza and Golden Eagle mines at Harqua Hala were worked by a succession of owners who realized from their operation varying degrees of profit and loss. Altogether the two properties, up to 1934, had produced around $2,500,000.

Meanwhile, Harqua Hala, the town, had enjoyed its little hour in the sun and had gradually withered away. Its stores had followed its newspaper into oblivion; its post office, established in 1891 and discontinued in 1918, had been revived in 1927 under the abbreviated name of Harqua, but had re-succumbed in the Depression year of 1932.

At the time of my first visit to Harqua Hala, in 1960, the old camp still had two residents—a man and wife named Johnson. Prowling over the townsite I found a couple of dozen wooden-frame and adobe buildings, some half fallen, others in fairly good repair. There were, also, the usual mine and mill ruins, old foundations, crumbling fireplaces, and rubble; and down on a wide, dry flat below town, the ground was half paved with fragments of crockery and broken glass, much of the glass darkly purpled by the zealous rays of Yuma County's desert sun . . .

[5]*Arizona Bureau of Mines Bulletin* No. 137.

[6]*Weekly Arizona Enterprise,* Jan. 3, 1891.

Five miles southeast of Salome, and about half a mile off the long, straight road that leads on to Tonopah and Buckeye, is the ghost town of Harrisburg, a contemporary of Harqua Hala.

Actually, the only reason Harrisburg came into being was because the site offered water in a region where water is almost scarcer than gold.

In 1849, years before the town was founded or any settlement had begun accumulating at that point, a covered-wagon party, enroute to the California gold fields, made camp one night at what was later known as the Harrisburg waterhole. Months afterward another band of emigrants found the burned and deserted wagon of the earlier party and in the whitening skeletons of its passengers read a mute but nonetheless eloquent story of Indian attack and massacre. When the victims of this early tragedy were interred on a desert hill slope near the waterhole, Harrisburg cemetery was inaugurated—though it still would be 37 years before the town was founded and given the name by Charles Harris, a Canadian, who had served with the Union Army in the Civil War and had risen to the rank of captain.

In partnership with Arizona's Territorial Governor Tritle, Harris moved a five-stamp mill from Prescott and set it up beside the desert waterhole, there to operate on ore from the Socorro and other nearby mines.

Harrisburg was not a spectacular camp and little news emanating from that point ever found its way into print. One exception was in August 1897.

"Harrisburg was visited by a severe wind-storm recently," reported *Arizona Weekly Journal-Miner*, on September 1. "The store of W. S. Moffatt, the post office building, and the dwelling house of Joseph Vickers, were blown down . . . Thomas Bouse, the butcher, lost the roof of his house. The Socorro G. Company office was unroofed, and O. B. Warner's house went down . . . No lives were lost."

But, except for this enthusiastic zephyr, Harrisburg lived and died virtually unsung—"no hits, no runs, no errors."

More enduring than the camp, itself, was old William Bear, its postmaster. When the Harrisburg post office was established in 1887, Bear assumed his official duties and was still going strong when the office was discontinued in 1906. Old Bill's wife, Mary, had died in 1891 and had been buried in the Harrisburg cemetery, alongside the nameless victims of the early Indian attack. At the time of the funeral—and numerous times afterwards—Bill told friends that he hoped, eventually, to be laid at Mary's side. But, for some reason, when Bill Bear died in 1920, he was not buried at Harrisburg—not immediately.

Time passed. Harrisburg and its cemetery, and even Bill Bear, were almost forgotten when, in 1936, a sheepherder stumbled upon the weed-swallowed grave of the massacred '49ers. The man reported his discovery at Salome where some of the old timers recalled the story of the early tragedy and everyone agreed that something should be done about marking the old grave. It was Jim Edwards, maintenance foreman for the Arizona State Highway Department, and his crew, who took the lead in translating that noble impulse into action. Working on their own, Edwards and his men cleared the cemetery of several decades' collection of brush and weeds, then erected there a fine, imposing monument of white quartz topped by a copper-and-silver silhouette of a covered wagon. On one side of the monument was mounted a copper plaque dedicating the marker "in memory of the pioneers who gave their lives to the development of the West."

During the course of this work someone remembered Old Bill Bear and his request to be laid beside his wife, Mary, by then 45 years in the grave. So Old Bill's body was exhumed, hauled to the little ghost town cemetery, and with a few old friends gathered at the graveside, the former postmaster of Harrisburg was lowered to rest in the grave where he should have been buried 16 years before.

One of those present at the belated burial was Fred Neff, a former resident of New York City, who had left there in the early 1920s, drifted West, and eventually located in this little Arizona ghost town.

Fred lived alone in a tiny unpainted shack, beneath a giant mesquite tree. He had a small garden, some chickens, and about 50 nondescript goats which he herded over the desert hills. When I first made his acquaintance he had lived at Harrisburg sufficiently long that he was, by then, known as "The Hermit." Winter or summer his clothing was simple. Above the waist there was only his tough, brown, wrinkled skin; below the waist only a pair of much-patched, faded blue overall-pants, bobbed off raggedly at the knees. His feet were shod in clumsy sandals contrived by himself from old auto tires and rawhide thongs.

Neff was a poet of sorts—not a second Robert Frost or Walt Whitman, but better than some. Sometimes he even set his poems to music. As he trudged over the dry, desert hills he would sing these homespun compositions to himself and to his goats; and in the evening, with his nannies bedded down for the night, he would sit in his little shack and sing to his old gray tiger cat.

Early in February 1942, Old Fred died of a heart attack and the next time I visited Harrisburg no one was living in his little cabin, nor anywhere in the town, and the old graveyard on the dust dry hillside seemed terribly quiet and alone.

Another later-day place in Yuma County that may merit inclusion here, is not a ghost town nor even a ghost village. However, if I have a correct understanding of phantoms and their favored haunts, I'm sure there must be at least two of them floating around these premises . . .

On the open desert, between Quartzsite and Hope, motorists on U.S. 60 pass a little way station called Brenda. Another mile or so farther east, and over a low hill, there's a roadside park with a thatched ramada and a table. Nearby stands a statue. It isn't I might say, an especially *good* statue, but the strange story behind it compensates for any lack of artistic skill.

There were several of the statues, originally. Mom and I first noticed them when we were driving through here about the time of World War II. There was no roadside park, then, nor picnic ramada. There was only the desert and these unusual figures of Indians, nude women, et cetera. Around the foot of each statue the desert growth had been cleared away and rocks and ores had been laid to make little winding paths—paths that didn't go anywhere. One statue, of a female nude, bore a sign designating it *The Yuma Nymph.*

The place was so unusual and so mystifying we wondered who might be responsible for it. Looking around, we noticed, for the first time, a small, gray-weathered, board cabin tucked away in the rocks about 200 yards distant.

The yard was neat and clean. There were bleached flour-sack curtains at the small windows, and a lone geranium was blooming in a red clay pot beside the door. If our rap had been answered by a white woman, or an Indian or Mexican, we wouldn't have been surprised; but it isn't a common thing to find a Negro woman living alone on the Arizona desert. She was a nice-looking old mammy. Not especially dark but still unmistakably African, she was comfortably plump, wide awake, and completely self-sufficient. We exchanged introductions, and before leaving, learned that the Negress was 67 years old and had been born in Missouri.

"I homesteaded this land in my own name and built this house myself," the woman declared proudly. "I sawed every board and hammered every nail. I gathered the rocks, mixed the cement, and built this fireplace. And I cleared my land and laid the walks you saw. Some of my kinfolk offered to help me, but I turned them down. After I'm dead and and gone I don't want them saying, 'Oh, yes! I built that house for mother!' so I just did all the work myself . . ."

When Mom inquired about *The Yuma Nymph* and the other strange figures, the woman's face went a little hard and blank.

"My son, Cecil, was a sculptor," she said, after a moment. "He made them . . ."

Inviting us inside her cabin and bidding us sit down, she told us a strange story—how she and her son had been separated soon after his birth in New York City, in 1900, and had gone through the years trying to locate each other but always failing to make connections. Finally, the boy had found her.

"When I asked him about his trade or profession—what he did for a living—Cecil said he was a sculptor; that he made figures on special order for fairs, carnivals, window displays, and things like that. He said he would make some figures for me, so I could see how good he was . . ."

"Does he live here with you?" asked Mom, the eternal newspaperwoman.

"No," said the Negro woman tonelessly. "He's dead. I had him with me only eleven months. He died in Blythe in 1937 . . . because of an accident."

"A car accident?" suggested Mom.

"No," said the woman, with the finality of closing a door. "I'd rather not talk about it . . ."

We spoke of other things for awhile, of the unseasonable heat, and how much the desert needed rain. Our hostess showed us some fancy-work she was making—an Afghan, I think; and then she served tea and homemade cookies.

"When you come back through here again you must stop and see me," she said, as we shook hands to leave. "I'd like to fix a nice meal for you . . ."

After we were rolling down the highway again, Mom said she wished we had asked the old lady's mailing address. "I'd like to send her a Christmas card," said Mom. I suggested that someone in the next small town would likely know where she got her mail.

"Why, yes!" said a pleasant-faced woman at the next little roadside store and eating place. "She gets her mail at Vicksburg . . ."

"Has her son been dead long?" asked Mom.

The woman thought a moment. "About eight or nine years," she said.

"He was a very fine man," she continued. "Tall, and coal black—very good looking. His character was excellent, and he was devoutly religious. Just one time he got in with some bad company. They went over across the river to Blythe. Several in the party had too much to drink, and all were arrested and taken to jail.

"Cecil was terribly ashamed of what had happened. I guess he figured he had disgraced the family and couldn't bear the thought of facing his mother. Anyhow, while he was still in jail, he killed himself . . ."

Soon after our visit the old Negress met death while crossing the highway near her cabin.

Each time since then, when Mom and I have traveled this stretch of desert highway, we have looked over to the little, sun-faded homestead shack and the sculptured figures, and have thought of the strange story behind them. Even as recently as last spring a couple of the statues still were standing and a few of the old lady's rock-bordered paths still were discernible beneath the encroaching desert brush.

Former Post Office, White Hills, 1947

Dry Gold Washer, White Hills, 1947

"Here is a real ghost town. No one is left. No stores or filling stations, or curio shops, or corner drug stores serving Cokes and postcards to tourists. No old codgers sitting around spinning their yarns of yesterday . . . White Hills is dead. It is the most ghostly kind of ghost town you could stumble upon anywhere. Only the sunburnt and windwhipped boards of forgotten shacks . . . A mill frame, an ore dump, a pile of empty whiskey bottles, a few tin cans. That is all. Nothing else. Nothing . . ."

—R. C. Carlson *in Arizona Highways, August 1942.*

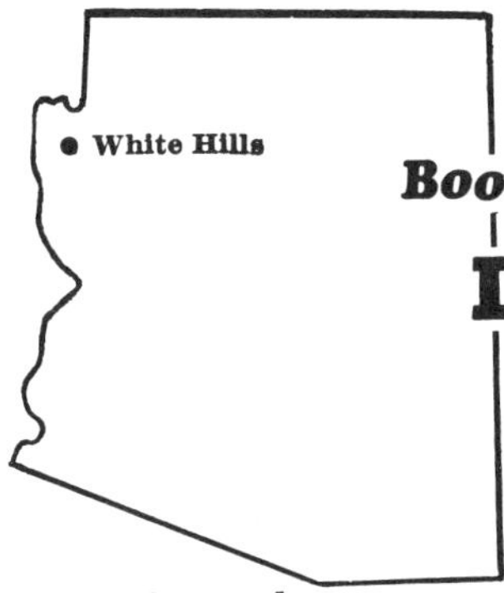

Boom Camp in the White Hills

In the eastern fringe of Detrital Valley, midway between Kingman and Hoover Dam, two mine-scarred ridges flank a desert wash. On the banks of this wash huddles all that remains of the ghost town of White Hills.

This is a place where no man lives, where noon sees no movement but gray lizards sunning in crumbling doorways, and night brings no sound but the empty wail of the desert wind.

Along two dusty streets stand the remnants of a few old buildings, the weathered brown of their wooden awnings and false fronts merging with the weathered brown of sun-scorched earth. Even in her flaming youth White Hills was not an attractive place, and today finds her as ugly and haggard as an old trollop.

And, like most old trollops, White Hills, too, has her story to tell.

Go there alone when the moon is swinging low over the barren hills to the east and all the gaunt canyons are welling with purple shadows. Stand in these empty streets—stand so quietly you can hear your own breathing, and the beating of your own heart—and, by and by, it will seem as if you can almost hear the old town speak.

"I wasn't always like this, you know," she seems to be saying. "Once I was proud and prosperous and, for a little while, I was one of the greatest of the great . . . ."

No man can say how much wealth has been produced in the vicinity of White Hills; neither does any man know how long the district has been mined. When white prospectors first entered this region they found Indians already taking out mineral—chiefly red iron oxide for decorating their faces and bodies. Because the redmen carefully hid the location of their workings, the area became known as the Indian Secret mining district. White miners were not received cordially in that district, and the area usually was given a wide berth.

It was Hualpai Jeff who sold out his tribe for a bottle of O Be Joyful, and to Henry Shaffer, a prospector at Gold Basin, revealed the long-guarded location of the Indian mines. The month was May; the year, 1892.

After Shaffer had made several locations he returned to Gold Basin to tip off his friends, John Burnett and John Sullivan, who also cut themselves in on the pie. When the three friends, working their separate mines, began shipping ore that averaged $1000 to the ton in silver, other men learned of the new bonanza and it wasn't long before a mining camp was taking form in the vicinity of the mines. Originally known as Silverado, the name of the camp had been changed to White Hills prior to the establishing there of a post office in October of that first year.

Like most old-time prospectors, Shaffer was interested in *finding* mines, not in *working* them, and he soon disposed of his holdings to Sullivan and Burnett who resold the property to R. T. Root, of Denver, one of the prominent promoters of that day. Associated with Root in the purchase—either as part owner or as financial backer—was D. H. Moffatt, Colorado railroad magnate. The White Hills Mining Company was organized with Root as president, a 10-stamp mill was built, mine shafts were deepened, and operations got under way. Not, of course, without certain difficulties. For one, the matter of fuel for operating the mill.

Nearest source of coal supply was at Kingman, 40 miles to the southeast, and there seemed to be no end of transportation difficulties. When freight teams and wagons were delayed and the coal pile reduced to dust—as happened, all too frequently—it was necessary to fire the mill with Joshua tree trunks and yucca stems, possibly the world's worst firewood but the nearest facsimile to fuel that the desert region provided.

Other mines came into production, and within two years of its founding the town of White Hills had grown into a full-blown drylands fleshpot replete with kerosene-lighted cafes, gambling houses and a dozen saloons—seven of them in a single row on Main Street. On the more constructive side of the ledger, the desert mining center also harbored several mercantile establishments, hotels, barber shops, assay offices, laundries, a post office and a school, and more than 1500 free-spending citizens tramped the dusty thoroughfares of the town.

In 1895 Root and Moffatt sold to an English syndicate for a reported $1,500,000. The new owners immediately increased the mill to 40 stamps, built an electric power plant to serve both the town and the mines, and laid seven miles of cypress-wood water pipe from a mountain spring to a huge concrete reservoir located on a hill above the town. But though some of the more affluent householders hastened to install bathrooms with cast-iron tubs and flush toilets, there was never enough water in the reservoir to operate such bizarre extravagances. Throughout the abbreviated life of the camp, its liquid requirements would be met with water and spirits freighted from Kingman by mule team—the water selling at "a buck a barrel," the whiskey at "two bits for two drinks."

"The Arizona Stage Company provided daily mail and passenger service, and the W. H. Taggart Mercantile Company, as well as Gaddis & Perry, ran beer and ice wagons to the desert metropolis," wrote Muriel Sibell Wolle.[1] "More buildings were erected, the Independent Order of Good Templars organized a lodge with 27 members, and Root announced that a railroad would soon reach the isolated paradise in which he so firmly believed."[2]

In August 1899, White Hills' perennial water shortage was temporarily relieved by a cloudburst in the nearby desert hills which sent a flash flood coursing through the camp's main street in a torrent as much as five feet in depth. Cabins were whirled end for end, the schoolhouse was swept off its foundation, two feet of muck was deposited on the street, and miners in the Grand Army Extension barely escaped with their lives when water poured into the mine, filling the shaft to the collar.

Reporting the flood in its issue of August 5, the *Mohave County Miner* noted that water customarily sold at a dollar a barrel in White Hills and wryly remarked, "We had a million dollar bath."

By 1900 the fine, big mill was running only half time and then dropping only half of its 40 stamps. Two years later, with $26,000 of the original purchase price still unpaid, Root and Moffatt sued the English syndicate and regained possession of the mine.

But White Hills' hour of glory was past. Ore that might be mined cheaply was no longer plentiful, and the price of silver was plummeting.

The mines and mill closed, workmen departed. Business establishments packed their more valuable merchandise, leaving the remainder on the shelves. Men who had arrived in the camp penniless left, for the main part, in the same condition. Scarcely had the dust settled behind the last departing stage until creosote brush was again sprouting in the empty streets and a camp that had added several million dollars to the nation's wealth was abandoned to the encroaching desert. As time went on and more settlers infiltrated Western Arizona, many of White Hills' business houses and homes were hauled away for the lumber they contained. Some fell prey to vandalism and fire. Others collapsed from sheer delapidation. One by one they disappeared—the barber shops, the bars, the brothels. And with the loss of each building, the restless ghosts seemed to swirl more densely around those remaining.

First time I visited White Hills was in 1946 when it had already stood virtually tenantless many years. Possibly a dozen of the original structures were then still standing. One of the best preserved buildings had served as a combination post office and mercantile. The pigeon-hole case for general delivery letters lay lopsidedly on the porch. Beside it

[1] *Bonanza Trail*, Bloomington: Indiana University Press, 1953.

[2] On September 14, 1897, E. G. McLennith filed with the Secretary of the Interior "notices of intention" covering the construction of three proposed rail lines: Kingman to White Hills, 50 miles; White Hills to northern boundary of Territory, 45 miles; and Hancock to White Hills, 60 miles. On the same day notice of intention was filed by Frederick Thomas covering the building of a railroad from Hancock to White Hills.

slumped an old time spice rack, and hanging from one corner of the building by a single nail was a battered tin sign advertising Star Chewing Tobacco. What sport the desert wind must have had with that clattering sign through all the 50-odd years it had been hanging there! In the lee of this building stood three old dry gold washers, a little shaky but still intact. Dora Tucker, of Las Vegas, who was with me on that trip, went through the motions of shoveling sand into one of the old washers while I took her picture—but we didn't find any gold.

Directly across the street from the post office stood another false-fronted structure that had housed a men's haberdashery. Rat-chewed rubble, ankle-deep over the floor, consisted mainly of yellowed style cards illustrating the *ne plus ultra* creations available from J. B. Price & Co., Chicago tailors. The cards were dated 1895, and attention of the potential customer was drawn to the advantage of "specially-tailored bicycle trousers" available in matching colors with all business suits.

Site of the White Hills Hotel was indicated by a mass of charcoal, heat-twisted iron bedsteads, and broken "thundermugs." Blackened crucibles and cupels marked the one-time location of an assay office evidently destroyed by fire. On a vacant lot, nearby, lay an old-style floor safe, its door blown off and missing. Except for wind-blown sand and a dead lizard the interior was bare. In the lee of an old livery stable lay a couple of big iron ore buckets, and on one wall of the stable hung an assortment of horse and mule-shoes. Dora and I wondered if this might be the livery stable where Jimmy Twiggs had held forth. Old Jimmy is said to have liked his nip, and when he had nipped a few nips too many his ringing cry allegedly could be heard the length and breadth of Main Street: "Hooray for Jesus Christ, George Washington and R. T. Root—the three best damned men in Mohave County!"

Jimmy, evidently, wasn't the only one in White Hills who had liked his nip. Of all souvenirs remaining in the town, the most plentiful were bottles. In every direction from the camp, a mile up the wash and a mile down, and well up the ridges on either side, were bottles without end! Green, brown, blue bottles; beer, whiskey, rum, and gin. Even champagne bottles—and every bottle as empty as a politician's promises!

In two old graveyards on a windswept flat below town, surrounded by yuccas and Joshua trees, Dora and I paid homage to those former residents who had met a faster gun or otherwise severed connections with temporal affairs. Many of the 100-or-more graves we located were unmarked save for a mound of desert rocks. Other burials were indicated by wooden headboards or crosses from which the inscriptions had completely disappeared.

We found only two markers on which the lettering was still legible. One of these, a marble slab, marked the grave of Robert Jackson, native of Tyrone, Ireland, who had departed this earth in December 1893. The other, a pine-board slab, indicated the burial place of L. S. DeBrecht, a native of Germany, who had died in November that same year.

Pitting the ridges and slopes on either side of town are scores of old mine tunnels and shafts, each with its dun-colored waste dump spilling down the hillside below. The Prince Albert, Great Diamond, Occident, Hidden Treasure, Chief of the Hills, West Treasure, Grand Central, Grand Army of the Republic, the Garfield, the Bryan—all stand as silent witnesses to those few great sparkling years when Silver was queen of the White Hills.

Desert Graveyard, White Hills, 1947

Portion of Jerome, 1952

Little Daisy Hotel, Jerome, 1948

"The history of Jerome is a story of tough men against a rough mountain. It's a story of hard rock, hard work, hard liquor, and hard play. Jerome's ups and downs have never been confined to its streets and houses, nor to its shafts and pits. Rises and falls in the price of copper forced ups and downs in employment and payrolls, in population and prosperity. Fortunes were made and lost. Hopes soared upward with a widening vein of high-grade ore and fell to new lows when the vein pinched out . . ."

—JAMES W. BREWER, JR.

Metropolis of Mingus Mountain

ON THE DRAFTY RUMP of Mingus Mountain, more than a mile above sea-level, the weary old skeleton of Jerome roosts in the sun.

Jerome is one of the most unusual ghost towns in America.

In the gunshot length between her north and south city limits lies an elevation range of more than a quarter mile. To this nearly-perpendicular plane cling her houses—stilted and shored and braced and blocked. There are houses whose tenants must climb three flights of stairs to enter their own basements; houses where children at play in their backyards may look down on the roofs of their own dwellings. There are homes below the level of the street, with the family garage built on top of the house; and homes, 'tis said, where a housewife may lean out her kitchen window and strike a match on her neighbor's chimney!

To further complicate the issue, this architectural miscellany, for the past 30-odd years, has been in the grip of a slow-moving landslide assertedly started in 1925 when an eighth of a ton of dynamite was fired in a single blast in the Black Pit. The unique circumstance of an entire city creeping downhill at the rate of several inches annually, once prompted the statement in a national magazine that another dozen months would see Jerome "totally destroyed."[1] Sixteen years have elapsed since the publication of that article—but stubborn old Jerome is still doing business

"Sure, we're slippin' downhill!" snorted Old Pop Clanton last time I talked with him. "Six or eight inches a year, they say. But you don't see nobody excited about it, do you?"

And no one was.

After meandering 300 feet down the mountainside, Jerome's former jail stands twisted and askew. Other vacant buildings are half up-ended or lie in heaped rubble. Buckled cement floors indicate where now-van-

[1]Rex Stanley, "Jerome—Ghost Town of Tomorrow, in *Westways,* March 1947.

ished buildings once stood, and many of the dwellings and business houses still occupied exhibit broken window panes and gaping cracks from cornice to foundation.

But, as Old Pop observed, nobody's excited. It is "just one of those things"—and Jerome is inured to playing the hand she is dealt. For 70 years the town's prosperity was gauged by the price of copper. When the red metal was "up" and times were booming, she ate tenderloin and tossed off her liquor at four-bits a shot. When copper was "down" she settled for Dago Red and beans without bacon.

Ample proof that Jerome's good years far exceeded the bad is found in the fact that her wonderful mines, prior to their closing in 1953, had produced close to $800,000,000 in copper, gold, and silver.

Nearly a *thousand million dollars* from one desert hill! Small wonder that mining men, when they speak the name of this old camp, speak softly, with pride lighting their eyes.

History of Jerome, the mining camp, had its beginning about 1884. The history of Mingus Mountain and its great copper deposit, however, goes back much farther. Even as long as 1000 years ago, Indians from the little hilltop pueblo of Tuzigoot, in the nearby Verde Valley, were scrambling up this same steep mountainside and with stone hammers and other crude tools were mining the oxidized surface outcroppings to be used as pigments for decorating their bodies and painting pottery. Before the abandonment of Tuzigoot, about A.D. 1400, her people had made a sizable hole in the mountain—a hole which later-come Indians would exhibit to the Spanish explorers, Espejo, in 1583, and Farfan in 1598.

But if Espejo and Farfan recognized the presence here of copper, they were not interested. They were seeking golden treasure rather than the lowly stuff of which pennies are made.

So the copper waited while another two-and-three-quarter centuries slipped by before the first mining claims were staked on Mingus Mountain in the spring of 1876.

Locator of the first claim is believed to have been Al Sieber, General Crook's able scout. Two other claims, the Eureka and Wade Hampton, were located about the same time by M. A. Ruffner and Angus McKinnon. With the aid of Angus's brother, George, the partners sunk a shaft about 45 feet deep; but because they knew themselves incapable of financing development of a copper mine and feared if they went any deeper the vein would pinch out entirely, Ruffner and McKinnon soon took steps to unload their property. When word of its potential reached Wall Street, certain financiers there sent Dr. James A. Douglas to investigate.

History, at times, can be very frustrating and exasperating—especially when facts which should be easily provable become a matter of controversy. In Arizona the story generally told is that Dr. Douglas investigated the incipient mine at Mingus Mountan and turned thumbs down. Yet,

James S. Douglas, son of Dr. Douglas, writing in *Mining and Metallurgy,* June 1935, states that his father spent three days examining the Eureka and Wade Hampton claims, that he thoroughly approved them, and upon returning to Maricopa telegraphed his principals to go ahead and exercise their option "with the understanding that the ore from the claims could not be profitably treated until the Atlantic and Pacific . . . then building westward from Albuquerque, should reach Ash Fork."

Whether Dr. Douglas gave or withheld his blessing, isn't of great importance. What matters is that the claims eventually came into possession of Eugene Jerome (later-to-be grandfather of Winston Churchill), who agreed to supply the money to finance further development of Ruffner and McKinnon's claims. In 1882 the United Verde Copper Company was formed, two small water-jacket blast furnaces were set up, and the camp that began accumulating in the environs of mine and furnace was named for the New Yorker who had evidenced his faith in the most concrete of all manners—cold cash.

Early-day mining at Jerome was a costly process, partly because coke for the smelter was imported from Wales, halfway around the world. After making the protracted journey around Cape Horn to San Francisco, the material was carried by rail to Ash Fork where it was loaded into wagons and hauled by long mule teams some 60 miles to Jerome. Despite this costly procedure everything was going fine by 1883, and the future looked unmistakably rosy. To the copper producers on Mingus Mountain it wasn't too important that Geronimo and his Apaches were on the warpath, or that the Pleasant Valley hostilities were growing unpleasantly warm only a couple of hills and hollows to the eastward. More important to them was the fact that their milling and smelting process was working so satisfactorily that they were saving almost enough gold and silver to pay all costs of the operation. The ore's copper content, its main value, was thus pure velvet—and copper was selling at 19 cents a pound . . .

Came 1885 and the United Verde sent a cabinet of its finest copper ore to the New Orleans Exposition where one of its interested viewers was William Andrews Clark of Montana—merchant, banker, and later-to-be copper tycoon and U.S. Senator. Clark knew copper, and he liked the appearance of these ore samples. Contacting Joseph L. Giroux, Montana mine superintendent, he asked him to have a look at the United Verde. When Giroux gave the nod, Clark set out to acquire the mine.

The United Verde, *per se,* was not for sale but its stock was on the open market—300,000 shares of it.

Before the Clarks quit they owned 299,000 of those shares. The program of procurement had cost them an asserted $60,000,000, but they had acquired a mine which soon would be rated the most valuable individually-owned mining property in the world.

In the spring of 1888, W. A. Clark moved to Jerome—all five feet four inches of him. The few remaining old timers whose memory reaches back to that event, recall that he wore a short beard that was parted in the middle and spread to either side of his face.

"He wore a cutaway and striped pants, and when he faced into the wind he resembled a tin rooster weather vane," recalled C. A. Brown, Jerome old timer who is now a resident of Merced, California. "In order to attract a better class of men to his employ he built a series of neat cottages for the married men, and for the single men he erected the Montana House. It was said to be, at that time, the largest stone structure in Arizona and provided comfortable quarters for 1000 men. We who worked for the United Verde had to live there—or else!

"The first time Mr. Clark came to Jerome after we had organized our brass band, we all turned out to give him a rootin', tootin' welcome," said Mr. Brown. "We rather hoped that such an enthusiastic salute would open his heart and purse-strings far enough that he would buy us some snazzy band uniforms. It didn't—but we reaped a lesser harvest: Caps from the company store . . ."

Many and varied are the tales of early-day Jerome. One of the more imperishable of these is that around the turn of the century water was being carried to Jerome by means of a 200-burro pack train owned and operated by the future revolutionist, Pancho Villa. The story may even be true—I know of nothing to disprove it.

There are stories of Jim Roberts, characterized as last of the old time shootin' sheriffs; and tales of Fred Hawkins, a later-day town marshal. Standing on the front porch of his dwelling in Jerome, it is said, Hawkins would sweep his telescope over the surrounding panorama of desert hills until spotting a lone campfire, perhaps many miles away. Mounting his horse he would start out in the direction of the fire and, several days later, would return with his prisoner—sometimes two or three of them.

Some of my favorite stories of Jerome are those told by my aforementioned friend, Mr. Brown, who landed in the camp in 1895.

"My older brother, Herb, met me at Jerome Junction," recalled this old timer when I asked him to dust-off his recollections of the mining town on Mingus Mountain. "The year before I arrived there, the United Verde had built a narrow gauge railroad to connect with the Santa Fe spur between Ash Fork and Prescott. The road was franchised under the name United Verde & Pacific, but was known locally as United Verde & Slag-pot. Herb was a car-whacker, stationed at the Junction. It was his job to keep the narrow-gauge rolling stock in repair.

"Lou Haskins, engineer, who tooled the Shae-gear engine around the impossible curves, pulling its long string of pygmy cars, assisted in the matter of upkeep by an unusual method. The new-fangled Triple Valve brakes sometimes were fouled with dust that swirled up from the track, so every wheel, on every car, had a spot of paint about the size of a camp biscuit near its outer rim. Riding in the engine, Lou could note if any of the wheels were stuck—that is, not turning. Lou said he could watch every wheel on the train except those on the engine. When the train slowed around the sharp curves even the wheels on the tender were visible at least three times in the 60-mile run . . .

"Jerome was flung like a skirt around the mountain below the smelter," recalled Mr. Brown. "The saloons were clustered mostly at a sort of flat. It was there we cashed our paychecks and rinsed our throats. There was always a Tillie or a Flossie to snuggle a bit until another guy came in with a check to cash. The Senate, The Fashion, and Tamborino's are the joints I remember best. Another of the saloons was O'Connor's. It was our habit to drop in there for a round of drinks with Andy Campbell, a canny old Scotsman, wrinkled as a prune. Once, on the waterfront in San Francisco, Andy had accepted a few free drinks from a stranger and had ended by being robbed of all his money. After that experience he devised sort of a 'possum belly pouch in his underwear, and carried his money in there. Whenever it came his turn to pay for the drinks, he would excuse himself and retire to the men's room where he would rob his bank in seclusion. Andy was an iron moulder and he had a trick that always brought down the house. He would fill his mouth with ordinary kerosene and exhale it onto the sand mould where a little blazing flare would cause the entire blanket of kerosene to burst into flame.

"The Fourth of July celebration at Jerome in 1895 included a 'Popularity Contest' between one of the local saloon keepers and the Catholic priest. Since it pitted the wicked against the pure the contest aroused a lot of interest. The prize was several hundred dollars, plus an ebony cane with a gold-plated knob, which we had bought through the T. F. Miller Company store for about $12.75. The priest won the contest and the cane; the money went to the church.

"Naturally, there was a drilling contest, too, and a full program of races. I remember that the Ladies of Light Love came to the races in a beribboned and betinseled super-spring wagon. They were very elegant. Their corseted figures were of the hourglass design then popular with the greatest actresses of the day. Their tiny colored parasols were held high and twirled jauntily in hands covered with gloves that reached nearly to their vaccination scars. My brother, Herb, who seemed to know most everyone, introduced me to these perfumed and gracious lovelies. They expressed their hope that we 'handsome young gentlemen' would attend their reception to be given that same evening. Oh, yes, they said, there would be dancing and refreshments. After shyly displaying their beauty patches on velvet cheeks, and giving us glimpses of sheer silk stockings, they waved goodbye and we men moved on to the long, improvised bar, stretching a full block. Big steins of foaming brew cost 10 cents, and there were viands of many kinds piled high on the end sections. Attendants in white aprons cut slices of meat and spooned salads into blue-decorated China plates.

"With all these pleasant distractions, I'm afraid I quite forgot the symbolism of a cracked Liberty Bell and John Hancock's king-size signature," grinned my friend, Brown.

"Among the women who roamed the West in those days were many who were quite picturesque in appearance and contour. As time went on almost every one of them acquired a nickname that seemed to become a part of her. There was one called Crazy Horse Nell. Sometimes she would have me boil a spoonful of powder over a stub of lighted candle, then blow on it till it was cool enough to fill the needles. At this point she always took over, for I could never find the muscles that could pump poison into a woman's arm. While she tied on her wire bustle she would declare I had saved her life and push half a dollar into my reluctant hand. Four bits for a service I abhor, even to this day . . .

"Belgian Jenny was a woman of means. She had a house of many rooms, and a front yard with a white fence and shrubs from foreign parts. She was given to periodic binges which seemed to have great influence on her vocal cords—but the only song she ever sang was the *Marseillaise,* first, last and always . . . Gold Tooth Mae was a real hoity-toity gal. She never leaned on a bar, she just sort of overcame it. Viewed from the sidelines she seemed to be devouring the mahogany slab . . . And there was Dirty Gert. She used to sew the buttons back on my shirt after the Chinese laundryman had popped them off with his charcoal iron. Everything that came from the Chink laundry smelled of opium . . .

"Tom Campbell, the postmaster, had his little stall in the store which faced the road leading down from the smelter. Tom had a bare electric-light globe hanging over his open-faced office. A Yavapai Indian slashed it off near the ceiling one night with the intention of lighting his wickiup down on the Verde River. I saw it later, hanging down through the mud and willow thatch.

"We had an organization of fishermen, and my first fishing license was made of leather. One year, with a companion, I spent my vacation in the sun-drenched watershed of Clear Creek. One of the saloon keepers at Jerome—the one who ran against the priest in the popularity contest—had recently married and he and his bride had their tent pitched not far from ours. The pretty little bride decided she wanted to make a pillow stuffed with the fluffy feathers of the wild pigeons that frequented the vicinity. Pigeon meat became sort of a by-product of our zeal. Since we had no storage bag for the feathers we all were saving, I gleefully surrendered a pair of long-handled underdrawers as a receptacle for the downy crop. With the ankle strings tied, my offering, at the last, resembled the bifurcated and nether portion of a field scarecrow.

"Those days, of course, weren't all happiness and horseplay," said the old timer. "My brother died in one of the fires that destroyed the town. The cemetery was located on what we called 'The Hogback.' A friend and I scooped out a hole in the rocky ridge, but all my efforts to get a 'man of the cloth' to preside were fruitless. A little Salvation Army lassie finally volunteered to help. She read some verses from the Bible while we all stood around in the hot sun with tears coursing down through the eternal red and yellow dust on our faces . . .

"When the newspaper reported the fire in which my brother had died, the story ended with this remark: 'The entire town burned, leaving at the upper end a church and at the bottom end the schoolhouse—outposts of Civilization . . .!"

The recurring fires alluded to by Mr. Brown were one of the chief hazards of life in early-day Jerome. Due to too many tents and ramshackle wooden structures, the camp was beset by frequent blazes and on three occasions between 1897 and 1899 was virtually wiped out of existence by sweeping conflagrations. One such fire prompted headlines in a Prescott newspaper: "JEROME BURNS AGAIN. ENTIRE BUSINESS DISTRICT OF 24 SALOONS AND 14 CHINK RESTAURANTS DESTROYED."

On September 11, 1898, fire again destroyed Jerome, leaving the camp with eight known dead, many more missing, and a loss calculated at more than half a million dollars.

In addition to devastating surface fires there was also the constant threat of fire in the mines. One such fire in the United Verde, presumably the result of friction caused by the caving of heavy sulphide ore, burned for many years and was never fully brought under control by underground methods. "For many years the mine employed a group of Spaniards who were recognized specialists in such work," states Charles H. Dunning.

"They became quite an institution in Jerome and were regarded with a bit of awe by everyone. Their art was practically handed down from father to son, and for years they were the only ones who would or could work within the fire area where temperatures were often over 150 degrees and rock temperatures very much higher."[2]

Despite these several flies in her pot of ointment, Jerome flourished and when she incorporated in 1899 she was the fifth largest city in the Territory of Arizona.

This same year George Hull, a miner at the town of Humboldt, also in Yavapai County, joined forces with J. J. Fisher, a mine surveyor, and the two went out hunting for "fractions." In mining parlance, a fraction is a small, irregular-shaped piece of ground which, through imperfect surveying at time of location, was not included within any of the legally established claims, therefore is open to filing.

As Hull had suspected, there was an unlocated fraction in the maze of mining claims surrounding the United Verde. Locating this fraction, situated just down the hill from the Verde, Hull and Fisher named it the Little Daisy and formed a company—the United Verde Extension Mining Company.

About ten years after location of the Little Daisy the ground was acquired by "Rawhide Jimmy" Douglas, son of Dr. Douglas, and his partners, Major A. J. Pickrell, and George Pickrell. Skilled geologists were employed and development begun. The first $200,000 expended by the company brought little if any encouragement. Another $100,000

[2]Dunning, *Rock to Riches,* Southwest Publishing Co., 1959.

was tossed into the pot. About this time a prominent Eastern mining "expert" visited the mine, shook his head disparagingly, and advised against any further work.

On the day following his visit, the drift broke into 45 per cent copper ore!

"The phenomenally rich strike on the UVX, one of the greatest ever made, brought a tremendous rush of promoters . . ." wrote Charles Dunning in *Rock to Riches*. "Almost every square foot of ground for miles around Jerome was acquired by someone . . . The streets were crowded with excited people day and night. Every nearby hill and canyon resounded to the roar and clang of legitimate, pseudo-legitimate, and strictly illegal attempts to strike another UVX . . ."

Before operations were concluded on the Little Daisy—the United Verde Extension—that fractional claim had produced more than 750-000,000 pounds of copper and had grossed a reported $125,000,000 of which some $42,000,000 had been paid to stockholders in the form of dividends.

During these prosperous pre-war years—in 1914, to be exact—there arrived in Jerome a fashionable young dressmaker, now Mrs. Madge Brissenden, of Long Beach, California.

"Walter C. Miller, manager of the T. F. Miller Company store, at Jerome, had asked J. W. Robinson's department store, in Los Angeles, to try to locate a dress designer who would come to Jerome to build up their yardage department and to alter the suits and dresses sold by Miller's," said Mrs. Brissenden. "The company agreed to provide a rent-free apartment, as well as a place of business next door to their store where such a seamstress could carry on her own dressmaking business on the side. When the job was offered me I accepted at once, packed my sewing machine and full-length mirror, and left for Jerome—a place I had never seen.

"Most of the American residents of Jerome at the time were young or middle-age married couples. Since the chief amusement of these ladies was going to card parties and trying to out dress each other, they welcomed me with open arms. In summer there remained in town only a few of us single and socially-acceptable young ladies, but in winter we had the competition of six pretty young school teachers. On Sundays, three or four carloads of couples, and all of us young ladies who wished to go, would drive down to the Verde Valley for quail hunting and picnicking; or we might go sightseeing to Montezuma's Castle, or boating on Montezuma's Lake. Sometimes we went swimming at a nearby ranch where there was a big soda spring. On such Sunday excursions the men did all the work, including cooking our dinner over a campfire. We usually had hashed brown potatoes, fried ham and scrambled eggs, with coffee boiled cowboy style and strong enough, as the saying goes, 'to float a rock.' All we ladies had to do was to go along and look beautiful . . ."

It was about this time Jerome experienced a burst of civic consciousness and set about building a recreation field on the least-perpendicular lot in the town's near-vertical environs. Mingus Mountain, rising abruptly on the south side of the field, provided a perfect gradient for stone seats and formed a grandstand guaranteed against any possibility of overloading or collapse. On this stadium were held Fourth of July and Labor Day ball games, parades, speeches and, most especially, drilling contests, the inevitable main event at any mining camp celebration in the old days when mining men were fiercely proud of their dexterity with drill steel and jackhammer.

Although conditions in her mines were far from perfect, Jerome seems to have had no organized labor trouble until 1907 when her miners struck for and won reduction of their 10-hour day to eight hours, plus a pay increase which brought their daily stipend to $2.75. Again, in 1917, when the camp was booming as result of World War I, the I.W.W.'s fomented a strike which backfired in a rather highhanded manner. Ousted from company property at gunpoint, several hundred striking miners and imported agitators were bundled into railroad cattle cars and hauled to a remote section of the Arizona desert where they were unloaded and abandoned to their own devices.

Mining by the open-pit method was begun at Jerome in 1920. Five years later the explosion of 250 pounds of dynamite at the Black Pit is presumed to have touched off the slide which has since held Jerome in its clutch and slowly but inevitably has carried its buildings and streets downhill. Undaunted even by this development, the old town on Mingus Mountain continued to do business as usual. With her pace and production gradually accelerating, Jerome's peak year of 1929 found her with a population of some 15,000, with 2345 men employed in her mines and nearly 2000 children enrolled in her several fine schools.

In this one banner year of 1929, alone, the United Verde produced more than $29,000,000 worth of copper!

But the great, gay sleigh ride was nearing its end. October 29, that year, brought the crash in Wall Street, one result of which was to send the price of copper hurtling downward to an eventual low of five cents a pound. With that development, in 1932, United Verde closed its mine and doused its smelter fires. Miners scattered to other jobs and relief rolls, and Jerome tightened its belt.

The United Verde was subsequently purchased by the great mining corporation of Phelps Dodge at a reported cost of $20,800,000, and both mine and smelter reopened in 1935. With outbreak of World War II there came a renewed demand for copper and the old camp on the mountain again hit her stride.

On my almost-yearly visits to Jerome through the later 1940s, Phelps Dodge was still working the United Verde and the town seemed prosperous. True, a few buildings were being destroyed by the creeping slide—the jail, J. C. Penney's store, and some dwellings were among the early casualties. Sundry other buildings were standing vacant on general principles. But, so far as a transient stranger could determine, Jerome's God was still in His heaven and all was more-or-less right with His world.

My annual visit to the town in 1950 found the picture greatly changed. Everyone was down in the dumps. Phelps Dodge, after operating the United Verde 15 years, during which time the company had realized on its investment some $40,500,000 after taxes, had closed the mine. For the first time in my memory of Jerome no plume of white smoke was rising from the tall stack at Clarkdale. The smelter had shut down for lack of ore.

"PeeDee is spending $100,000 for exploratory drilling . . . but things don't look so good," I was told by at least half a dozen different townspeople.

My visit the following year found the gloom even thicker. The hospital had closed. So had many of the stores, and it looked as if that would be the last year Jerome's school system would operate. With beginning of the 1952 term, Jerome's few remaining youngsters would be hauled down the mountain to the school at Clarkdale.

Jerome, in other words, was joining the ranks of the has-beens—the mining camps with lots of past but no foreseeable future. The town was dying. Her people were moving away. Everyone was discouraged.

Such was the state of affairs, one day in 1952, when Jerome's city council met in one of its regular sessions. It sometimes seemed as if there wasn't much point in holding such meetings for about all the councilmen discussed was the moribund condition of their city and what could be done to save it from complete downfall. Unfortunately, no one ever seemed able to contribute any concrete suggestions.

The present meeting was following its usual trend when the door opened to admit Jim Brewer, superintendent of nearby Tuzigoot National Monument. After listening awhile to the dispirited discussion, Jim offered a suggestion.

"Why not make Jerome into a ghost town?"

Most of the councilmen thought Jim was joking. Being a ghost town was the very thing they were trying to avoid.

"It's already a ghost town," muttered one of the members.

But Jim Brewer was not joking. Before leaving the council chambers that day he had convinced the city fathers that Jerome's best future lay in "tourism," and its best chance to attract tourist dollars was to become the liveliest ghost town in America.

Jerome has done exactly that.

The first step, in June 1952, was to organize the Jerome Historical Society, which immediately set up a work program. With a rare combination of humor and business acumen, everyone got into the act. A former saloon was cleaned up, remodeled, and turned into a museum to be operated by the Society. A call was sent out for exhibit material interpretative of Jerome's bonanza years. In poured jackhammers, miner's lamps, old ore cars, photographs, paintings, ore samples—everything under the sun. The two little blast furnaces placed in operation by United Verde in 1882 later had been removed to the smelter yard at Clarkdale. One of these furnaces was trundled back up the hill to occupy a place of honor across the street from the Mine Museum. A pamphlet covering the highlights of Jerome's history was written by Jim Brewer, published by Southwestern Monuments Association, and placed on sale at ten cents. Street signs that had read "Pedestrian Crossing" were changed to read "Spooks' Crossing," and the spectral likeness of a cartoon character, "Luke the Spook," began to appear on signboards bidding visitors welcome to this newest and most fascinating of Arizona's ghost towns.

The Jerome of today is not the bonanza town of 1929—but neither is it the disheartened place I visited in 1950 soon after the mine and smelter had closed down and those who peered into the future saw nothing there that might be mistaken for hope.

Today, Jerome is a "fun place"—not only fun for its visitors, who number into the tens of thousands every year, but fun for its hundred-or-so inhabitants, as well. Fortunately, it has learned one of the most valuable lessons any city or man can ever learn—how to laugh at itself.

On the more serious side, Jerome has been characterized "a great outdoor museum dedicated to its own marvelous story of mines, men and money." Like a somewhat disheveled old eagle, the town roosts high on the barren shoulder of Mingus Mountain, midway between the cottonwoods and green fields of Verde Valley and the ponderosa pines of the loftier levels. Its main street—U.S. 89A—loops upward through the residential and business districts in a series of sharp switchbacks offering grades of 9 to 14 per cent. Along these hairpin turns rise Jerome's high, narrow, cadaverous buildings, their paint peeling and cracking, windows staring sightlessly. Yonder squats the perambulating concrete jail that has traveled some 300 feet downhill in the past 38 years. Visitors grin and point to a fading sign on one side of the old bastile: *Danger—Keep Out!* Here is a shop named "Spooks' Retreat," another called "Spooks' Roost." On the hill slope, below the level of the main business section but visible from almost any part of town, is the once elaborate United Verde Extension Hotel—"the Little Daisy,"—with its vaulted corridors and archways and multiple fireplaces, its fancy inlaid ceramic tiles.

Whenever work or play takes me anywhere near Jerome, I pay a visit to the old town. One of the more recent times I was through there I parked in the center of the business district on a square of tiled paving that had been the floor of a big bank building. Over at one edge of the tiled square a couple of shaggy gray burros were drowsing under pack-saddles bearing similated loads of prospecting gear. When I paused to speak to the little fellows and rub them behind their ears, I was greeted pleasantly by an old man whom I had not seen before. I asked him what had become of Old Pop Clanton who had been operating the "burro concession" when I was there the previous summer? He said Pop had died that winter. I was sorry to hear it. I had liked Old Pop. He claimed to be a nephew of Ike Clanton of Tombstone fame. Pop had lived down at Cottonwood, I believe, but each morning had driven up the long grade to Jerome where he regaled visitors with stories of the old days, added local color to the streets, and posed with his burros in exchange for small gratuities offered by tourist photographers.

Crossing the street toward the post office, I dropped in at Sally's Rock Shop to say hello to Sally's brother, Claude Parker, another supplier of local color. I found him contributing to the general effect by sporting a handsome white beard that made him look quite a lot like Moses, and long white hair that fell below his shoulders. He was wearing a red scarf tied around his forehead, and three big finger-rings with gigantic turquoise sets, and a turquoise belt buckle large as a saucer. Claude gave me an introduction to a Hopi Indian—likewise loaded with turquoise—and the three of us lounged across Sally's counter and talked almost an hour.

Again I started on toward the post office but soon stopped to inspect a display window filled with pioneer relics and antiques. A couple of old timers were sitting on a pine bench at the edge of the sidewalk, soaking in the warm mountain sunshine and talking of "hanging walls" and "foot walls." Another old whiskerino came clomping along the walk puffing furiously on an ancient cob pipe. Glancing at the occupants of the bench the passerby snarled frigidly, "If I've ever said anything to ye I oughta be sorry fur—I'm glad of it!" Both of the men on the bench were grinning as the other clomped on up the sidewalk . . . and I'm sure that somewhere under his white shag, Old Whiskers was grinning, too.

I like Jerome. Her big mines and fat payrolls may be gone forever; but so long as the old town holds fast to her sense of humor, she'll get by.

"Sacred to the memory of these pioneers; They came when only the brave dared come. They stayed where only the valiant could stay . . . They carved a home from the wilderness. With the warp of labor and the woof of dreams they wove a pattern of life as beautiful as the sunsets, and as enduring as the mountains they loved so well."

Inscription on gate of Erickson family cemetery, Chiricahua Mountains, Cochise County, Arizona.

The Miners Called It Paradise

MOM AND I had been camping for nearly a week in the tall pines and high, clean air of Rustler's Park, close to the summit of the Chiricahua Mountains, in Cochise County. In all that while we had not seen another person. It was October, a wonderful time of year for drowsing and dreaming, and we had been content to spend most of our days in complete indolence.

One morning, feeling more ambitious, we took a couple of sandwiches and canteen of water and hiked to Barfoot Lookout, almost 9000 feet above sea-level. From that lofty eminence we could sweep our eyes over tier after tier of dark mountains spreading away toward New Mexico and Sonora and on to infinity. All the nearer ranges were clothed in conifers, tall and slim and past counting. Ranges more distant were dark with shadows and mystery. A few wild flowers—the last gallant stragglers of autumn—still bloomed among the rocks and along the trail; but deer already were leaving their summer feeding grounds, and the cold fingers of winter soon would grip the high country.

Next morning found a film of ice on our waterbucket, and white frost-diamonds sparkled on the grass.

"Maybe we should move down the mountain to a warmer elevation," suggested Mom. Then, by the way of inducement, "We might find a ghost town to explore . . ."

"There's Paradise and Galeyville," I said. "I don't know exactly where they're located, but we might be able to find them."

"Let's try!" shivered Mom.

Minutes later our camp gear was packed and we had said goodbye to Rustler's Park and the high elevations for another year. Following a road that clung to the mountainside like a nervous pencil mark, we started down the easterly face of the Chiricahuas. Soon after crossing Turkey Creek, where cold, sparkling water danced over the rocks as brightly as any brook in the freshness of April, I turned the car into a side road. A friendly little sun-freckled way, it nosed in and out of ravines, ducked under red canopies of wild grapevines, and politely turned aside to by-pass queenly old sycamores, their smooth white arms aflame with autumn's gold.

We hadn't traveled far on the little side road until we began seeing old ruins of rock masonry and adobe and wooden frame. Here sat an ore car, brown with rust and accumulated years. Yonder, a set of cement steps led upward to nothing—the building they had served, had vanished. An old wooden cabin, lacking in both windows and doors and time-weathered to the soft brown of pine bark, was strewn with faded letters and mouse-chewed magazines and newspapers of 30 years past.

One neat cottage bore the stamp of occupancy but my rap at the door and my called greeting brought only the saucy scolding of an Arizona jay. Beyond the screen door I could glimpse an immaculate living room where a plush-cushioned rocking chair wore crocheted tidies on its back and arms, and old-fashioned pictures in gilt frames hung on the walls.

No greeting came from the house across the street, and its neighboring house exhibited the unmistakable footprints of neglect and disuse. Weeds grew high around the front step; several window panes were broken.

Cruising on down the quiet street we eventually came to a pleasant little home with a lot of well-tended flowers and vines. Two gray-haired women were standing on the porch chatting in neighborly fashion. When I asked if this was Paradise, the ladies laughed.

"We think so!" said one.

"At least," said the other, "it's Paradise, Arizona."

Explaining Mom's and my interest in old mining camps I asked which of the local buildings was historically significant. The ladies went into conference.

"They might be interested in the old dancehall and store," suggested one.

"And they might like to see the old jail," said the other.

When the ladies tried to tell me how to locate and identify these buildings, their instructions seemed rather complicated and I had to ask further particulars. Finally, one of the ladies said it was such a nice morning for a walk, why didn't they just go along, too? Thereupon, they introduced themselves as Mrs. Potter and Mrs. Sanders, and the four of us started back down the sunny road.

Mrs. Potter identified the wooden building, with the old magazines on the floor, as one of the town's early post offices. This precipitated, a friendly discussion concerning location of the town's first post office. Mrs. Potter said she wasn't sure about the site of the *first* post office, but she certainly knew the location of the *last* one because it had been in her home and she had been postmaster until the office was discontinued in 1943.

Our guides pointed out a rock wall which had been part of a big general store; and the cement steps we had noted, they said, had led into the dancehall.

"I surely remember the first dance I ever attended in that hall," said Mrs. Sanders. "I was only a young girl, and thrilled half to death!"

"Yonder across the ravine," said one of the ladies, "was the redlight district. In only one block, according to my husband's Dad, there were seven saloons and 42 whores!"

"I thought it was 42 saloons and seven whores," put in the other.

"No," declared the first lady. "I'm quite sure Dad said 'seven saloons and 42 whores.'"

"The jail's over this way," said Mrs. Potter. Following her lead, we traipsed across a ravine, through burrs and brush and a tangled thatch of weeds.

It was a very substantial jail, its single window barely large enough for a man to stick his arm through—and even it was barred. The cell doors had been removed and the roof was gone, but it was still quite a building. Before this jail had been built, we were told, the town had used an open-air establishment consisting of a log chain stretched between two oak trees. Prisoners had been shackled to the chain.

There are at least three explanations why this Cochise County mining camp was named Paradise (though its natural surroundings, alone, certainly would entitle it to that designation)!

According to Judge J. C. Hancock, Paradise pioneer, prospectors in this vicinity, around the turn of the century, had to cross the valley to San Simon for their supplies and mail—a trip of about 28 miles with no water and little shade on the way, and terribly hot and dry during summer months.

"When they would get back up here," recalled the old judge, "they would take a drink of spring water and lay back in the shade and say, 'Well, this is Paradise, all right!' When George Walker and George Myers went to shipping ore, it made a camp large enough for a post office. When the Department asked what name, they all said 'Paradise.'"[1]

That's one version of the naming. Another was given by Frank C. Lockwood, dean of Arizona historians. Writing in the magazine, *Progressive Arizona,* Lockwood stated: "The name Paradise, and its beginnings, date back to about 1900 when George C. Walker and George Meyers, prospectors from Pearce, staked down a small lead-silver mining claim and made their camp at a fine cold spring where several madrone trees were growing, called by prospectors, "Trees of Paradise.'"[2]

Still more recently, Byrd Granger came up with a third explanation: "One of the Reed daughters (of Cave Creek) married George A. Walker. The young couple moved to an isolated location which, because of their happiness, they named Paradise . . ."

[1]Barnes.

[2]Issue, January 1932.

With the new camp of Paradise named—not once, but thrice—we can now get on with the boom, which, according to Walter Noble Burns, "went up like a rocket and came down like a stick."[3]

Some lead was taken out by Walker and Myers, but the boom at Paradise did not get under way until about 1903 when Capt. Thomas Burns and Capt. J. A. Duncan became interested. They, in turn, secured the financial backing of Calumet & Arizona, and the Chiricahua Development Company was formed, with C. & A. floating the shares. Work continued over several years and some ore was shipped, though not in sufficient quantity to compensate the company for the near half million dollars allegedly sunk in the venture prior to 1907.

Paradise was never a large town. At the crest of its boom, about 1904, the camp had around 300 inhabitants and a business district consisting of a few stores, Judge Hancock's hotel (built in 1903), and 13 saloons—an average of one for each 20-or-so citizens, which should have been quite ample. Although the boom had petered out some 45 years before Mom's and my first visit there, in 1951, the little settlement was still a delightful spot and we were glad to make the acquaintance of its ghost. The town then had a population of six—our guides and their respective husbands, and a Mr. and Mrs. Epley.

When we inquired for Galeyville we were told it was only about three miles from Paradise. The ladies gave us minute directions but, at the same time, warned we would likely have trouble locating the place.

"I'd love to go with you and show you the way," said Mrs. Potter. "But I'm sure my bread must be about ready to put in the oven . . ."

Driving down the canyon to what we interpreted as the proper turnoff, we headed up a steep, rocky trail to the left. When the trail ended in a barnyard, we knew our interpretation had been wrong.

"Pardon me," I said to the old lady who answered my knock at the back door of the house. "We're looking for Galeyville. It is near here?"

"Eh?" said the ancient one, cupping a hand behind her right ear.

"GALEYVILLE!" I shouted. "The old mining town. We want to go there. Where is it?"

"Oh yes . . . Galeyville," the old lady nodded. "I rec'lect some of th' youngn's used to go to Galeyville for good times. But I don't know how they got there. Reckon you'll have to ask somebody besides me. I was there once, when I was a young'n . . . but I'm too old to remember. I don't think anybody lives at Galeyville, now. No, I don't think so. They're all gone . . . all dead . . ."

As the mountain woman melted back into her cabin her lips were still forming the words, ". . . all gone . . ."

Returning to the Paradise-Portal road, Mom and I soon found another left-trending side trail which carried us up a ravine to a small farmhouse. A saddled cow-pony drowsed in the yard. There were a few cows in a pole corral. A woman, framed in the open doorway of the cabin, regarded our approach indifferently.

[3]*Tombstone.*

"Maybe we'd better ask this lady," said Mom.

Even before our wheels stopped turning, dogs had erupted from every corner of cabin and yard, some barking in high soprano, some howling in baritone, others just yapping without any regard for artistry or finesse.

When Mom halted at the gate and called across the yard that she was afraid to come any closer because of the dogs, the woman ambled forward to meet her.

"They won't hurt you," she said. "They're lion dogs. My boys use 'em for hunting . . ."

The woman was a tiny mite of a thing. An old sunbonnet, faded by multiple tubbings and the strong rays of the mountain sun, covered her thin gray hair. Her dress was almost hidden beneath a checked-gingham apron.

When Mom inquired for Galeyville, the woman said to go past their corral and turn left up the hill.

"How will we know when we get there?" asked Mom. "Are there any buildings?"

"No," the mistress of the hounds shook her head. "No buildings. Most of those that survived the big fire in the '80s were torn down and moved to Paradise at the time of its boom. But you'll know when you get there, all right. Just drive up the mountain as far as you can go in your car and you'll be in Galeyville!"

As the woman had said, there were no buildings, not even one crumbling wall. Still, there was no mistaking that a town had occupied this site at the road's end. Shards of sun-purpled glass, scraps of rusted iron and age-stiffened leather, occasional square-cut nails, were adequate proof of man's one-time presence.

Galeyville was a contemporary of Tombstone, Charleston, Contention City, and Fairbank, all in Cochise County. Except for Tombstone, Galeyville had a larger population than any of her sister cities; yet, ironically, had less valid excuse for being, and was shorter-lived than any of the others.

John H. Galey, pioneer oil and gas producer of Pennsylvania, arrived in Arizona Territory when the magic name of Tombstone was like sweet music on the lips of the multitude. In that frontier city of promise, only then taking form, Galey was told of a wonderful silver prospect in the Turkey Creek Mountains, less than 60 crow-flight miles to the northeast. Visualizing another millionaire-making mine of the calibre discovered by Ed Schieffelin, Galey communicated his enthusiasm to the McKinney brothers, of Titusville, Pennsylvania, already rich from the proceeds of oil and gas, but not adverse to adding more millions through the medium of Arizona silver.

Solidly backed by McKinney money, plus several thousand dollars of his own and his family's money, John Galey opened the Texas mine and in November 1880, laid out the new city of Galeyville.

"The continued clatter of the hammer and saw is a convincing argument that the boom is getting more and more substantial and there is not the shadow of a doubt that the near future will show up Galeyville as one of the best mining towns in Arizona," declared *Arizona Weekly Star,* on February 17, 1881. "The new smelter of the Texas Mining Company is expected daily, and as soon as running will be the means of giving a further impetus to mining matters . . ."

On the social side it was reported that Mr. Weasles' new and commodious lodging-house tent had been the setting for "another of the delightful hops for which Galeyville is getting famous.

"For its size," declared the *Citizen,* Galeyville can more than hold its own for the number and beauty of the fair sex that adorn its society. The dance and lunch, which was gotten up in style by the Cosmopolitan Hotel, was greatly enjoyed by all . . ."

Upon blowing-in the smelter it was found that ore of the Texas mine was not self-fluxing, but after a brief period of uneasiness a suitable flux was found in the Granite Mountains, just across the line in New Mexico. By using one part of this ore with two parts of the Galeyville ore, the reduction process was completely successful. Galey's company bonded the Granite Gap property, put a crew of men to work, and began hauling the New Mexico ore across the line and up the mountain to Galeyville. By the time everything was operating satisfactorily, the company had in its employ at mine and smelter close to 100 men who were drawing from five to six dollars a day.

Table-board, at this same time, was $1.50 a day and drinks at the town's eleven saloons were two bits apiece. In addition to this generous supply of barrooms and bistros, Galeyville then embraced 42 business houses and professional men. These included a short-lived newspaper, the *Galeyville Bulletin;* six general mercantiles, two hotels, two restaurants, two butcher shops, a dairy, two blacksmith and wagon shops, a laundry, three lumber yards, a boot and shoemaker, jewelry store, three lawyers, three notaries public, an assayer, three physicians, a sign maker, post office, and Wells Fargo office.[4]

The Galeyville Militia, bristling with 20 guns and 5000 rounds of ammunition, also was organized that same year of 1881.

But the goose that swung so high for a little while, soon came tumbling down.

Why the Texas company so quickly began drifting into straightened circumstances isn't quite clear. Even though the ore was deficient—as, of course, it was—it would seem that the McKinney's oil and gas money should have financed a longer run than Galeyville enjoyed.

Within two years of the town's founding, the wages of miners and millmen already were going unpaid and red ink was gathering on the ledgers of boarding houses and merchants.

After first moving to the comparative safety of Tucson, Manager John Galey sent back word to his foreman that the company had decided to

[4]*Disturnell's Gazetteer,* 1881.

shut down operations. With the receipt of this announcement, irate miners and smeltermen declared their intention of taking over the property forcibly and operating it themselves, at least long enough to recover the back wages due them. But the Granite Gap company refused to recognize the freebooters and without the New Mexico fluxing ore the big Galeyville smelter was as useless as one of the big brown rocks on the mountainside.

"Uncle John came home badly in debt," wrote John Galey's nephew, Titus Galey, of Owensboro, Kentucky. "He had lost $180,000 of his own and the McKinney brothers' money. He had skipped out of Galeyville leaving debts. He began to receive letters threatening him with death unless he paid up. He borrowed $1500 from father, and with some additional cash he was able to scare up, he returned to Arizona and satisfied his creditors.

"What disposition was made of the mine by Uncle John's creditors, I do not know. Presumably it was always a total failure. When the mine was closed down, the smelter was moved to Benson and the town of Galeyville emptied like a circus tent after a ring show."[5]

Between the time when the miners and smeltermen moved out of Galeyville and the ghosts moved in, the camp passed through the Era of the Outlaws. Many a captivating story has been written of those days and the old town's most notorious outlaw citizen, Curly Bill Brocius. These anecdotes, without exception, are entertaining, fascinating, and delightful.

A few of them, it is possible, may even be true.

Curly Bill is pictured as a merry murderer who lived dangerously but happily. Neither in appearance nor performance could he have been classified as a penny ante gunslinger. As powerfully built as a Brahma bull, he loomed half-a-head taller than the average man, even in this land where men were big and bold. According to the Legacy of Legend that has descended unto this third generation, Curly Bill's undulating black hair was habitually topped with a wide-brimmed white hat, and around the swarthy neck that two nations were eager to measure for a hangman's noose was invariably knotted a flaming red cravat. Except for these excusable vanities the Outlaw King of Galeyville effected the same faded garb as might have draped the frame of any $30-a-month saddle tramp.

Curly Bill romped through life, exacting tribute, impartially, from stage coaches, trains, cattle ranchers, and Mexican smugglers. As a bandit he was not only a very bad boy but, also, a very busy boy.

While all forms of outlawry appealed to Curly Bill, the participant sport he favored over all others seems to have been cattle rustling—an avocation marked by both profit and peril.

In the late 1870s and early '80s, when Curly's star was at its zenith, Uncle Sam's army was in the market for vast quantities of beef to be used in feeding its reservation-interned Indians. Although Army officers

[5]Letter to Arizona Pioneers' Historical Society, 1942.

didn't actually go so far as to suggest cattle rustling as one of the quickest and most efficient methods of producing beef, neither did they require that every animal offered them be accompanied by its birth certificate and bill of sale. Furthermore, the prices they offered were inordinately good.

Virtually over the hill from this lucrative sales outlet lay the baronial ranches of Chihuahua and Sonora, ranged by uncounted millions of longhorn cattle. The fact that Mexico's *bandito*-harassed stockmen had earned a reputation for shooting fast and accurately only whetted Curly's interest in their animals.

In addition to a few bullets and, possibly, a couple of corpses, one night's South-of-the-Border foray oftimes netted the Brocius gang as many as 1000 stolen beeves. As leader of the gang, Curly Bill of Galeyville attained the status of an international menace. Not only were his cross-border deeds a subject for discussion in the Halls of Congress but, likewise, the subject of copious correspondence between the government of Mexico and the U. S. Department of State.

With big premiums on his head, both in the United States and Mexico, Curly Bill one day vanished from Galeyville like the smoke of yesterday's campfire, and for 80-odd years his disappearance has been a matter for conjecture in the region where he formerly roamed. Wyatt Earp, frontier marshal of Tombstone, claimed to have shot and killed him in a remote section of the mountains—but could offer no proof to substantiate his claim. One Tombstone newspaper of that day posted $1000 reward for evidence of the merry murderer's death; a rival newspaper countered with $2000 reward for proof that he still lived.

Neither reward was ever claimed.

Even though Curly Bill's disappearance went unsolved and unremunerated, it had the effect of breaking the back of the outlaw element. With the desperados scattered, Galeyville soon slipped into the status of a ghost town, its last remaining residents Seward Smith, former justice of the peace there, and Rube Hadden.

"For 20 years these two lived in the empty town alone, and then Smith moved away," wrote Titus Galey. "Rube Hadden remained as Galeyville's last and only inhabitant. He did a little mining and assessment work in the hills, and when he sat in the evening at his front door he could see from one end of the main street to the other. The old saloons where Curly Bill and his outlaws had once caroused were falling into decay, the bars still standing and the mirrors behind the bars reflecting only cobwebs and empty darkness. The dance hall, where the rustlers used to shoot the French heels off the slippers of the girls, whirling in the quadrille, was tumbling down. The rotting wooden sidewalks were caving in, and grass and weeds grew in the streets. But Rube didn't seem to mind the silence and loneliness. He grew old and gray as Galeyville's lone citizen. He might be there yet if it hadn't been for the new silver strike in Silver Creek Canyon. Then he moved a little closer to Paradise, where he died in his little log cabin, aged nearly 100 years.[6]

"What was left of Galeyville was finally torn down and the lumber used for building construction in Paradise. Now, to find Galeyville's location one must be guided to the spot where he may see where the earth has been worked, and he may even find broken glass here and there about. The silence is in vivid contrast to the time when Galeyville was at the height of its glory, when it is recorded that every five minutes, day or night, somebody seemed to be firing a gun and several doctors had a fine practice constantly patching people together . . ."

With our pilgrimage to Galeyville concluded, Mom and I drove back down the hill to Paradise and on to Portal and Cave Creek Canyon, where we made camp in a pleasant nook bordered by sheer cliffs and tall rock spires.

At this lower elevation the temperature was much softer and warmer than on the mountaintop where we had been camping for a week past. Even late afternoon found no chill in the air, and we were eating supper out under the trees when the sun slipped from sight behind the high, cave-pitted walls and the sky began coloring with rose and saffron. Against this radiant curtain, separating light from night, the canyon's tall spires and grotesque rock formations stood silhouetted like a convocation of giant priests and priestesses clad in robes of black velvet.

With supper eaten and camp made shipshape for the night, we sat under the trees and talked until the sky darkened and a few stars appeared, the pleasant sounds of approaching dusk faded away, and even the rustling leaves of the golden sycamores grew still.

It had been a full day, and our sleeping bags looked wonderfully inviting.

[6]According to the reminiscences of Hadden's close friend, Judge John C. Hancock, who located at Galeyville early in 1881 and remained in those precincts most of his life, Old Rube died in the Pioneers' Home in Prescott, in February 1930.

Early Post Office, Paradise, 1951

"Where is there a desert so barren or a mountain so rugged that it is not traversed or explored by the irrepressible Prospector? . . . He is the most sanguine of men, the most persistent of explorers."

—J. Ross Browne, *Adventures in the Apache Country, 1863.*

Two Golden Ghosts and a Fig Tree

This chapter is concerned, primarily, with the ghost mining towns of Oatman and Goldroad, in western Mohave County. But it is also the story—the obituary—of a tree.

It is more than a dozen years since Marion Johnson and I were first told of Fig Springs. According to our informant the tree for which the springs were named had been planted by Spanish mission fathers in the 16th or 17th century, and had grown to be the largest fig tree in Arizona. Or was it the largest in the world? I'm not quite sure on that point; but I do remember that we were skeptical about the tale—especially that part concerning the padres and the 17th century. Not that it was especially important. To Marion and me, a wildgoose chase can be thoroughly enjoyable—especially when the chasing is done in Arizona's desert hills, in early May! Even if no fig tree nor any springs rewarded our search, we knew that the time spent hunting them would not be without compensation.

We had been told to take the Oatman road (old U.S. 66) southwest out of Kingman, and to watch on the right for a ruined building. Just beyond the old hulk we were supposed to see a trail leading off toward the Black Mountains. It was all very plain; and yet, we were more than a little surprised when we actually *found* an old building fitting the description given us and, just beyond that building, discovered a pair of dim wheel tracks that led westerly across a wide, dry valley, toward a distant range of mountains.

Slowly and cautiously—almost like a canny horse that tests the planks of a weak bridge before trusting his full weight upon it—my old red Mercury followed the sinuous trail across the desert flat. It had been a dry winter and except for a sparse scattering of perennials—creosote brush, yuccas, catsclaw, cacti and snake weed—the flat was virtually bare. In the tawny heat haze, a mile away, we could see a couple of small whirlwinds, each lifting into the warm air its wavering spiral of dead thistles, and bits of grass, and dust. But for those twin spirals, the afternoon air lay like a dead thing on the parched flat, and the powdery

Opposite Page: Fire Bell and Main Street, Oatman, 1953

white dust cloud slowly lengthening along the trail behind us seemed too lethargic either to drift away or to settle back to earth.

"Look!" cried Marion, when we were about half way across the wide valley. "Aren't those *trees*?"

Tracing the line of her pointing finger I saw a faint green smudge at the base of a high cliff, still a long way distant. It was true! That green smudge simply *had* to be trees; and where there were trees, in this dry-as-dust land, there *had* to be a spring. Could it be Fig Springs? I looked at Marion; she looked at me. Doubt, obviously, was still in the saddle and riding hard—but Hope was gaining! Thereafter, whenever our trail divided and we were uncertain which fork to take, we invariably chose the branch leading more directly toward the little smudge of green. As we approached nearer, the oasis seemed to expand in scope, the trees began to assume individual forms, the air grew a bit cooler and moister, and our nostrils identified the delicious perfume of green grass!

Then, at last, with the old car boiling hard but still pulling gallantly, we toiled up the last steep pitch and rolled to a halt in the welcome shade of the grove. Our first eager glance had identified the trees as commonplace mesquites and palo verde—a bit of a disappointment to a couple of gals hoping to find "the biggest fig tree in the world;" but the grove, if unoccupied, would at least be a pleasant place to camp.

For a minute or two we remained in the car, just looking around us, and listening. A little deeper in the grove we could see an old rock house and we half expected that someone would emerge from it, either to bid us welcome or to challenge our presence.

"I think the place is deserted," said Marion.

We got out of the car and walked over to the house. I rapped at the door.

"Anybody home?" called Marion.

There was no answer. Slowly I turned the doorknob and pushed the door open a few inches and called again, "Anybody home?"

The interior of the house was bare except for a rusty, wood burning cookstove, a couple of apple-box cupboards, and bushels of packrat litter. Back in the yard, exploring more boldly now, we found several peach trees set with small, fuzzy, green fruits. There was a little worn-out bed of purple iris, just coming into bloom, and a clump of rhubarb.

We walked on, back through the oasis, rejoicing in its moist coolness after the heat of the valley.

Then, suddenly, there *it* was before us—the great fig tree of Fig Springs!

"Oh, no!" cried Marion. "It's *dead!*"

Even that season new leaves had sprouted on a few of the younger boughs; but the old tree no longer had possessed the vitality necessary to feed them and the forming leaves had shriveled and died. There was no question but that the gallant old tree had put forth its last effort.

It had been a veritable giant with a spread of more than 100 feet with several of its larger branches each measuring as much as five or six feet in circumference. Even in death, its trunk and limbs were smooth and white and cool to the touch, like those of a sycamore.

Water, emerging from a spring somewhere above, was seeping past the old tree in a marshy swath a dozen feet wide. Here the spongy ground was grown to watercress and false mimulus, and a big green frog eyed us reproachfully and muttered something under his breath before disappearing in the lush grass.

Further exploration of the oasis revealed the presence of a cement-lined reservoir, 54 by 30 feet in area and nearly ten feet at its greatest depth. Since fairly large trees and brush had grown up through cracks in the cement bottom, it seemed likely the tank had been dry for many years. A fine spring, rocked-up and roofed, held a pool of water a dozen feet across and three feet deep. The water smelled pure and fresh, and was so clear that even small pebbles on the bottom of the pool were clearly visible.

We made camp not far from the big fig tree. Marion cooked a good supper for us, over the campfire. Afterward we sat and talked until the flames had died away to embers and even the embers were growing cold.

Though the moon lacked a couple of days of being full, the night had a strange, bright, faery-like quality.

Looking back over the wide desert flat we had crossed from the highway, we could see trains of the Santa Fe line as they wound their way downgrade from Kingman toward Yucca. Lights of an occasional auto moved slowly along the highway, about eight miles distant, and the aerial beacon at Kingman winked on and off as regularly—and as silently—as the planets in their courses.

During the earlier part of the night a sprinkling of bats dipped and swooped over our camp in the grove. Nighthawks made merry, and a few crickets tuned up their fiddles. Once, about midnight, I was awakened by a strange, long-drawn whistle, almost like a ghost-train might make. I was on the point of wakening Marion to hear it when, suddenly, the whistle developed into the chilling bray of a wild jackass somewhere in the surrounding hills. Later, a desert fox barked sharply, and the low, mournful hoot of an owl drifted through the grove.

Dawn found a blaze of red spread across the eastern sky and it seemed as if all the songbirds in creation had gathered to serenade us! We took our time about breakfasting and breaking camp, and it was nearly noon before we drove back across the valley to the highway and continued on toward Oatman. At the summit of Sitgreaves Pass there was a small station selling a few staple groceries, beer, and gasoline. We stopped for a Coke, and Marion inquired about Fig Springs—who owned it, and whether they might sell it?

"I don't really know who owns the place," said the woman station-keeper. "Some of the Bonellis, I suppose . . . As far back as I can remember, the place has belonged to the Bonellis.

"When Mr. Frank Bonelli owned the springs, we young folks would go up there to picnic under the trees," continued the woman—and, for just a moment, her memory of those gone days seemed to light her lined face and make it almost young again. "Mr. Frank was a fine man, and we always had lots of fun when we went to the Springs. After I began running this station, more than 30 years ago, he would bring in boxes of figs and trade them to me for merchandise. They were the best-flavored figs I've ever tasted—great big black fellows!"

Marion asked the woman if she knew the tree was dead?

"*Dead*?" she said, in surprise. "No, I didn't know that . . . It's been years since I have been out there. Folks said the road was washed out and the place all in ruins. I didn't want to see it like that . . . so I just didn't go back."

Suddenly turning away from us, the woman grabbed a towel and began a great flurry of wiping the back bar.

"It seems a great pity," she said, at last. "It was such a pretty place when I was a girl . . ."

From the lonely little way station at Sitgreaves Pass, Marion and I coasted down the Oatman Grade into the ghost town of Goldroad, not too long ago one of Arizona's great gold mining camps. This was 1951 and there was still one business house in part-time operation—a combination grocery, gasoline station, bar, and blacksmith shop. Our visit happened to fall on one of the days when the place was closed, and though we spent more than two hours in the old town we didn't see even one person.

Most of the frame dwellings had been torn down two years before, chiefly to reduce taxes on the no-longer productive property. The townsite was strewn with mangled and tangled litter, lengths of wall paper, broken shingles, fragments of dishes, swatches of tarpaper, heaps of shattered bottles and rusted tin cans. In some of the yards a few pitiful remnants of flowers still struggled for life—hollyhocks and bachelor buttons, and other species that can accommodate themselves to little water and less care. A few poor old trees still clung tenaciously to life. Most of the adobe and rock buildings had been de-roofed and their empty shells stood hollow and ghostly. In all the time we spent rummaging the townsite the only building we saw that gave evidence of being inhabited was the combination store and bar. Several times we heard the braying of an unseen burro; but no dog barked, no roosters crowed, no child's laugh rang through the ruined streets.

Goldroad and Oatman, three miles downgrade, are unique among the great gold mining centers of Arizona in that their birth, boom, and decadence, all took place within the 20th century.

The discovery that triggered the boom at Goldroad was made about 1900 by José Jerez,[1] a Mexican prospector, working on a $12.50 grubstake supplied by Henry Lovin, owner of a general mercantile at Kingman. The original samples brought to Lovin by Jerez are said to have assayed

[1]Contemporary newspapers and records spelled this name in a variety of ways: Jerez, Jeres, Jerise; even Jeneres.

40 ounces of gold to the ton, and the wonder is not that Jerez found the outcrop but that no one had found it sooner. The Moss ledge, Mohave County's first successful gold mine, located in 1863 by John Moss and party, was situated only four miles to the northwest, and the first wagon road built through this district had paralleled the Goldroad lode for two miles at a distance of no more than 1000 feet.[2]

Within only a few months Jerez and Lovin had sold their claim for $50,000. After another change of ownership and considerable development work the mine was incorporated in 1902 at $1,500,000, and by February 1903, 150 men were employed in the mine and in building a 150-ton cyanide plant. Soon after this plant went into action the working force was increased to 300 men and between $12,000 and $15,000 worth of gold bullion was being shipped monthly. By the end of 1907 the mine is credited with a gross yield of $2,250,000. During the next four years production was somewhat curtailed, and in 1911 the property was sold to Needles Mining and Smelting Company for $1,200,000.[3] Followed, then, the greatest period of prosperity known to Goldroad and its mine, and by 1931 the mine had grossed, officially, $7,300,000.

Despite its original showing of 40-ounce ore, and its 30-year record as a consistent money-maker, the Goldroad mine was, generally speaking, a low-grade proposition. Throughout the productive life of the mine its ore averaged only about seven dollars to the ton.

Three miles downgrade from Goldroad was its sister mining camp of Oatman, *nee* Vivian. Here the first discovery was made about 1902 by Ben Paddock, a halfbreed Indian, who was assertedly riding along the trail when he noticed free gold sparkling on the ground and stopped long enough to locate a claim. In 1903 Paddock sold this property to Judge E. M. Ross and Col. Thomas Ewing. These men, in turn, sold to the Vivian Mining Co., of Los Angeles, and the post office of Vivian was established early in 1904. In the next three years considerable gold and silver was mined in the vicinity and the town of Vivian grew to embrace a dozen stores, two banks, a hotel or two, a Chamber of Commerce, and a depot serving the Mohave & Milltown Railroad.

In 1909 the name of Vivian post office was changed to Oatman, probably after Olive Oatman, an emigrant girl kidnapped by Indians in 1851 and assertedly hidden for awhile at what came to be known as Ollie Oatman Spring, about half a mile north of the townsite. (According to one authority, however, the town may have been named for John Oatman, a prominent local mining man, who claimed to be a son of Ollie and a Mohave Indian.)

In either case the camp was named and began to grow into the sort of place that miners call "A good town to tie-to." Work was steady, wages were good. Between 1908 and 1932 the Bluebird Mine—now more commonly known as the Tom Reed—produced around $13,000,000 ac-

[2]*Mohave County Miner,* Sept. 15, 1900.

[3]Dr. G. M. Butler in *Engineering and Mining Journal,* Oct. 30, 1928.

Tom Reed Mine and Mill, Oatman, 1949

Honolulu Club, Oatman, 1953

cording to the Arizona Bureau of Mines, and was sunk to a depth of 1400 feet, making it the deepest mine in the district. Another Oatman mine, the United Eastern, between 1913 and 1926, produced nearly $15,000,000. Both mines were later operated by lessees, with varying degrees of financial success, until 1942 when working gold mines were closed by order of the War Production Board.

Oatman's period of greatest popularity and prosperity lay in the double-decade between 1914 and 1934, when the town's population approached the 10,000 mark, and its business district was one of the most extensive in Mohave County.

I am just a little proud to be able to say that I knew Oatman—briefly—during this period. It was in 1928, back in the Model-T days when the rocky, crooked, Oatman Grade, ascending to Sitgreaves Pass, was one of the most feared mountain crossings on the entire route of U.S. 66. (The only grade on Sixty-Six possibly more dreaded was La Bajada, between Santa Fe and Albuquerque.) Motorists descending from Sitgreaves to Oatman burned their brake linings, and motorists climbing the grade ordered passengers to get out and push. Eastbound travelers were advised to lay over at Oatman so that they might cross "the hump" in the cool of morning, with a cool engine.

If they heeded that advice they saw a busy, booming town—but not one designed by a city planning board. Its main business street, which did double duty as U.S. 66, followed the bottom of the twisting canyon. Bordering the unpaved street, on both sides, were wooden stores and shops and cafes, built cheek-to-cheek, rarely with a firewall between. In front of these buildings ran a wide wooden sidewalk, its outer edge supported by stilts, and its surface reached by splintered wooden steps. Many of the business houses had false fronts—the so-called "battlement" front—and wooden awnings. These provided a bit of shade where Old Men of the town could sit on pine benches and talk in the heat of afternoon, as they would have sat under shade trees in the park, if there had been any park, or any trees. Oatman didn't have many trees. Water was too much of a luxury to expend it on anything larger than a sunflower.

Branching away from the main street of the town, dusty side streets, some of them little more than footpaths, struggled up the canyon sides to the homes of miners and millmen. Most of these building sites commanded a wonderful view of the Tom Reed Mine and spreading tiers of tawny mill tailings.

Holidays, such as the Fourth of July and Labor Day, drew attendance from all over the country. Highlighting such days were the mucking and drilling contests; boxing and wrestling matches held in the pavilion on Main Street; tugs-o'-war, women's nail-driving contests, foot races for both sexes and all ages, and other homespun events enlivened by band music and assorted spirits, and always concluded with a big dance.

After my first experience with Oatman, in 1928, when everything was still going strong, I didn't get back there until after World War II.

Driving along the main street of this formerly-active town I saw one business house after another boarded and closed. Homes were vacant. The mines and mill were standing idle. Everywhere were visible the hallmarks of neglect and discouragement.

Parking in front of the Arizona Hotel I was starting up the street when a man appeared, as if on cue. He grinned at me—a nice, open, good-natured grin.

"Howdy!" said the man. "You lookin' for somebody?"

"Not exactly!" I stammered. "But, what happened here? Where is everyone?"

The man shrugged his shoulders. "Gone."

"Did the mines play out?"

"Play out, hell! The ore was still running $14 to the ton and there was plenty of it in sight. It was the damned WPB. Uncle Sam figured that closing all the gold mines would force the miners to go to work in copper and zinc mines and other industries more necessary in winning the war. But it didn't work that way. All it did was wreck the gold mining industry, without any gain to anybody.

"Miners streamed out of here like rats leaving a sinking ship," continued the man. "The first day after the mines closed, 60 men, some of them with families, left Oatman. Others left as soon as they could make arrangements.

"They really cut us off at the pockets . . ."

After we had visited awhile the fellow introduced himself as John Voynich. He said he had been mining in Arizona about 30 years but had been at Oatman only since 1940. He had a little money saved up, he said, and eventually would get Social Security, so he figured he would just stay on at Oatman.

"It's been a good town to live in," he said. "I reckon it'll be as good as any to die in . . ."

Re-visiting Oatman in 1949, and with Marion Johnson in 1951, I could notice little change; but my return there, in 1953, found the town completely demoralized. Scarcely any of the hillside dwellings appeared to be occupied, and not a man was visible on Main Street. Inquiry at the little fourth-class post office brought the good tidings that John Voynich was still a resident of the town.

"Oh, yes!" said the postmaster. "Old John and a few more of us are the rear-guard, just stickin' around to see that everybody else gets out safely!"

After John and I had greeted each other and set down on the edge of the sidewalk to visit, I asked him what had happened to bring such a drastic change in the town, so quickly?

"They took our road away from us, that's what!" said the old timer. I remembered, then, reading about the 46-mile cutoff from Topoc to Kingman by way of Yucca Station—a road that would by-pass Oatman and leave it without further excuse for existence.

"I didn't know it was to be finished so soon," I said.

John grinned wryly. "The damned contractor finished it ahead of schedule! One morning the traffic stopped coming through—like that!" he snapped his fingers. "We thought maybe there had been a bad slide or something. When we phoned to Kingman to ask about it, we learned the new highway had been opened. It sure knocked the props from under us. In the next few days nine service stations closed. All the businessmen got out except the post office and one small bar. Even the school wouldn't have opened this year except that a fellow with seven kids decided he could live at Oatman and drive to work cheaper than he could live at Kingman . . ."

Later, John Voynich and I walked down the empty street to the elegant Honolulu Club, a two-story building that was still in operation at the time of my last previous visit. Now it was closed and the interior was bare except for the miscellaneous litter that follows departure. Painted jungle scenes, supporting the club's name, still decorated its facade, and fading signs advertised liquors and beer, dine & dance, sandwiches deluxe, and set forth the fact that the place was air cooled. Next door, at the Shell Garage, workmen were moving out the last remaining equipment and loading it in a truck backed up to the entrance. In one of the dusty front windows of the garage were several ore samples.

"I'd like to have those specimens," I said. "I wonder if anyone would care?"

For answer, John Voynich stepped inside the building, collected all the samples and piled them in my arms. "These buildings have already been sold," he said. "They're going to be torn down and trucked to Las Vegas. Nobody cares about a few ore samples . . ."

We walked on past the drugstore where no clerks remained behind the counters; nor was there any goods on the shelves. The two-story Arizona Hotel and cafe also was vacant. The big glass-fronted office of Triumph Gold Mines, Inc., was empty except for some more ore samples in the window—these securely protected by a padlock on the door. Other business houses showed no more activity. Across the street, in front of the firehouse, the big old firebell still hung on its iron-pipe standard.

"Here's something you might like to see," said John, indicating a stout plank door with strap-iron hinges and hasp. The door swung from a cement slab used to block off a niche in the rock face of the cliff. "There was so much danger of fire here, and so little water to fight it, many of the merchants had these hillside vaults back of their place of business. Each night they would put their money in the vault to protect it from fire; or a jeweler would store his best watches and other items."

When Mr. Voynich and I completed our tour of Main Street and were back sitting on the sidewalk again, he called my attention to a group of jagged rock promontories that formed the skyline to the northeast.

"That largest one is what we call 'The Elephant's Tooth,'" he said. "In the evening, when a man was sitting on his porch after a hard day's work in the mine, there wasn't a grander sight anywhere in the world than to look off across town and see that Tooth in the last rays of the setting sun. It looked like it was painted with gold—brighter than any gold we had mined that day. The picture it makes at sunrise is just as beautiful, but entirely different. Then the Tooth is rough and black, like something a kid might have cut out of black paper—except that it looks terribly big and tall, and sort of mysterious.

"Yes," said John Voynich, nodding as if to himself, "If the time ever comes when I have to move away from Oatman and live somewhere else, I suspect I'm going to miss that old Elephant's Tooth more than all the rest of it . . ."

Devastation, Goldroad, 1951

"Life in a mining camp . . . is a strange life, unnatural, unwholesome, leading to no good . . . The worst of it is that the life is as fascinating as it is unwholesome. 'Once a miner, always a miner,' is a proverb which is little less than an extract of truth. The life is simply a gamester's life with the wide earth for a hazard table, and the instances are rare in which a person who has once come under its spell ever breaks away."

—*Atlantic Monthly, December, 1883.*

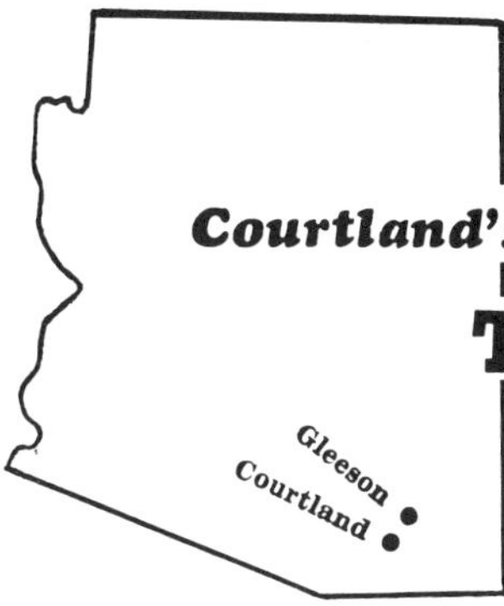

Courtland's Boom Was Copper Plated

THE BIRTH of Courtland preceded my own by several months and was attended by much greater fanfare; but I have had my revenge.

Courtland, today, is a ghost. I am not.

Almost the entire story of this Cochise County copper town—particularly the high hopes and exhilaration prevalent at the time of the camp's founding and throughout its youthful days—is faithfully set forth in the weekly news columns of the *Courtland Arizonan,* first issue of which went into the mails under date of Feb. 13, 1909, one month before Courtland's post office was officially established. It was a large, lively, good looking sheet, with black banners and skyline streaming across top and center of Page 1, and diminishing banks of black subheads, like miniature tornadic funnels, introducing each of its several main stories. As befitted a brand new mining camp where opportunity appeared to have taken a long-time lease on the guest room, practically the entire front page of the journal's Vol. 1 No. 1 was devoted to singing the praises of Courtland's several mines.

Heralded by a page-wide streamer, followed by three deep subheads bristling with exclamation points, appeared the story destined to excite mining circles from Wall Street to Waukegan.

"That the development work in the Germania shaft of the Calumet & Arizona, and by the Great Western Company, is responsible for the devilish haste with which the Southern Pacific is building south, and the El Paso & Southwestern north from Douglas, into the new camp, is not to be questioned," exulted the *Arizonan.* "It was learned Friday from an official of the Great Western Company that sufficient ore has been blocked out to produce from the Germania and Great Western shafts 2,000,000 pounds of copper monthly for the next four years!

" 'There has never been another camp in the history of Arizona which has developed so great a tonnage of high-grade ore in the same length of time,' said the official quoted. Since the station was cut on the Germania and Great Western, on August 1 last, over 8000 feet of development work has been done, better than 90 per cent of which runs through ore running 7½ per cent or better . . . At present there are four companies operating here—The Great Western, Calumet & Arizona, Copper Queen, and Leadville . . .

"Six weeks ago the present townsite was practically barren. Today, within the neighborhood of 2000 people are living in tents and shacks. . . . Many are doing a lively business in all branches of trade. There are restaurants, groceries, general merchandise stores, a barber shop, poolroom, hardware store, and a number of hotels where one has the rare privilege these formation days of jumping at the chance to get a good bed, over a dust floor, for from 75 cents to $1.50 a night. The short-order houses, located in the various townsites, are running at full capacity day and night. Assurance has been given that within six months . . . from 1500 to 2000 men will be put to work in the shafts of the Great Western and Calumet & Arizona properties . . .

"It is entirely within the bounds of reason to expect a population of 10,000 in Courtland in the next year or 18 months."

So new and unfledged was the town that this initial edition of the *Arizonan*, printed in the office of the *Miner*, at Bisbee, contained numerous advertisements for business concerns in Bisbee and Tombstone, but only ten from Courtland itself. These included the Southwestern Mercantile Company which had opened for business in Courtland less than two weeks before the newspaper made its own debut and now announced its ability to supply "hardware, lumber, fuel, oils, paints, and everything a hardware store should have." There was an ad for the Citizens Bank and Trust Company, whose Courtland branch had been opened January 22—"Capitalization $50,000." W. H. Gardner was advertising his Courtland Cafe, and Carter and Woods claimed to have the best short-order house in Courtland. John Rock, Tombstone dealer in gents', ladies' and childrens' furnishings, wished it known that he had opened a branch in this new copper town. There also were ads for Courtland Meat Market, George Land, Broker; Tester and Hicks, Brokers ("Copper Stocks Our Specialty;") and J. A. Hatley, contractor and builder.

In addition to these modest bids for trade, all of Page 6 was occupied by an advertisement for J. N. McFate's Courtland Townsite.

The pulling power of that highly optimistic announcement was greater than anyone had imagined it would be.

First day sale of building lots in the McFate Townsite aggregated over $60,000, stated the *Arizonan* in its issue of February 28. The line of prospective purchasers had begun forming in the late afternoon previous to the sale (scheduled for 11 o'clock the next morning). The cold, windy darkness of that February midnight had found many persons shivering in line, including a number of women, and by breakfast time, next morning, 146 lot seekers were standing in line.

"For hours before the opening of the sale, speculators with active partners on the outside worked energetically to sell their places to those who were eager to buy," stated the *Arizonan*. "Quite a number places in the line were sold for prices exceeding $100, and wild rumors went up and down the line of as high as $500 and $1000 being paid for places near the head of the division . . ."

"Over 1000 men are on the payrolls of Courtland today and the building of a city has barely begun," reported the local news medium in its next issue following the land sale. "These 1000 men are drawing on an average of $5.00 a day . . . Within a radius of six miles from Courtland are Southern Pacific and El Paso & Southwestern Railroad camps employing at least 700 more men . . ."

On March 14, 1909, readers of the *Arizonan* learned that President Taft had, that week, formally appointed Harry Locke as Courtland's first postmaster. "The appointment was made directly by the president, an action unprecedented, as appointments have always been made through the head of the Postal Department," exulted the boomtown journal, as if that VIP treatment made the Courtland post office something extra special. Tucked into the same issue was a news item to the effect that notice had been served "on the two-score or more women who occupy the cribs adjoining the dancehall . . ."

During the first 14 weeks of its journalistic career the *Arizonan* was put together in the printshop of the *Bisbee Miner*. Along in mid-April the editor of the *Arizonan* proudly announced that a fine new structure for its eventual use was nearing completion, and that a freight carload of newspaper and job office equipment had been purchased from the American Type Founders in Los Angeles, that week.

"The car is due in Pearce, Sunday, and shipments will begin from that point just as soon as the building is ready for the machinery," it was stated. "The news and job plants will be superior to those of any printing plant in Cochise County, and rivalled only by the big dailies of Tucson, Phoenix and Prescott." The *Arizonan*, it was noted, had already filed its bond for an Associated Press report, and daily publication was foreseen.

"The Courtland *Arizonan* has the honor of having raised the first American flag over Courtland," boasted the editor on April 25. "Old Glory was floated to the breeze over the new home of the *Arizonan* on Third Street, in the upper end of the main town, Saturday morning. Across the street, in front of the schoolhouse, the members of the Women's Club of Courtland stood reverently while the flag was being raised by business manager W. A. Sherwood, assisted by Supt. Ted Healey of the mechanical department, and sang *The Star Spangled Banner*. The flag was furnished for the occasion by Mrs. Lee Hunt, the patriotic proprietor of the Germania Hotel . . . It is prized by the Hunt family far beyond material worth, for it was one of the keepsakes left behind by a little daughter who died some years ago in Colorado. The precious patriotic flag will wave until the arrival of an eight-foot banner ordered by Manager Sherwood for the *Arizonan* building."

In the same issue appeared a lengthy story concerning construction progress of the Courtland Hotel, on which 15 carpenters and numerous other artisans were then employed. "The building is to have a foundation of stone and concrete, with frame superstructure. There will be 28 large rooms. The entire building is to be designed along the most approved modern lines and will be a strictly first class and up-to-date hotel . . . It will be in the center of the city, almost next door to the Southwestern depot . . ."

Another of the lead stories in this issue of April 25, 1909, concerned "Clovis Day" at the Courtland Townsite.

"The New Mexico town was well represented in Courtland and over $25,000 passed hands in realty transactions," it was set forth. "Dr. Bailey, owner of the Courtland Drug Store, invested in Great Western lots at the opening of the townsite some months past. He is the original Clovis man on the ground . . . The influx of Clovis capital is the result of his visit to that city since his purchase . . ."

On May 9 the *Arizonan* announced that the Great Western Hotel would be completed May 15. "The interior finishing and decorating of the hotel will be the best that money can obtain or skill devise (with) the wood in Flemish Oak and Mission. Electric fixtures and electric illumination will be used throughout."

By the end of June, that first year, two railroads had built into Courtland—the Mexico & Colorado Branch of the El Paso & Southwestern, and the Arizona & Colorado Branch of the Southern Pacific—and trains were arriving and departing daily. Before the year's close there had been organized at Courtland a lively Chamber of Commerce and Mines, with 36 charter members, and the Courtland Water and Ice Company was laying five miles of mains to assure the town "a liberal supply of good water."[1]

In the years that followed, Courtland had its recurring revivals and slumps. Outbreak of World War I in Europe sent the camp into a mild upswing (a copper camp has to be on its last legs, indeed, if it fails to respond to the stimulus of a major war).

"Once again Courtland is on the upgrade," declared the loyal and persistent *Arizonan* on February 5, 1916. "Since the first of the year the number of men engaged in and about the camp has nearly doubled . . . Mrs. Brown at the post office reports that some 20 lock boxes have been rented . . . (and) Mrs. R. G. McBride, of Webb, was in Courtland to answer inquiries relative to the renting of her Turquoise Hotel that has been vacant two years . . ."

It was this period of renewed activity that brought to Courtland a 19-year-old young woman, Viva J. Johnson (now Mrs. Viva Taylor, of Pomona, Calif.). Actually, it was a return visit for Miss Johnson who, in

[1]*Courtland Arizonan,* Feb. 12, 1910.

1904, had come West to Arizona Territory with her father and mother, Mr. and Mrs. Isaac F. Johnson, and their three other children.

The Johnsons left the train at Cochise where they were met by Viva's uncle, Daniel W. Brown, who had come for them with a team and wagon.

"We children rode on blankets in the back of the wagon," recalled Mrs. Taylor. "I was eight years old. I still remember the mystery of the mirage we saw between Cochise and Courtland. Actually, there was no Courtland at that time, and the only house on the site where the town later would be was the one my Uncle Dan had built. A few Mexicans lived in tents and worked around his mine. Soon we also had two tents—one for our kitchen, the other for sleeping quarters. All supplies were hauled in from Tombstone. We children made playhouses in abandoned mine tunnels, got lost climbing Brown's Peak, dodged rattlesnakes, and had an all around wonderful time! That fall, when school opened, we moved to Tombstone, later to Phoenix, and I did not return to Courtland until October, 1915.

"During this interval Uncle Dan had persuaded his three brothers Mate, Charlie, and Will, to join him in his mining enterprises. Together with Len Shattuck, of Bisbee, they had organized the Great Western Mining Company and Courtland had begun to boom.

"Uncle Charlie was a widower with one son, Arthur. One day, when Arthur was about eight years old, he and his father went to Tombstone, with team and wagon, for supplies. As they planned to shoot some rabbits on the way home, Uncle Charlie's shotgun was leaning against the wagon seat between them. As they rode along, Uncle Charlie playfully began tickling Arthur in the ribs; Arthur accidentally kicked the trigger of the shotgun, and the entire blast entered his father's chest. Little Arthur snatched the reins and lashed the tired horses into a gallop, but before they reached Courtland Uncle Charlie had bled to death . . .

"On November 2, 1915, we received word that Pancho Villa and his troops were threatening Douglas," recalled Mrs. Taylor. "Uncle Dan's son and his family lived there, so Uncle Dan sent his nephew to Douglas to bring them to Courtland, away from danger. That night we climbed a small rise and watched the firing at Douglas, as General Pershing drove Villa's troops back into Mexico.

"But these two incidents were exceptions. For me, a 19-year-old girl, life in Courtland was mostly a round of parties and dances. Until my uncle bought his first Motel T Ford car, we went to such affairs on horseback and in hayracks.

"In March, 1916, the Guggenheim Mining interests opened a mine near Leadville, just over the hill north of Courtland, under the name of Needles Mining Company. I got a job with them as secretary and bookkeeper. As there wasn't enough office work to keep me busy, I persuaded the assayer to teach me his trade. I used to go down in the mine to get samples of ore to assay, and because I found blueprints interesting

Post Office, Courtland, 1914 —*Courtesy Viva L. Ta*

Western Hotel, Courtland, 1914 —*Courtesy Viva L. Tay*

the mine surveyor taught me how to draw them for him. In February, 1917, the Needles Company shut down their mines—out of ore. I took a job as court reporter for Judge Bolton, at Courtland. I remember, especially, one trial for claim jumping. Most of the testimony was in Spanish. I worked six hours taking notes and many more hours transcribing them. My fee was $50 . . .

"One day a cowboy in the valley was thrown from his horse and killed. His widow, Eva Shearer, was left with two small girls to support. As there were no good restaurants in Courtland at this time, and many of the miners were single men, Eva suggested that she and I start a boarding house. She would do the cooking; I could set the tables, wash dishes, etc. We were to divide the profits . . . but there never seemed to be any.

"When Eva got her husband's life insurance money, about $1000, she bought a new Maxwell automobile. One day it was parked in front of her house which was built on the side of the hill above the road leading to North Courtland. Her two little girls, then aged five and seven, got in the car to play. I was sweeping the front porch when I saw the car start to roll down hill! Running down the 12 steps to the street, I managed to jump onto the running board and grabbed the emergency brake just as the front wheels went over the bank. Jumping free from the car, I watched in terror as it traveled all the way to the bottom of the hill—miraculously without turning over! Except for two terribly frightened little girls, there was no damage . . .

"After a few months a beer parlor in North Courtland began serving meals and Eva and I lost most of our customers. When the Needles Mining Company reopened their mine in September 1917, I went back to work there at a salary of $75 a month, and everything looked rosy again . . ."

But the rosy glow was only skin deep. Despite the incontestable enterprise of her people, the enthusiasm of her newspapers, and the original promise of her mines, Courtland was not destined to take her place with Butte and Bisbee and Bingham, and other Western centers made rich and famous by copper.

"Courtland looked so good that the Phelps Dodge Company and the Calumet & Arizona almost fought for claims," reported Ira B. Joralemon in 1934. "The Calumet & Arizona wasted a few hundred thousand dollars on the Bill Holmes Leadville claims and then found a continuation of the Mary ore in the Germania. The upper levels of the Mary and Germania had as sweet a high-grade orebody as an engineer ever dreamed of. The Southern Pacific and the El Paso & Southwestern rushed in to grade branch lines, each trying to block the other from entering the wonderful new district. Three townsites sprang up to house the thousands who joined the rush . . . The saloons and dancehalls were like the good old days in Tombstone and Bisbee.

"One brief flash of greatness and it was all over . . . As soon as the Mary and Germania sank 300 feet they . . . encountered the hardest and barrenest limestone that ever dulled a drill. The orebodies lasted only a few brief years. Young's Great Western Mining Company made a profit of a few hundred dollars out of the Mary, and lost it all hunting for more ore. The Calumet & Arizona got back most of the money it spent on the Germania. The rest of the investments in Courtland were a complete loss . . ."[2]

As a consequence, even the loyal and enduring *Arizonan* threw in the towel with its issue of December 25, 1920. The Courtland post office struggled manfully until 1942 when it finally gave up the ghost; and Courtland, after 30-odd years of ups and downs, dwindled to a ghost town with one or two die-hard residents.

Meanwhile, four miles to the south of Courtland had flowered—and was likewise fading—the town of Gleeson, named for John Gleeson, an Irish emigrant, who had come from Colorado to the mining boom at Pearce in the 1890s. Gleeson was located on the site of, or near, the still older mining camp of Turquoise, which was a lively center in 1890 when its five hotels and three saloons were doing a very satisfactory volume of business.[3]

No one knows who may have been the original miner at Turquoise. Long before the Spanish explorers appeared in this vicinity, Indians were mining the cherished blue gemstone from the desert hill that would become known to white men as Turquoise Mountain. Later, Tiffany & Company, of New York, acquired ownership of these same mines and the camp of Turquoise was established. Its life—the life of its post office, at least—was a brief one: from October 1890, to September 1894. Six years later the moribund town was reborn as Gleeson, and a new post office established under that name on October 15, 1900.

During the first nine years of her life, Gleeson failed to distinguish herself; but in 1909, with all the whoop-de-doo at neighboring Courtland it was impossible that some of the excitement should not rub off on the neighboring town.

"Not since the institution of this old mining camp . . . has Gleeson enjoyed the promise of future greatness which is hers now," declared the *Courtland Arizonan*, March 7, 1909. "The railroad builders are abroad in the land . . . the big Herriman line, pushing fast down the valley toward Douglas, is already at work on its Gleeson Spur . . . and has all the men and teams available working early and late . . . With two railroads to carry the product of the Gleeson mines only 35 miles to the Douglas smelters, the interests here are arguing that the future of Gleeson is assured beyond a doubt. That Courtland and Gleeson will eventually be connected by a trolley line . . . is the belief of people here . . . Progress is in the air! People are listening for the assured whistle of the locomotive, and things do move."

[2]Ira B. Joralemon, *Romantic Copper; Its Lure and Lore* (N.Y.: Appleton-Century Co., 1934).

[3]*Graham County Bulletin,* September 12, 1890.

"For 25 years[4] ores have been freighted out of Gleeson to Pearce and Cochise," stated the *Arizonan,* June 20, 1909. "Its mines were kept in operation when camps with twice the notoriety were idle, because of the valuable smelting properties of the Gleeson sulphides. More ore has been mined at Gleeson, to the amount of development in the mines, than in any camp in the world. The whole mountainside opposite the new railroad townsite is a blow-out of copper . . . The two railroads, the terminals, the agricultural and mining environment, and the mines themselves, with prospects for the establishment of the county seat, point to a population of 5000 to 10,000 people in five years . . ."

But Gleeson never got the courthouse, or its street trolley to Courtland, and the envisioned "5000 or 10,000 people" fell several thousand short of that mark. The situation wasn't helped by the fact that the town, on several occasions, was virtually destroyed by fire.

Of these blazes, none was more devastating than the conflagration that hit Gleeson in June 1912, when 28 buildings and their contents went up in smoke. One of the town's leading businessmen, B. A. Taylor, announced himself loser to the extent of $85,000 with only $10,000 insurance coverage. (This, incidentally, was the only insurance carried on the section of the camp that was destroyed.)

Starting in one of the Taylor warehouses, the fire had spread rapidly, taking every building on both sides of the street for a considerable distance. Taylor's first efforts were directed toward the removal of two drums of gasoline and a quantity of powder stored in one of his warehouses. Otherwise, all that could be saved from the 17 buildings Taylor lost in the fire, including seven warehouses, were the cash register and account books from one store, an automobile, and the moving picture machine and roller skates from the amusement hall. Included in the stock destroyed was a carload of coal and one and a half carloads of flour which were still burning briskly nearly 24 hours later.

"Gleeson had no fire-fighting facilities other than by means of a bucket brigade . . . for though there was a water main through the street, recently laid by Mr. Van Meter, the preparations have not been completed for the handling of water and the supply was limited," reported the *Courtland Arizonan,* June 8, 1912. "Mr. Van Meter started to pump as soon as he realized the situation, helping greatly in supplying the buckets by means of which several buildings were saved. Dynamite was also used to remove some buildings, but they flamed up almost immediately after being wrecked."

It was typical of mining camp courage that both Taylor and Van Meter announced their intention to rebuild at once, using concrete, "which," commented *The Arizonan,* "will result in a much improved appearance in the business portion of the main camp."

Fortunes of Gleeson paralleled those of Courtland for a few good years, but Gleeson was to prove the more enduring of the two. World

[4]In using this figure the *Arizonan* was either in error, or was including shipments made from the earlier town of Turquoise.

War I found her mines still producing copper, and with America's entry in that international holocaust, the little town on the mountain put forth a reasonable facsimile of a boom.

It was that year of 1917 that brought to Gleeson Miss Ethel Prosser, who had joined the camp's schoolteaching staff of four young women.

Though I never met Miss Prosser I feel as if I had known her personally. She first contacted me in the summer of 1960 and for several months we enjoyed an exchange of correspondence concerning her recollections of life in the Gleeson of 43 years earlier. It was always a joy to receive one of her long letters, flawlessly typed, with every name and initial and date complete, and every page sparkling with enthusiasm and a wonderful zest for living.

"Gleeson was quite a busy town when I went there in 1917, as the United States had entered World War I and copper was very much in demand," wrote Miss Prosser in her first letter to me. "The resultant increase in population demanded that a new four-room school be built as quickly as possible. While it was in construction—we moved into it in the early spring of 1918—three teachers taught in the cramped quarters of the old school, while I held forth with 35 pupils of the 4th, 5th, and 6th grades, in an old store building on Main Street.

"We four teachers lived in what was called 'The Teacherage'—a very small three-room house with the usual Chic Sale in the back yard. This Teacherage, we were told, had been the original Gleeson house. It was so tiny that the woman principal asked the school board to give her sleeping quarters at the mine hospital, but she continued to share meals with us at The Teacherage.

"The highlight of that 1917-18 school year was our tin can gathering campaign. The tin, it seemed, was needed in the war effort. Every recess the youngsters were turned loose to search through the washes, back yards, and along the streets, to collect the tossed-away cans of this small community where garbage pick-up and service disposal units were non-existent.

"By close of the campaign we had gathered from the Gleeson environs the astounding total of *12 tons of cans!* These were hauled to the railroad station in a Studebaker wagon loaned for the purpose by one of the town families, and were loaded into a freight car provided by the mine company. Culmination of such a highly successful drive naturally called for a parade, so every child in the school marched sturdily up Main Street and down to the railroad station, each carrying one last can on the end of a stick, hobo fashion. Bringing up the rear of the parade was the Studebaker wagon with our school's reed organ setting in the bed, with several big boys to hold it steady, and with me playing the campaign songs, and all the children singing them lustily. (We had written parodies on a number of the then currently-popular war songs such as *Tipperary*, and *Over There*, all our parodies being to the effect that Gleeson's tin cans were on their way to the Bisbee can pile.) As we jolted down the rough street, with me pumping away at the organ, I felt like the calliope player in a circus!

"Our campaign netted $180 which was used for the next year's Junior Red Cross activities. We felt justified, however, in spending enough of this money to give the hard-working youngsters an ice cream feed—really a rare treat for them.

"In the early decades of this century, every mining town had its teachers who, like me, were seeking experience which would lead eventually, to better jobs. Believe me, we got it in some pretty rugged forms, for the 'call to duty' came, sometimes, in most unusual ways.

"In the interest of cleanliness and health we teachers were asked by the school board to issue orders, in both Spanish and English, for home shampoos, after which we were to inspect our urchins for head lice.

"Armed with fine-tooth combs and basins holding a solution of vinegar and kerosene, we lined up the lower-grade pupils (the older ones were subjected to only an urgent appeal to their personal sense of pride) and grimly inspected heads. The children, and mothers, too, had been warned in advance that any head which didn't pass inspection would be shaved clean to the skin. Before the day ended there were several shining bald pates for the unfortunate ones whose home treatment had failed. The youngsters gave a somewhat amused cooperation and I'm afraid that a bald head was considered more or less a mark of distinction, which was worn above a broad grin!

"Tactful discretion won a reprieve for one little Mexican boy with long, black braids. It seemed that his mother-to-be, while carrying the child, had been so desirous of a son that she had made a vow to the Virgin Mary that if a male child were sent her she would never cut his hair until he was a grown man. The little fellow didn't pass inspection but out of respect for the mother's religious vow she was encouraged to 'try again.' She was known to be a very good mother who was doing the best she could to rear her family properly.

"Possibly I shouldn't have even mentioned such a garish creature as a louse . . . but I'm in good company with poet Robert Burns and his vivid picture of that 'blasted' creature crawling on the lady's hat, in church . . ."

After our delightful exchange of letters during the summer of 1960, I wrote Miss Prosser that I was leaving to spend the winter in Mexico and hoped we might resume our correspondence upon my return in the spring. But I never heard from her again. I eventually learned that she had passed away April 7, 1961. The word came to me from her close friend and one-time roommate, Dorothy C. Smith, of La Mesa, California.

"I know how thoroughly delighted Ethel was in the correspondence which had grown up between you during the past year or so, giving her the opportunity to renew her recollections and share her vivid memories of teaching experiences 40-odd years ago in Arizona . . . It was one of the most stimulating features in her recent years, curtailed as her activities have been by crippling arthritis, to have this little association with historical research. You must have come to realize through her letters, painfully written though they were, something of the zestful spirit which she possessed. On her behalf, and that of her brother, I wish to express appreciation for the delightful contacts that you renewed for her . . ."

May I say, somewhat tritely, the pleasure was all mine . . .

The first I heard of Gleeson, I believe, was in 1933 when Paramount Pictures was in Arizona filming *The Mysterious Rider,* an adaptation of a Zane Grey novel. For the story's climaxing sequence Producer Harry Sherman settled upon the almost-ghost town of Gleeson as an ideal place to double for the frontier settlement in which the action supposedly was laid. The old Liberty Hotel in Gleeson was transformed into what appeared to be a two-story structure; false fronts were built on two blocks of adobe houses. Hitching racks were set up, cacti and other desert growths planted in the yards, and a company of 75 Paramount actors, actresses, technicians, and cameramen, plus 125 extras from Tucson and Tombstone, made a jolting 90-mile trip over the rough roads of that day to the old copper town on the mountain. Despite the general fakery of the stage props, the Gleeson jail was filmed exactly as its stood without any face-lifting whatever. Producer Sherman declared it "perfect."

By 1940 Gleeson was very much a ghost town. Her fourth class post office had been discontinued the year before; roofs of many buildings were falling in, and most of the structures along her once-busy Main Street stood silent and empty. Even the fine, modern school, where Edith Prosser and her three colleagues had taught in 1918, was showing the blight of the years, and it would have been possible to fire a shotgun down almost any street in the town without hitting a living creature, man or beast.

Courtland, too, was on its last legs and sinking fast. A visitor there in 1942 described it as "an old camp, with just a few weather-worn houses set down between bleak hills. Scores of mine and prospect holes, like bleary eyes, peered at us out of the mists of a productive past—and beyond, into the uncertainty of an empty future . . ."[5]

Fourteen years later I read in *Sunset Magazine*: "Though you may be tempted to stop for a look around the ghost town of Courtland, we don't recommend the social climate. Our reporter found the only inhabitant to be a blind, gun-toting hermit, named Yokum, who didn't approve of sightseers . . ."

Until reading that brief paragraph I had cherished no particular desire to re-visit either Courtland or Gleeson. I had seen both towns a couple of times and couldn't remember that I had lost anything in either. The perversity of human nature, however, is a peculiar thing. Soon as I read that warning in *Sunset Magazine* I developed an overwhelming desire to return to the old towns on the mountain! Just knowing there

[5]Bertha G. Brown, "Fairy Crystals from an Old Mine Dump, *Desert Magazine,* September 1942.

was someone who might object to my presence there made all the difference in the world.

At this time I was bending all my efforts toward completing my book, *Sovereigns of the Sage*, and I didn't get back to Cochise County until the fall of 1959 when Mom and I were spending a leisurely month hunting ghost towns in the Arizona back country. We had camped one night at Apache Pass and, next morning, were planning to head down into the Huachuchas.

"Instead of going around by the paved highway, let's cut over the Dragoons through Courtland and Gleeson," I suggested. "I'd like to have a try at 'softening up' that hermit who objects to visitors."

Mom said I'd better leave the hermit alone; but I knew, of course, that she didn't mean a word of it.

We halted for a look around the little mining town of Dos Cabezas, then rambled down the Sulphur Springs Valley, through Kansas Settlement, where the fall lettuce harvest was in full swing, and got to Pearce in time to eat our noon lunch beside the tumbled-down office of the old Commonwealth Mine.

Leaving Pearce our road began climbing gradually through rolling hills and well-grassed plains, and after about 11 miles we were seeing ahead of us gray mine dumps and tumbled terrain. A few old foundations and ruins made their appearance, and it wasn't long until a big two-story building came into view. It looked as if its bottom half once might have held a couple of stores, with maybe a hotel overhead, and there was a faded barber pole beside the door. The building looked as if it were completely deserted. Its shutters were hanging awry, windows were broken, some of the walls were partly fallen, and rubble lay everywhere.

Turning the car off the road, I was about to park in the lee of the old ruin when I became conscious of a great clutter of crudely lettered signs and placards. Pressing my foot irresolutely on the brake pedal, I peered through our dusty windshield to read such warnings as No Parking—Keep Out—This is Not Snooperville—Snoopers Not Welcome—Trespassers Will be Prosecuted—Rockhounds and Tourists Not Welcome—This is Not a State Highway or County Road . . .

Even the warning paragraph in *Sunset* hadn't prepared us for any reception such as this and we didn't know what to make of it. It just didn't figure! When Mom and I stop at an out-of-the-way place in the desert or mountains, both clad decently and conducting ourselves in a manner thoroughly ladylike, with no axes to grind, nothing to sell, no ulterior motives, offering nothing but brotherly love and goodwill toward all men and even the beasts of the field and the snakes in their burrows, it is fundamentally impossible for us to imagine ourselves not welcome. Even here I couldn't believe the signs actually meant what they said.

Easing my foot from the brake, I slowly coasted the car into parking position in front of the big old building. Switching off the ignition, I set the emergency brake and slowly raised my left hand to the doorlatch. At the same moment I felt Mom nudge me on the leg and saw her give an almost imperceptible nod toward the shadowy outline of a man half hidden within the dusky interior of the building.

"Hi!" called Mom cheerily. "Is anybody home?"

After a full minute the sagging screen door began to push open, slowly and creakingly. For a moment an old man stood framed in the doorway; then he was shuffling across the splintered threshold and the screen door had closed wearily behind him. Halting halfway across the broken porch, the man stood regarding us hostilely. He was thin to the point of emaciation; his trousers were threadbare, his shirt torn. An old gray felt hat was pushed back from his forehead and one knotted hand held a folded newspaper.

"Who are you?" It was not so much a question as a challenge.

"We were coming up this way so we thought we would drop in and say 'hello' and bring you some fresh fruit," said Mom brightly, at the same time reaching over into the back seat for a sack of peaches and oranges we'd bought the previous day at Willcox. "This is a nice old building you have here," she remarked, as she extended the sack of fruit in the man's direction. "Was it a hotel?"

The man made no reply, nor did he make any effort to take the fruit from Mom's outstretched hand. Then, without further introduction or provocation, he remarked, "I been readin' an article in this newspaper some Tucson folks left here last Sunday. I cain't read much. Blind in one eye and got a bad cataract on the other, so it's taken me three days to puzzle-out this one story. It's about the credit unions. Them things are gonna take over the country. Here,"—shoving the paper through the open window of the car toward Mom—"you read it!"

"I can't see much better than you," said Mom.

"Then *you* read it!" commanded Courtland's hermit, pushing the paper over to me.

I read the indicated article aloud and tried to exhibit a polite interest —which was more than I felt, because I have no war with the credit unions. As a matter of fact I scarcely would know one from a hole in the ground, and if they'll keep out of my hair I'll certainly keep out of theirs.

"See what I mean?" said The Hermit. "They're gonna take over the country!"

For the next quarter hour Mom and I tried desperately to shift the conversation to Courtland, Pearce, the possibility of rain, and how far was it to Tombstone? But the old man would have none of it. All he wanted to talk about was credit unions.

Finally, I said casually, "While you and Mom are visiting, I guess you don't care if I take a picture of your building, do you?"

The man whirled on me like a maniac. "No, by God! Nobody's going to take a picture of this building!"

"But *why?*" I said. "What harm would it do?"

"Some woman took a pitcher of it once and put it in th' paper, and called Courtland a ghost town. And then city folks began comin' up here to see th' place. Well, I won't have it, I tell you! I won't have snoopers comin' around here pickin' up rocks an' everything that ain't nailed down!"

"Why do you care about the rocks?" asked Mom. "You've certainly got lots of them around here!"

"Yeah!" growled the Hermit of Courtland. "And if one of them snoopers falls and hurts hisself, he can sue me for $10,000!"

"Tell them they can pick rocks, but you won't be responsible if they get hurt," said Mom. "Make them sign a waiver, or something . . ."

"Yeah . . . and then they'll take rocks when I ain't lookin' and I'll fall in one of th' holes they leave, and *I'll* get hurt!" muttered the old man.

"But I don't want any rocks," I pleaded. "Jeepers! All I want is just one picture of this old building—and then we'll go away and never bother you again. I have pictures of old mining camps all over the West, and I certainly want a camp as famous as Courtland to be included with the others. It's just a hobby of mine," I explained. "I'll give you my word I won't allow the picture to be published in any book or magazine . . ."

When the old hermit made no reply I hopefully interpreted his silence for consent and again started to ease the car door open.

"Don't you get out of that car!" screamed the man, shrilly. With a swiftness almost unbelievable in the face of his apparent decrepitude, he leaped to a pile of stovewood on the porch, snatched a piece about as long and hefty as a baseball bat and raised it menacingly to shoulder height.

"I won't have you taking pictures of this building! I won't have it, I tell you!" Shaking the weapon menacingly, he took a couple of quick steps toward my side of the car. "You get out of here!" he screamed. "Get out!"

I have always held there are certain times in the lives of men—and roving reporters—when it is advisable to run up the yellow flag of cowardice. This, in my considered judgment, was such a time.

Mom said afterward she had not known it was possible to get a car started and under way so quickly.

With the engine of the old Mercury at full throttle, we roared on up the hill, past the other ruins of North Courtland, never slackening speed and scarcely looking to right or left until we topped the divide and dropped down to the one-time suburb of South Courtland. Since it seemed likely that this was out of the old man's jurisdiction, we stopped here long enough to snap pictures of some old cement buildings, then went on to Gleeson, three miles south. Here we found ruins enough to delight the heart of any ghost town prowler.

Only a big, impressive arch, some half-fallen walls and tons of cement and-stone rubble, marked the site of the fine "new" schoolhouse built in 1917. A long adobe building, divided into several rooms, looked as if it may have been a rooming house; and a square-fronted building, bearing the sign, *General Merchandise,* was padlocked and clearly inoperative. The windows and doors of some dwellings were boarded shut; others gaped open to the wind and dust.

In the north part of town we found a rather nice looking adobe house, with a smaller adobe on the rear of the same lot. A sign in the front yard read: "This House For Sale, Furnished. Two Lots, 50 by 120 Ft. $600. Squatter's Title." I tried to convince Mom we should buy the place, but she said she wouldn't live this close to the Hermit of Courtland.

On a hill that overlooked the townsite from a mile or two distant, we could see the remains of the old mines which once had held forth such promise and, indeed, had operated profitably so many years. Sweeping the hillside with our binoculars we could see that the dumps bore the medley of color that so frequently tints the residue of copper mines—bluish and yellow, white, and gray. Machinery was visible around some of the mines, but the glasses revealed no actual activity in progress.

Before leaving town we stopped to pay our respects to the cemetery, a rather unkempt little burying ground, with the usual fallen pickets, crosses and headboards leaning awry, rusty old fences, and graves heaped high with desert stones—a typical ghost town graveyard.

It was nearly sundown when we reached Tombstone. As we had visited this interesting town many times in the past and would be there again, a few days later, we stopped only long enough to buy a pound of ground beef and a loaf of fresh bread. After driving out of town several miles on the road to Charleston we turned off into a remembered grove of mesquites and made camp. In jig time we had our sleeping bags unrolled, our small campfire was burning brightly, and the ground beef—seasoned with chopped onion—was sizzling in the skillet.

Supper over, I barely had time to type a page of notes before the soft desert darkness closed in upon our world.

In 1959 this impressive arch still marked site of Gleeson's "new" school

"Hanging on the wall was a large map of Calabazas. Upon it was depicted a flourishing city. Great hotels towered up from the business thoroughfares . . . Church steeples pierced the air . . . Handsome private residences graced the streets . . . Palatial public edifices, and monuments, surrounded by parks, occupied squares . . . and double rows of black lines indicated street railways . . ."

—JAMES CABELL BROWN, *Calabazas, or Amusing Recollection of an Arizona City, 1892.*

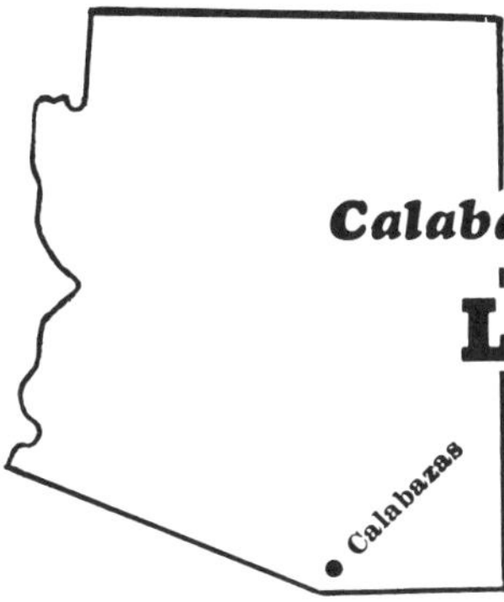

Calabazas, City of the Pumpkins

LOST GHOST TOWNS are simple to classify. Like billiard balls plunking into their pockets, they slip into their prescribed slots—mining or milling town, farming settlement, shipping port, stage station, army camp. But a few towns are neither fish nor fowl. They can't be classified because there is no slot to fit them. They are a little of everything, and not much of anything at all.

One of these mavericks is Calabazas in the pleasant, green valley of the Santa Cruz River.

Settled two centuries ago, the townsite of Calabazas is, today, one of the oldest in Arizona. Its ghosts speak Indian, Latin, Castilian Spanish, Mexican, and Anglo-American—and at least a few of them speak German. They wear the sombre robes of Jesuit priests and chant the Mass; they speak in the dulcet voices of dark-eyed señoritas, and blister the air with the salty jargon of army muleskinners. Though Calabazas, as a town, was neither fish nor fowl, she was a lot of other things—Indian ranchería, Spanish mission settlement, the site of army garrisons manned, respectively, by both Mexico and Uncle Sam, a mining town, farming settlement and, last of all, the focal point of a terrific railroad promotion deal that, like a designing gold digger, promised everything and gave nothing.

And now, Calabazas can't even be called a memory, since probably no man now living has a personal recollection of those days when she was an up-and-coming town, a budding metropolis . . . .

No one knows, with certainty, when the later-to-be site of Calabazas first was settled. Father Kino, arriving in Arizona in the late 1600s, found already established here an Indian ranchería which is indicated on the Kino map of 1701 as San Gaetan. About this same time the place also was being called San Cayetano de Calabazas and, as such, functioned throughout the later 1700s as a *visita* of Guevavi mission. In 1807 the King of Spain patented the townsite to the Indian people resident there. Several years later, for some reason, the site was abandoned and

Calabazas became a ghost town. Followed, then, the overthrow of Spanish domination and the birth of Mexico as a free nation, and in 1844 the National Treasury of Mexico sold the Calabazas property at public auction to Don Francisco Aguilar. The site was thereupon developed as a fine ranch, but in 1851 Commissioner Richard Bartlett of the Boundary Survey party found the hacienda deserted and in ruins. With the Gadsden Purchase, of 1853, the townsite came into possession of the United States and on August 11, 1855, *San Francisco Weekly Chronicle* published a brief description of Calabazas and its valley, as translated from the *German Democrat.*

"The valley is about 36 miles long and, like all the valleys in that vicinity, is rich in soil and timber," read this report. "It contains a number of ruins of deserted ranches. Herds of sheep, horses, mules, and horned cattle, are to be seen grazing in the open timber and amidst the dense brushwood. The chief building at Calabasas[1] is a citadel with walls about 20 feet high, unbroken except by a solitary doorway provided with very heavy doors. A bell is placed near it for the purpose of giving the alarm when hostile Indians are seen approaching. The place is now occupied by a couple of Germans, with some Mexicans and some tame Apaches in the neighborhood. These settlers have to be continually on the lookout for the wild Apaches who frequently attempt to drive off the stock. If the valley were protected by the presence of a detachment of American soldiers, the land near Calabasas would soon be well-tilled; besides, there are rich copper and silver mines in the vicinity which ought to be wrought, and will be, at no distant time . . ."

Within the next year following that publication, Calabazas became headquarters of a contingent of Uncle Sam's First Dragoons who, with Major Enoch Steen in command, took up residence in the old Calabazas hacienda.

For two decades thereafter the fortunes of the town vacillated like changes of the moon. Soldiers came, served out their tour of duty, and were replaced by others. Predatory Indians raided the valley herds and killed its settlers, or were themselves slain. At rare intervals a baby was added to the town's population, or a new adobe building was erected. But for every child born, each new wall laid, two oldsters cashed in their chips, and two walls crumbled back to the earth whence they were formed.

How long this program of progressive retrogression might have continued before Calabazas would have vanished in the dust of oblivion is hard to say, since an event of July 24, 1877, had the effect of changing the picture completely.

When Don Manuel Maria Gandara, ex-governor of Sonora, on that day sold his Calabazas land grant to Colonel C. P. Sykes the stage was set for a gigantic promotion scheme that would fill the old veins of the

[1]The spelling of Calabazas is a wayward thing, the 7th letter vacillating between "s" and "z." The Spanish word, *calabaza,* meaning gourd, pumpkin, squash—or an ignorant person—is correctly spelled with a "z."

town with vigorous new blood and would dwarf anything that settlement had ever known or, likely, will ever know again.

Almost before the ink had dried on Colonel Sykes' new deed, Plans–with a capital P–were being drafted and made public. Plans for a railroad–The Calabazas, Tucson, and Northwestern; plans to develop Calabazas as Arizona's chief port of entry from Mexico, and to build a swank resort city with an elegant hotel, parks and tree-shaded avenues and street railways and tall office buildings. Plans for an in-pouring of prospective settlers and moneyed visitors to be drawn hence from all less-favored parts of God's green earth.

"The townsite of Calabazas will soon be the scene of much interest," reported *Arizona Daily Star* on February 9, 1879. "Already the bricks are being made for a hotel, which will be erected during the coming season. When complete it will be the finest in the Territory, with all modern improvements. Ground will be broken for the foundation in about two weeks . . . It is the object of the company[2] to make the town of Calabasas the finest place of resort in the Southwest."

The booming city of Tombstone, 50-odd miles to the northeast of Calabazas, took a dim view of these prosperous and promising developments–a not-uncommon trait among neighbors slightly green-eyed.

"Possibly there is no spot in the United States where the spirit of deviltry and pure cussedness has so much play as in the extreme southern wing of Pima County, hugging the State of Sonora," muttered the *Tombstone Epitaph* on August 12, 1882. "Ever since the building of the Arizona and New Mexico railroad began, a wild, worthless pack of outlaws from every quarter of the globe has been gathering along its line. The Sonoita Valley, made beautiful by nature, has been irrigated with blood from end to end. Men, who were as ready to be killed as they were to kill, and women whose faces were not shrouded with a blush for years, made Tarfist, Crittenden, and adjoining camps their headquarters. Riot, robbery, and bloodshed was the order of the day, murder an amusement, and deviltry that would puzzle hell to reproduce or Omnipotence to duplicate, a mere pastime.

"Within a few miles from the line of the railroad, a few speculators have laid out a town on the banks of the 'raging Santa Cruz.' The townsite is owned by a company, and prospectuses glittering with generalities and radiant with metaphor were stricken off. There were three articles of this kind–one for circulation in Arizona, another for distribution in the Eastern States, while the third was to enlighten the European mind on the grandeur of the enterprise. This European prospectus was a grand affair. It was filled with magnificent drawings representing the raging Santa Cruz at high tide, with steamboats and heavy vessels at the docks, unloading, while yonder in the distance could be seen the smoke of a dozen furnaces, with lively locomotives hauling ore to the numerous mills. A little further south was the Grand Promenade, bounded on one side by stately mansions, and on the other by the placid waters of the

[2]Çalabazas Land and Mining Co.

Santa Cruz, with the tide rolling and ebbing along. The spires of a dozen churches and public buildings, the campus grounds of colleges and schoolhouses, with troops of merry children gamboling on the greensward, added another enchantment to the view. This paradise was called Calabasas.

"Calabasas, in Mexican Spanish, means pumpkin; so, in clear English, the town should be called Pumpkinville. It was, indeed, a heavenly place, if extreme quietness and tranquility could make it so, before the railroad struck its environs. As it is, between leaden pills and malaria, it requires an everlasting boom to keep its population from falling off . . ."

Only two months later to the day—October 12, 1882—*Arizona Weekly Star* headlined the opening of Colonel Sykes' Hotel Santa Rita, at Calabazas. Dated October 6, the release read:

"What was but only a few years ago a comparative wilderness, best known only to the savages, was yesterday and last night the scene of one of the most enjoyable receptions ever attended in the Territory. For completeness of every necessary detail, or hospitality of the host, it could not be excelled. In short, the formal opening of the Hotel Santa Rita marks an important era in Arizona's progress.

"At noon, yesterday, Colonel Sykes hoisted the Stars and Stripes over the hotel . . . and the town band joined by playing the national anthem. At 1:30 p.m. your correspondent, with the other visitors from distant places, arrived after a delightful trip over the S.P. and the AT&SF railroad. They marched to the hotel in military style, headed by the Tucson band, and were cordially received by Colonel Sykes and the guests who had already arrived. After a half hour spent in preparing toilettes and admiring the picturesque scenery surrounding the handsome hotel, the guests were ushered into the dining room where 100 hungry palates were tickled by the choice viands and the following menu:

Soup—Mock turtle, tomato.

Entrees: Boiled leg of mutton, caper sauce; chicken, cream sauce; tongue, ham, roast beef, pig, chicken.

Lobster salad, salmon salad.

Game: Wild pigeon, potted roast teal duck, wild turkey, mallard duck, curlew, plover, quail, English snipes.

Vegetables: potatoes, mashed and fried; sweet potatoes, sweet corn, succotash, tomatoes, relishes, Worcester sauce, tomato ketchup, mixed pickles, olives, pepper hash.

Pastry: Plumb [*sic*] pudding, apple pie, peach pie, custard pudding, rhubarb pie.

Fruit: Apples, peaches, pears, grapes, pomegranates, nuts, and raisins.

Crackers, cheese, coffee, tea.

"The delighted diners lingered over the dainties long after the stars brightened the unclouded heavens; and when all the toasts had been offered and responses returned, the host complimented over the substantial benefits he was showering on that vicinity, and the cook flattered for her culinary abilities, the guests flew to the airy verandas only to be recalled at 10:30 to the terpischorean enchantment also on the bill of entertainment. It was early morning when the ballroom lights were extinguished, and the pleased participants had retired to rest . . .

"The Hotel Santa Rita, which graces one of the prettiest valleys in Arizona, especially deserves a brief description . . . The building contains 30 bedrooms, together with dining halls, smoking, bar, and billiard rooms. The parlor is elegantly furnished. Handsome oil paintings grace the walls, while the furniture is of the latest pattern and most expensive finish. The dining room is large and well lighted, the windows opening upon broad piazzas that completely encircle the house. Every bed chamber is completely furnished and has a stationary washstand with running water, hot and cold. For comfortable apartments, excellent food, the choicest wines and liquors, pure water and polite attention, the Hotel Santa Rita can compete with any other beween San Francisco and the East . . ."

Echoing this journalistic tribute was an advertisement carried in subsequent issues of *The Star*:

"Situated at Calabasas, Arizona, is the finest appointed house, in every particular, between San Francisco and St. Louis," read that appeal for patronage. "The table is always supplied with the best the market affords. It is the regular meal station for the A.T. & S. F. Railroad, and being situated in one of the picturesque valleys of the Pacific Coast, having wonderful salubrity of climate, not excelled by any part of the United States, and obtaining an abundance of wild game, it offers special inducements to tourists and pleasure seekers. Good stable in connection with the hotel. Terms, $3.00 per day. A reasonable reduction made to families, or parties, by the week or month.

C. P. SYKES, *proprietor*."

But Colonel Sykes' Big Plans stopped short of fruition. A new town called Nogales, located directly on the Mexican border, undermined the hopes of Calabazas to become port of entry for Mexico's West Coast. Other things, too, went wrong, and before the big hotel was a year old it (together with most of the town of Calabazas), was sold to the Territory of Arizona for delinquent taxes.

Several years after The City of the Pumpkins had resumed its briefly broken status of ghost town, a lengthy feature story concerning it was published in the *St. Louis Democrat* and subsequently reprinted in the *Arizona Citizen* of August 10, 1895.

"An old Mexican land grant with a strange history is the Calabasas," wrote the special correspondent to the *Democrat*. "It has just been wiped off the map of Arizona by a decision of the Court of Private Land Claims. When the news officially reaches Boston counting rooms, there will be $75,000 and upward to credit to profit and loss . . . One of the largest and best appointed hotels in Arizona, a handsome brick structure, stands on the Calabasas grant, a mile from any other house. Why it was put there is one of the mysteries to be charged up to the mental vagaries of the boomer . . .

"The Santa Cruz River flows through the grant. It is a very respectable stream, in the eyes of the irrigator; but in the mind of the navigator it would inspire only contempt. Yet, when the grant was unloaded on Boston moneyed men the deal was accompanied by pictures showing steamboats puffing along on the bosom of the Santa Cruz. There was once an Arkansas commodore who vowed his sternwheeler was so light of draft she could run on a heavy dew. That was more probable than the pictures of water transportation on the Calabasas Grant.

"The lonesome hotel and some other improvements are accounted for by a story. It is said to have been the purpose of the rich men of Boston to make a profitable plaything of the Calabasas Grant. The cattle investments were expected to furnish the profits. The hotel was to be part of a great hacienda on the Mexican plan, where the capitalists and their families could go to enjoy the finest climate in the world, and where, by irrigation, they could feast on the products of two zones.

"The walls of the old mission church of Calabasas still stand, but the roof is gone. In the cement floor excavations have been made to some depth and a fairly well authenticated legend belongs to them. Knowing their danger from Apaches, the priests of the early period of mission work among the Pimas buried the gold and silver which the Indians brought them as offerings. The hiding places were beneath the floor of the church. When a deposit was made in this primitive bank, a diagram was drawn and the secret place was shown on this diagram by measurements from the altar. A copy of the diagram was forwarded to Rome so that, if the mission was wiped out by the Apaches at any time, the church authorities would have a guide as to location of the treasure. This forethought resulted just as intended. Forty years after, the Apaches came down like a wolf on the fold and destroyed the mission. A Roman Catholic bishop, with an escort, traveled from Santa Fe to Calabasas. He carried papers sent to him from Rome. The ruins of the old church were found. The measurements from the altar, as shown on the old map, were taken; the cement floor was cut, and the buried treasure was found exactly as described in the report sent in by the mission fathers. As much gold and silver as would load two of the two-wheeled carts common to that period, was taken from beneath the floor of the old mission. The holes are there today as evidence that the search was made . . ."

Little remains to add to the Calabazas story. Its post office, established in 1866 and discontinued in '68, was re-established with the outset of the land boom in 1880 and endured until 1913 when it succumbed due to lack of patrons and patronage. The Calabazas school continued to function several years longer.

Oddly enough, almost the only memorial there is to this strange boomtown is a book, *Calabazas, or Amusing Recollections of an Arizona City,* privately published in 1892 by its author, James Cabell Brown,

Long out of print and almost non-existent except in the libraries of a few collectors of Western Americana, the book was reprinted in a photo off-set edition in 1961 by Arizona Silhouettes, of Tucson.

In the closing paragraphs of that book, penned more than 70 years ago, James C. Brown wrote the obituary of Calabazas—a better obit than has been written since, or is likely to be written in the future.

"Of Calabazas, the name only remains," wrote Brown. "Where is Handsome George, Curly Pete, or Bottle Bob, and Casino Harry, the leaders of fashion and arbiters of that burg? Where is Drinkwater, the Judge and Custom Officer? Where is Crandall, the good-natured and generous? Where the Calabazas Baby and Widow? the hurdy girls? the terrified Kid and the energetic Preacher? the 'Bad men'? the vigilant Constable, and all that went to make that town lively?

"If living, they are no doubt in just such another town, undergoing such other experiences and dangers, for their life could not be lived in a less reckless or quieter place. Their changing moods and generous acts, their impulsive sympathy, their cold-blooded nerve, bloodthirsty quickness, profanity, dissipation, and horse-play, remain as a memory only . . . It may be that an old-timer, visiting the deserted site of the once lively town, can hear the ghostly yells of the mob, the sounds of ghostly music, swearing and dancing from the hurdy houses, or the whistling of ghostly bullets through the blistering air."

Those words of James Cabell Brown were written more than two generations ago. With each passing decade the ghostly sounds of Calabazas have grown fewer and fainter, and the old gray ruins, each year, are a little harder to find.

Ruins of Burfind Hotel, Gillett, 1959

"To the brave and rugged prospector, with his cheap outfit of mining tools, his empty stomach and hopeful heart, is due the rapid development of the West. He disappears in the rocky fastnesses of the mountains and with no companion but his pick and frying pan, is lost to memory. Suddenly comes the news of another discovery, and a city rises like a pillar of flame in the wilderness . . ."

—SAM DAVIS, *in Carson (Nevada) Appeal, 1897.*

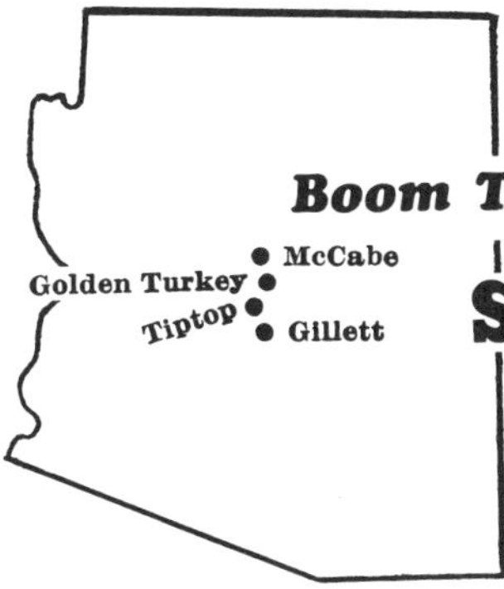

Boom Towns of the Black Canyon

SHOULD the equanimity of Phoenix ever be threatened by the International Bad Boys, its residents will have their choice of several escape hatches. One of these so-called Evacuation Routes, plainly marked with round, blue-and-white signs, is the Black Canyon Highway, leading from Arizona's capital city north to Prescott. Wide and smooth, and paved throughout, with no sharp curves or steep grades, this present road through the canyon—officially known as Highway 69—follows, in a general way, the route of the original Black Canyon road, built nearly a century ago. Over the rocks and sand of this earlier road, horse-drawn stages were rumbling and clattering in the 1860s, and along that road were spotted such well-known stage stations as King Woolsey's ranch, Lord Duppa's, Capt. Calderwood's, and others no less notable. The same road also served various mining and milling camps, long ago reduced to ghost towns.

With the idea of spending a week in the Black Canyon country, visiting all these historical spots possible, I left Phoenix on a bright April morning in 1959.

After passing through Lower and Upper Deer Valleys, Happy Valley, Ocotillo Flats and Ocotillo Hills, and Village Meadows, all pleasant little farming-suburban communities between the city and the foothills, it seemed good to be getting into a more-sparsely-settled region where the land was strewn with black lava. Spilling down from the hill slopes to the edge of the right-of-way were a lot of my old desert friends—the giant saguaros, ocotillos, palo verde, the belligerent cholla clan, and the feathery-branched creosote brush. Tissue-petaled prickly poppies bloomed beside the road, and the long white trumpets of *Datura* would be opening within a few days.

Two of the old Black Canyon towns I especially wanted to visit were Gillett and Tiptop. If only I could locate Gillett, I could puzzle out the way from there to Tiptop—I was sure of it. Unfortunately, I wasn't at all sure I could find Gillett. Back in 1950, before this Black Canyon Highway was even black-topped, I had spent two days trying to find Gillett and finally had to give up the search. Several years later, I had come armed with a sketch map and specific directions for finding the town. But the penciled map had left me confused and confounded, and the directions had led to a sandy wash where the trail had petered out with sickening completeness.

This time I had been told to see Bill Hostetter, owner of a rockhound shop and motel in the small town of Black Canyon. Bill, like most desert folk, was very friendly.

"Sure!" he said. "I can tell you how to find Gillett—if you are real good at following directions! Go back about five miles to Moore Gulch. Just before you cross the bridge, turn off the highway to your right and go through the gate. Be sure to close it behind you. The road soon forks. Take the right-hand branch—the left goes to an old deserted goat ranch. At the next fork, keep to your left. Follow the road to the top of the hill. Before it starts down the other side, park your car. You'll have to walk the rest of the way—and when you get to the bottom of the hill you'll have to wade the Agua Fria River because Gillett is on the other side."

Having had some experience with Arizona rivers I asked if there was any danger of quicksand in the Agua Fria?

"Oh, sure!" said Bill. "But not at this time of year!"

I hadn't known that quicksand was choosy about its season; but, as long as Bill said it was safe, I figured I could take a chance.

Following the directions given, it was only a little while until I was parking my car at the top of the hill, a few yards from the point where the road tipped over the edge and started sharply down into Agua Fria Canyon. From the hilltop eminence there wasn't a ghost town anywhere in sight. Neither was there a house nor any other sign of man—except, of course, the little road and my own car.

Laden with camera, extra films, a two-quart canteen of water, binoculars, and a small packet of food, I started down the steep, rocky trail. After winding downward a considerable way, the road climbed another rise and so brought me in sight of the greenery fringing the Agua Fria, still some distance below. Stirred by the great sweep of Arizona spread out before me, I raised the binoculars to my eyes—and looked straight into the midst of a bunch of stone ruins on the opposite side of the river! The structures were still half masked by the dense shrubbery of the river flat, but just knowing that this time I had actually found the old ghost milling town was enough to lend wings to my feet.

Hurrying down the road to the foot of the hill I floundered across a width of hot, seemingly-bottomless sand, and halted, a little breathless, at the edge of the river. Moving lazily over its pebble-and-sand bottom, its water did not appear to be anywhere more than a foot deep. Removing my shoes and stockings I waded experimentally into the stream—still a bit apprehensive that quicksand might have changed its "season." Moments later I was scrambling through the copse of thorny brush fringing the river and knew that my destination, at last, was at hand!

My first close view of Gillett's ruins thrilled me to the bottoms of my feet! Delighted as I am to find even the bare site of a historical town, my cup of joy almost spilled over as I emerged from the thicket and looked upon a large building, comprising seven units, arranged around three sides of a patio or small plaza.

Either these units had been built at different times, or by different owners, as several types of construction were represented. Largest of the several structures was built of white stone, its squared edges laid to the wall's outer face. Frames of the doors and windows once had been painted blue, and the building held two corner fireplaces. From descriptions read, and one picture I had seen, it was easy to recognize this as the former Burfind Hotel. Another of the ruins—this fabricated of adobe—contained hand-hewn ceiling beams decorated in orange and blue paint. Still another unit was built of random stone; a fourth, of large cobbles from the nearby river. Some of the adobe walls were plastered, others not. Most of the units contained at least one fireplace. Desert brush was crowding in jealously from all sides, and one large saguaro stood overlooking the ruins. In one corner of the patio, or plaza, quite a hole had been dug at sometime in the past—likely by treasure seekers.

From this first group of buildings I ranged over a desert flat to the rear. Here were numerous foundations, caved cellars, low walls, broken glass and assorted castoffs—and the beginning of a rutted road leading to Tiptop.

The Tiptop Mine, which supplied the lifeblood to Gillett's mill, was discovered in 1875 by Jack Moore and Bill Corning, jackass prospectors, while on their way to Castle Hot Springs from Prescott. When word got around that even surface rock from the new discovery assayed as high as 1000 ounces of silver to the ton, a small army of prospectors moved down from the higher Bradshaws and many locations were made. A few of these, including the Eldorado, Silver Museum, the Lane, 76, Foy, and possibly others, eventually developed into fairly good mines—but none to equal the Tiptop. In 1877, Corning and Moore sold to San Francisco interests for a reported $85,000 and, later that year, a new milling town began to take form on the Agua Fria River, eight miles from the mine.

"I arrived at the Tiptop millsite after a long and tedious journey," wrote a reporter for *Arizona Weekly Miner* in late January 1878. "Tender smiles greeted me from the corner of every saloon—and they are quite numerous, I assure you. Messrs. Ganz & Co., have opened out their liquor department . . . where, in a couple of weeks, the choicest wines and liquors will be lavishly dealt out.

"The townsite has been located, surveyed, and lots are being rapidly disposed of. They vary from $100 to $250 each, according to location. The principal streets in this town are California, Mill, Main, Pine, and Market . . . Anders and Rowe are hauling in lumber by the thousands . . . The real estate men held a grand mass meeting yesterday, Jack Swilling acting as chairman. After several hours . . . it was decided to christen the new-born city Gillett in honor of one of the most affable gentlemen in the place . . ."[1]

"The mill at this place is about completed," announced the Gillett correspondence in the *Miner* of February 8, 1878. "The stamps will begin pounding about the 15th. The miners are in great glee, betting that Tiptop will go up to $50 a share in 60 days . . . The Tiptop saloon, kept by Brooks and Collins, late of Broadway, New York, is a secure place. Moss, the post office man, has a log house and dispenses fluids. I don't suppose the Postmaster General would object . . ."

Gillett got its sealegs fast. By the end of March 1878, three men already had been murdered there, and on April 12 the *Miner* reported that the new and thriving town contained several stores carrying large stocks of goods "that will compare favorably with the mercantile houses of our own town, Prescott."

"Anders & Rowe, C. P. Head & Co., and C. T. Hayden, are the owners of these mammoth establishments," it was set forth. "A brewery has been established by Mr. Peter Arnold . . . Saloons dot the town . . . Mr. Edwards has erected a meat market . . . Building is progressing all over town and the place presents a lively appearance . . . Wood choppers and haulers are numerous and help fill the place with an industrious class of people. Corrals are going up and are being filled with hay for the accommodation of the freighters and others who are arriving continually from the capital and other points throughout the Territory. Messrs. Clark and Crooks have a thriving blacksmith shop in full blast . . ."

In the same issue of the *Miner* as the foregoing report was published a letter, half a column long, signed Eliza A. Edwards. After giving details of another tragic shooting (that of William H. Fulkerson, accidentally killed by his brother) Mrs. Edwards concluded:

"Only one thing I regret in Gillett, and that is the little regard in which many people here hold God's sacred day. It is nothing but hammer and saw all day on the Sabbath, from morning until night. O, Mr. Editor, how can we expect to flourish if we stand in defiance of His sacred commandments?"

Before another week had rolled around, a feed, seed, and livery stable had been opened at Gillett by Tom Cusack, and the Gillett Boarding House had been opened by Mrs. J. E. Brown. "The table will, at all times, be furnished with everything the market affords," announced the paid advertisement of that good lady. "Board, per day, $2.00. Single meals, 75 cents." Not to be outdone, Martin Kopsecher, owner of the Tiptop Boarding House, was accommodating 75 regular customers at $11 per week. (Early in May, that year, the Kopsecher establishment would be purchased by Mrs. A. L. Ayers.)

Most prominent family in the Tiptop-Gillett vicinity, was, undoubtedly, the Wagers.

Ed Wager was a storekeeper, saloonkeeper, and postmaster, and with his brother, George, alternated as justice of the peace. Another

[1]D. B. Gillett was superintendent of the Tiptop mill.

brother, Henry, was constable; a fourth brother, Frank, ran the assay office. Oscar Wager, a son of George, climbed on the municipal bandwagon at the age of ten years when he began carrying the mail, horseback, from Gillett to Tiptop, and before he was 20 he was serving with Bucky O'Neill's famous Rough Riders. All the Wagers owned mines in the vicinity and, eventually, became owners of the Tiptop.

It was Clarence Wager, a younger son of George, who supplied much of the historical information contained in this chapter.

According to this former resident of Tiptop, the camp was divided into two parts. In Lower Town was located the original Tiptop Mine and, later, the mill. In Upper Town were the business houses, school, several less important mines, and a spring of water that supplied the domestic needs of the camp. A short way above Upper Town, near the bottom of the gulch, was the cemetery.

At the time of my visit, in 1959, ruins of Tiptop still straggled up and down the canyon for more than a quarter of a mile. In addition to several buildings in near-habitable condition, there were numerous foundations of weathered stone, many crumbled adobe walls, and dozens of foundations where buildings formerly had stood.

Next day, when I returned to Black Canyon, I found Bill Hostetter shooting the breeze with another old timer. Bill produced a round of cold drinks and the three of us went outside and sat down on a bench in the shade and talked—mostly of Gillett and Tiptop, and men and women who had called it home. Our talk veered to gold and silver mining and, eventually, to uranium.

"Guess you've heard the story of Maggie Baker?" said Bill. I nodded. Everyone in the mining country has heard of Maggie Baker who, with her husband Riley, took a homestead on Badger Creek in Houserock Valley, Coconino County, in 1932. Almost all her life Maggie had been a rockhound. She was especially interested in petrified wood and liked to make long horseback rides over the desert looking for it. One day, in 1947, she found some especially pretty wood in a canyon breaking out of the Vermilion Cliffs, west of Navajo Bridge. This particular wood was a strange greenish color, inlaid with yellow, and Maggie sent some of it away to be made into buttons. The yellow streakings proved to be carnotite—one of the uranium ores.

Instead of a mere deposit of petrified wood, Maggie Baker had discovered what would be one of the first uranium mines to be developed commercially in the United States.

"Yes," I said. "I remember the story of Maggie. What about her?"

"After Maggie discovered the carnotite, I went into partnership with her and Riley," said Bill Hostetter. "They dug out the petrified wood and I packed it to the highway on seven burros, making two trips daily. It wasn't long before some promoters learned of Maggie's discovery and offered her $40 000 for it. Maggie asked me what I thought we should do?

" 'Sell it quick!' I told her.

" 'I don't know!' protested Maggie. 'That petrified wood's awful purty!'

" 'It sure is, Maggie,' I said. 'But $40,000 in nice crisp greenbacks is a hell of a lot *purtier*!' "

A few miles beyond Bill Hostetter's place I turned off on the Horsethief Basin road. Although it was too early in the year to visit the high, frosty country around The Basin and Crown King, where the road would still be buried in winter's snow, other camps in the warm, dry foothills, merited attention.

First of these was Gloriana, about a mile from the highway. It was never a large camp, and the remaining ruins are somewhat less than spectacular. Prowling about the site I found a cement reservoir, dry and cracked; two walls of a rock-and-adobe cabin, foundations, the remains of a small mill, and assorted rubble. So far as I know the little camp never had a post office or store—but still its site was not an unpleasant place to spend a few moments on a pleasant morning.

From the site of Gloriana I dropped down to the floor of the canyon—and Bumblebee. Originally a mining camp, later a ghost town, Bumblebee had recently shown an indication of returning life, and the April day of my visit found in operation there a small trading post, a fourth-class post office, and a restaurant. A few summer cabins also were being offered for rent. In addition to these signs of reincarnation, the townsite still was spotted with old cement foundations, empty buildings, and other relics testifying to a livelier past.

Some of the first mining to be done in the Bradshaws was in this vicinity where prospectors were at work as early as 1863. When these men assertedly found a bumblebee's nest filled with honey they named the nearby stream Bumblebee Creek. A stage station later adopted the name, and when a post office was established here, in 1879, it followed suit.

Clarence Wager, of the Tiptop Wagers, recalls a trip made through this section with his uncle and a young Texan, remembered only as Lafe. After passing the night and eating breakfast with owners of the Bumblebee stage station, Clarence and his uncle had gone outside and were sitting in their buckboard waiting for Lafe to finish eating. Suddenly, the station door flew open and the young Texan came flying out, a half eaten biscuit in his hand, and closely pursued by the station keeper's wife who was threatening him with a broom and delivering a stinging tirade. Leaping to the driver's seat, Lafe snatched the lines and slapped them hard against the horses' rumps.

"What in the world did you do to start that rumpus?" asked the elder Wager when the trio was safely out of the station yard and headed down the road.

"Nothin,' darn it!" said Lafe. "I just said, 'Golly, I shore like your biscuits, Miz Bumblebee, and she riz right up and tore into me!"

The older man laughed. "She thought you were making fun of her," he said. "Bumblebee is the name of the station—not the lady!"

Lafe pondered the matter for a moment. "Well," he said, at last, "I'd ruther tangle with a bumblebee . . ."

Continuing along the unpaved road leading northwesterly, I was soon seeing the ruins of another mining camp. There was a big old shaft house, the foundations of a large mill, some cabins, and a huge pile of mill tailings. Near the mill stood a two-story building in good repair—evidently a one-time boarding house. There were petunias and other flowers in the yard, as well as a nice vegetable garden, and a few peach trees were just breaking into bloom.

My rap at the screen door was answered by a pleasant looking, middle-aged man. After explaining my interest in ghost towns I said I would like to look around the mill and take some pictures, if it would be all right?

"Why, yes! Of course!" said the man. And then he asked if I had been to Cleater?

"*Cleater*?" I said, in surprise. "I thought *this* was Cleater!"

"Oh, no!" said the man. "This is Golden Turkey!"

After introducing himself as Charles H. Manly, my host invited me in the house to meet his wife, and the two of them gave me a run-down on history of the camp.

The Golden Turkey Mine, according to the Manlys, was discovered early in the present century but saw little production until about 1933 when the increased price of gold made it profitable to work lower-grade ores. In the next eight or nine years, said Manly, the mine produced close to $3,000,000 in gold and silver, with some copper and lead, but was forced to close in 1942 by the WPB order barring gold mines.

Like most Western gold mines where work was interrupted by that order, the Golden Turkey had never reopened.

After operating the store at Crown King from 1934 to 1945, the Manlys, in 1955, acquired the Golden Turkey property—including 16 claims, the townsite, and 275,000 tons of mill tailings. Remodeling the old boarding house for living quarters, they had been there ever since.

"If you are going to write a story about the Golden Turkey, you should go through the mine," said Manly. When I said I would like that, very much, if he had time to show me through, my host produced a pair of battered carbide lamps and the two of us climbed to the portal of the incline shaft about 400 feet from the house.

From the portal we started downward at an angle of about 30 degrees. It was pleasantly cool in the mine; the air was fresh and sweet. Many stopes led off from the main shaft into great chambers where the ore had been mined out, occasional pillars of rock being left to help support the ceiling. Timbering was at a minimum. Each stope we turned into sloped downward until we were halted, at last, by a pool of pure, fresh water, ice cold and crystal clear, and extending from wall to wall.

To my surprise, we eventually emerged from the lowest level of the mine and stepped out into the sunshine, having walked all the way through the underground workings!

Returning to the house we found Mrs. Manly waiting with tall glasses of cold lemonade. We sat down on the porch to drink it, and talk. Mr. Manly said they would like to sell the property—if they could get $50,000 for it.

"By why?" I asked. "Where could you possibly find a nicer place to live?"

Manly nodded. "I know . . ." he said. "It's wonderful here. But I'd like to get about 50 acres nearer some town . . ."

"Fifty acres is too much land to take care of," said Mrs. Manly. I agreed that it did seem like quite a lot of land.

"Why do you want such a big place?" I asked.

My host grinned, a little sheepishly, I thought. "I'd like to keep some goats . . . and maybe a paint pony," he said.

"Fifty acres of *goats*?" I exclaimed.

"That's what I say," echoed Mrs. Manly. "*Fifty acres of goats!*"

By the time the second round of lemonade was gone it was too late to resume traveling, so I camped that night at Golden Turkey. Next morning took me on up the road to Cleater, two miles deeper in the hills.

To unscramble the genealogy of Turkey, Turkey Creek, and Cleater, would require a better historian than I.

The Turkey Creek mining district was organized in 1864 and the original name of its post office, established in 1869, apparently was Turkey Creek. This post office was short lived. In 1902, however, when the Prescott & Middleton branch of AT&SF was being built from Mayer to Crown King, Turkey Siding was established at or near the old town of Turkey Creek, and the new post office at this point was called Turkey. Levrette P. Nellis soon started a store and saloon here, and later in 1904, or early in 1905, sold a half-interest in the business to a prospector friend, James Patrick Cleater, a native of the Isle of Man. Cleater, in time, became postmaster and, in 1915, acquired his partner's interest in the store and saloon. In 1919 he married and began accumulating a family, and about 1920 the Manxman had the town's name changed to Cleater. With mines developing in the surrounding mountains and the district becoming steadily more prosperous and important, Jimmie Cleater enlarged his business and began building cabins to rent.

Eventually, however, the mines began turning off men, the mills were closing. About 1934 the branch railroad ceased operations, the rails were pulled, and Cleater became a ghost town, with Jimmie Cleater its last remaining merchant. In 1954 the post office was discontinued, and the time of my visit found only about a dozen buildings remaining on the townsite—one of these a handsome stone schoolhouse built during WPA days and long unused. Two old timers—Jimmie Cleater's widow and Homer Snyder, each crowding the 80-year mark—were still holding down the fort. Completing the population were Mr. and Mrs. Phil Reasoner who had located in Cleater about five years earlier and were operating a rock shop.

Back on the highway I drove north to Humboldt and again turned left toward the hills. This time I hoped to locate the old town of McCabe, existence of which had been first brought to my attention several weeks earlier by Mrs. Breckinridge, assistant custodian of the Sharlot Hall Memorial Museum, in Prescott. Mrs. B. said that her grandmother, Jane Lane, had kept one of McCabe's boarding houses, feeding 125 men regularly. In addition, I was told, there had been two other large boarding houses and 11 saloons, which suggested a mining town of more than average scope.

Snaking through the dump of the huge Iron King mine which had up to that time, produced some 3,000,000 tons of ore and was the largest lead and zinc producer in the state of Arizona, I picked up a typical ghost town road and followed it back into the hills. Twice the road forked, leaving me faced with the necessity of choosing between the two branches, both equally good—or bad—and neither signposted. A little dubious whether I had chosen the proper fork, I was gratified to come upon an old cemetery quite extensive enough to have served a town with three boarding houses "and 11 saloons." It was impossible to determine how many graves the plot embraced. Only five headstones carried inscriptions that could still be read. Three of these were for persons named McCarty—Martha Ellen, 1845-1916; Thomas H., 1844-1910; and Jack C., 1879-1925. A third stone indicated the grave of one D. M. Thomas (1888-1906); and a tall, white obelisk marked the burial place of Lena Reichelt "and baby," died 1906. No legible inscription remained on any of the wooden markers I saw, and many other graves carried no markers, whatever, nor were they even mounded.

Half a mile beyond the cemetery the road again forked. Since there was no indication which way might lead to McCabe, I took the binoculars and climbed to the top of the highest peak in that immediate vicinity, an eminence that afforded a view for 360 degrees around.

Sweeping the binoculars slowly over the immense spread of country, inspecting it canyon by canyon and ridge by ridge, I caught, at last, the glint of sunlight on broken glass! There seemed to be literally acres of it, and I knew this must be the old townsite.

Driving back to the cemetery I turned upon a dim sideroad I had overlooked earlier. It was badly washed into deep ditches and strewn with boulders that would have been disastrous to a lower-slung car. Within half a mile, when even my 1946 model was beginning to brush alarmingly against high centers, I parked in the shade of a sparse-foliaged tree and continued afoot.

All along the road, now, was an abundance of rubble—old tins, broken glass, fragments of crockery dishes, an old cook stove, the twisted chassis of a baby cab, splintered boards, dented powder cans, and rusty horseshoes. Upon reaching the hill slope seen from the distant crest, I found it hard to believe that any mining camp could have left in its wake such a bounty of bottles! Beer bottles, champagne bottles, wine and whiskey—all empty, and virtually every one broken. Glass originally clear had turned darkly purple, or amber, or had taken on a milky iridescence that caused the glass to gleam rainbow-like as it was turned in the sun. This, evidently, had been "Whiskey Row."

Continuing along the rutted road I would occasionally see a foundation where some building had stood, or would skirt a caving cellar or an open mine shaft. Topping the hill, I looked down into the canyon ahead and saw with amazement a mill tailings pond that appeared to cover at least 50 acres. Above the tailings stood the ruins of an old mill surrounded by a jumbled chaos of tangled cables, old castings, and crumpled sheet iron. On the opposite hillside, not far above the trough of the valley, were seven huge cement cyaniding tanks (subsequent measurement showed each to be 40 feet in diameter); and half hidden in the hillside brush, back of the tanks, laid an old steam boiler.

Elsewhere in the canyon I discovered the ruins of what had been a large stone store building. Mangled boards, bricks, and shingles, littered the site. One gable end, in falling, had struck the ground in an up-tilted manner that made it look curiously like one of the ultra modern buildings architectured by Frank Lloyd Wright. An old steel safe lay in the midst of the rubble, its interior gaping empty.

Climbing up the hillside a little way above the old building, I found a sheltered spot in the lee of a big rock where the sun fell pleasantly warm and the chill April breeze was broken. Settling myself comfortably on a cushion of dry, brown leaves, I leaned back against the rock and from one of the pockets in my camera bag drew forth a letter. Several pages long, it had been written to me by my earlier-mentioned friend, Clarence Wager, who had known McCabe in its more or-less giddy youth of 60 years earlier.

"Next time you are ghost-towning in Yavapai County you might enjoy a visit to the old camp of McCabe, about five miles west of Humboldt," Mr. Wager had written.

"Although there are numerous mines in the vicinity, McCabe owed its existence, mainly, to two of them: the McCabe, and the Gladstone, situated in Galena Gulch, above and westerly from the camp." (I looked to the west where a sombre dump spilled down from the entrance of a silent mine . . .) "Both shafts eventually reached a depth of 900 feet, with horizontal drifts at each 100-foot level," continued the letter. "The McCabe had a steam-driven hoist, the boilers being fired by wood hauled from Mount Elloit, several miles to the west. The mine shaft equipment was very crude. An iron bucket attached to a cable slid up and down the 900-foot shaft on wooden skids. The ore was hoisted in this bucket, and so were the men. One man crouched inside the bucket, and three or four men stood on the bucket's rim, continually fending the spinning bucket away from the sidewalls of the shaft. A hemp rope, usually wet and slick, reached from the shaft bottom to the surface where it was attached to a signal bell. Occasionally this rope broke and hundreds of feet of it poured down the shaft in wet, writhing coils, leaving the men to ride the bucket with no means of signalling to the hoist man—and the bucket traveled on. Strangely, there were few serious accidents.

"In the year 1900, the population of McCabe was about 300, including men, women, and children. All dwellings were flimsy frame houses and tents, with exception of one brick store building. In the earlier days of the camp an itinerant peddler had made regular visits there (as well as to other camps in the vicinity), driving a team of horses and a light wagon, and selling drygoods, clothing, etc. Eventually this man settled at McCabe and later built the only brick building in the town. This was George Flammer's General Store. McCabe also supported nine saloons. Six of these were built wall-to-wall at the upper edge of town, and on payday nights it might be said that joy was somewhat unconfined. Free-for-all fighting and shooting scrapes were more or less frequent.

"Like most mining camps, McCabe was a gathering place for odd characters, especially tinhorn gamblers. Many of these were known by expressive nicknames such as Mickey the Mike, Corduroy Bill, and Big Turk—a hulking, raw-boned, pock-marked man. All the tin-horns worked in the mines between paydays. Payday night they began their poker games and often stayed at the tables two or three days and nights—or until their money was gone. Then, back to the mines for another stake . . .

"McCabe never had a school larger than one room, and one teacher had to handle all grades from the first to the eighth, and all subjects. In McCabe's earlier years this required as much fortitude as it did brains. The first school was a small building located in the scrub oak brush northwest of camp, near the old Triangle mine. During the term of 1898-99 three successive teachers were required. The first two, Mr. States and Mr. Alexander, both subscribed to the belief that sparing the rod was inclined to spoil the boy—and each, in his turn, was overcome by physical combat in the schoolroom. The third teacher employed that term, a Miss Kerbaugh, was more successful. As long as this building stood, the wall behind the teacher's desk wore a huge, dark splotch, a reminder of the poor aim of one of the larger pupils who once threw a pint bottle of ink at Mr. Alexander . . .

"In a mining camp such as McCabe there are lots of little episodes that scarcely merit the telling, in after years, but are funny at the time. Such as the incident of Frank Beard and the rooster.

"Frank was one of the miners, a droll, humorous, likeable fellow, who lived in a cabin beside the road leading out to the cemetery. He was tall and angular in build, with a long mustache that drooped at the tips. A short way from Frank's cabin lived a family that owned a few chickens, including a large and boisterous rooster. When Frank worked the night shift and tried to sleep during the day, it seemed that this rooster always chose the doorsill of Frank's cabin for his morning crowing place. In consequence of this habit there was a long-standing feud between the two.

"One day there was a mine accident in which a man was killed. His funeral procession was scheduled to pass Frank's cabin early in the morning. This was one of the mornings when the rooster seemed determined to stand on Frank's doorsill crowing vociferously. Frank chased it away. In a few minutes the bird was back again. This time Frank hurled a hobnailed shoe, and the rooster retreated—briefly. Frank had just got to sleep when the rooster returned with an even more resounding blast. This was too much. Frank threw back the bedclothes and in bare feet, and wielding the other shoe, went after the bird.

"Back in those days only city fellers and dudes wore pajamas or nightshirts. The miners of McCabe slept in their long-handled underwear, with full southern exposure, and on Frank this style didn't look so good.

"As Frank, in his Long Johns, busted out through the door, the funeral procession was approaching on its way to the cemetery—but Frank saw only that cussed rooster. Still wielding the shoe he chased the bird all the way to its owner's place, where the rooster ducked into the chicken house. Satisfied, Frank turned to go back to his cabin—only to find himself facing the funeral procession, in which were a number of women. Since there was no place else he could hide, Frank dived into the chicken house along with the rooster. After the procession had passed, the woman of the house and one of her lady friends stood in the yard and talked half the forenoon, about everything under the sun, while Frank huddled in the chicken house, shivering and barefooted, without any place to sit down, and with that diabolic rooster crowing all the while at the top of his voice!

"I think I should tell you, too, about Biddy Foran and her guitar. Biddy was a very fat lady who owned several buildings not far from Whiskey Row. She rented these buildings to some of the unmarried miners. They all loved Biddy, and to express that love they often made her the butt of their practical jokes.

"Biddy's most cherished possession was a guitar. One day she loaned this instrument to Hank Sauer, foreman of the Redman mine. Hank took the guitar into one of the saloons and began accompanying a fiddle played by one of the saloon men. There was a small audience of off-shift miners, among whom was one of the camp's practical jokers. Noticing a tangled ball of strings in the violin case, this fellow surreptitiously pocketed them and strolled out. Hurrying to the porch where Biddy sat fanning herself, he showed the mass of violin strings to her and told her that Sauer had completely wrecked her guitar.

"Biddy went wild. She was too fat to run, but she waddled as fast as she could up the road to the saloon and barged inside. Ladies didn't go into or around saloons in those days—but Biddy didn't care about that. Not with Sauer wrecking her beloved guitar.

"The music makers were playing with their backs toward the door and didn't see her coming. The miners grinned and watched, and said nothing. Waddling up behind Sauer, Biddy smacked one big first down on his bald head, snatched her guitar, delivered a few blistering remarks, and left him speechless.

"Sauer, needless to say, was called upon to buy drinks for the crowd . . .

"Most of the incidents I have related in this letter took place prior to 1905. By 1907 the mines were failing and the exodus had begun.

"The old graveyard, incidentally, is more extensive than it now appears," my old timer friend had written. "Even 20 years ago, when I last saw the place, it was difficult to identify many of the graves. As I walked through it on that occasion, my foot struck a loose board buried in the grass and thistles. Drawing it forth I saw it was the headboard of Alex Denny, a man I had known well. It was impossible to know which grave might be his; but I stood the board up, anyhow. Even if I put it on the wrong grave, I had a feeling that Alex would understand . . ."

Folding the letter and returning it to its envelope, I sat for several minutes looking out over the silent mill tailings where the April breeze was whipping up little curls of fine, yellow dust. My eyes moved on to the ruined mill, the huge cyanide tanks, the jumbled wreck of the old store building—and on up the hill toward the silent sepulchre of Whiskey Row.

Even to one long inured to the desolation of abandoned cities and towns, the scene was a little saddening. For nearly a week I had been moving through a strange, quiet land where ghost towns far outnumbered towns still active, and the living population is vastly exceeded by the dead. Gillett, Tiptop, Gloriana, Golden Turkey, Cleater, and McCabe, as well as a dozen other towns in the vicinity I had not visited, all had known briefly their place in the sun; each, in turn, had watched its sun go down.

From the ruins of this old town of McCabe, where so many dreams and hopes lay buried, I lifted my gaze to the bright, cloudless blue of the wide Arizona sky.

An eagle was circling over the ridge to the west, soaring, soaring endlessly. On the hill slope across the canyon a couple of red-and-white calves was engaged in a lively game of tag, their thin tails held stiffly aloft like little flagstaffs. In the oak brush behind the old store a pair of quail was visiting in low tones—probably discussing the best place to build their nest—and an Arizona jay was scolding in the tree overhead.

There was no doubting it longer, Spring was just around the corner. All the lower levels of the Bradshaws were greening with new grass. The first wild flowers of the season were showing their faces on slopes warmed by the sun, and each day saw the snowline creeping higher on the shoulders of the range. Every part of the whole wonderful drama of Springtime was slipping into place, smoothly, unerringly. Every big and little player from the budding aspens to the tiniest Johnny-jump-up seemed to be waiting impatiently in the wings, ready to appear promptly on cue, exactly as it had appeared the year before and all the years before that since the dawning of Creation.

Suddenly, my melancholy mood of moments earlier was gone, as if whisked away on the fresh winds of April, and a gay, glad song seemed to be bursting inside me. What mattered worked out mines and ruined towns, even the death of man's plans, little or great? What matter that man's works are as transitory as man himself? These are not tragedies, not really; not so long as the great miracle of Spring is eternal and everlasting.

ROSTER OF THE GHOSTS

INCLUDING FORMER TOWNS, MINING CAMPS, STAGE STATIONS, ARMY FORTS, AND RANCHES

(Unless specifically noted "by road," all mileages quoted are based on approximate airline distance between geographical points mentioned. In many cases no direct road exists between place described and point of location cited in the text. Persons planning to visit any off-the-beaten-track destination should first make local inquiry concerning best routes, road conditions, and other factors.)

ADAMSVILLE, Pinal Co., N. side Gila R., 4 m. W. of Florence. Founded 1866 by Chas. H. Adams (b. Ohio, 1834.) In 1868 or '69 Nick and Wm. Bichard moved to this place their modern steam flour mill, earlier erected at Casa Blanca. By 1870 Adamsville had 400 residents. P.O. estab. 1871 as Sanford after Capt. Geo. B. Sanford, U.S. Cavalry, commander Fort McDowell, 1867. Due to popular protest, original name soon restored. Over a 10-year period Adamsville was scene of considerable violence, including the lynching of John Willis for murder of Robert L. Swope, but by 1881 the town had shriveled to "a cluster of tumble-down adobe and an unsavory memory."

AGUA FRIA, Maricopa Co., E. side Agua Fria R., nr. Calderwood Butte; approx. 6 m. from present town of Marinette. Station on Black Canyon (Prescott-to-Phoenix) stage road. Covered in text.

ALEXANDRA, Yavapai Co., 10 m. S.W. Mayer, on Highway 69. Early mining camp. Covered in text.

ALLEN CITY, Pima Co., Papago Indian Reservation, T. 14 S., R. 4 E. (Barnes). After John Brackett ("Pie") Allen, b. Maine c. 1819. In 1868 Allen built the first steam flour mill at Altar, Sonora. A pioneer in the business life of early Tucson, where he resided 22 years, Allen became an inveterate follower of mining booms and was among the first hotelkeepers and restaurateurs to set up shop in Tombstone, Bisbee, Gunsight, Quijotoa, and other Arizona camps. See. GUNSIGHT.

AMERICAN FLAG, Mohave Co., W. side Hualpai Mtns., nr. summit. Small mining camp named for rich American Flag silver mine originally located by Cienega Shoulders who sold it to four "Cousin Jacks"—Monte and Jim Penberthy, Jack Coran, and Billy Richards. These men operated the mine several years, sorting and sacking the rich ore and packing it on burros 14 miles to the foot of the mountains where it was picked up by freight wagons and hauled to the Colorado River to begin its long water journey around Cape Horn to Swansea, Wales, for smelting. Due to the great distance involved, and slow methods of transportation, a full year was required for each shipment to reach its destination and only one shipment was made annually. Upon receiving returns for their 12-months' labor, the partners would journey to San Francisco for a terrific celebration. Before its conclusion they would have spent almost the entire proceeds of the shipment, withholding only sufficient funds to finance another year's work in the mine.

AMERICAN FLAG, Pinal Co., small mining camp in Catalina Mtns., S. of Oracle. American Flag mine at this point was located by Isaac Lorraine, of Martinique, in late 1870s. P.O. 1880-90.

AMERICAN RANCH, Yavapai Co., early station on Prescott-Ehrenberg stage road. Situated 100 yds. S.W. of Prescott-Williamson Valley road at a point 12 m. W. of Prescott. Adobe ruins, old fireplaces, and rubble, in 1961.

ARIVACA, Pima Co., 24 m. (by road) S. W. of U.S. 89 at a point 28 m. N. of

Nogales. Small settlement, old, historic, and interesting. Not a ghost town. Covered in text.

ARIZONA CITY, Yuma Co., original name of Yuma. Covered in text.

AUBREY CITY (or LANDING), Mohave Co., early river port and smelting town on N. bank Bill Williams R. at its confluence with Colorado R. Covered in text.

AUSTERLITZ MINE, Santa Cruz Co., S. E. 1/4 Sec. 36, T. 22 S., R. 10 E. On Arivaca-Peña Lake road between Oro Blanco and Ruby. Ruins. Covered in text.

BAKERVILLE, Cochise Co. "A Dutch baker named Jake Perung settled here in the 1880s and built a rooming house and hotel." Granger.

BATES WELL, Pima Co., T. 14 S., R. 6 W. In Growler Pass between Bates and Growler Mtns., 13 m. S. of Ajo, 11 m. W. of Rte. 85. Site of a mine, store, and dwellings dating from 1890s. Abandoned 1920.

BAYARD, Yavapai Co. Camp of Oro Bella Mining Co., on Humbug Ck., near Mt. Wasson. P.O. 1887-1888.

BEALE SPRING, Mohave Co. Named for Lieut. Edward Fitzgerald Beale who camped here Oct. 8, 1857. Due to skirmishes with Hualpai Indians of vicinity there was established here Mar. 25, 1871, a "tent fort" commanded by Capt. Thomas Bryne, 12th Infantry. On complaint of Dr. J. A. Tonner, agent at Colorado River Indian Reservation, the "fort" was abandoned April 6, 1874. Later, a way-stop for travelers, with hotel, eating house, bar, and corral operated by A. F. Simonds, and a general store kept by Charles Cohn. "Beale Springs promises to become a town of importance,"—*Alta Arizona,* Jan. 14, 1882.

BELLEVUE, Gila Co., T. 15 S., R. 14 E. About 5 m. S. W. of Miami. P.O. 1906-1927.

BELLE AIR, Pinal Co. "Belle Air is the name of the new city seven miles southerly from Pinal . . . The Silver Belle, Columbia, Martinez, and many other mines center there. Goldman & Co. have a large store with a well-assorted stock of goods . . . There is a hotel and butcher shop, blacksmith shop, saloons, and all the et ceteras . . ."—*Pinal Drill,* Nov. 26, 1881.

"Another saloon is going to be opened (at Belle Air) which is urgently called for, as there are at present only five saloons for more than 80 persons."—*Pinal Drill,* Dec. 24, 1881.

"It seems we were in error last week in stating that the $130 found on the person of Wm. Perkins, who was killed at Belle Air, were confiscated by the crowd, Judge Benson having shown us the county treasurer's receipt for the amount less necessary expenses. Mr. Benson seems to fear that the public may understand that he was mixed up in the riotous proceedings which accompanied and followed the burial, but the idea is altogether too absurd . . ."—*Pinal Drill,* Jan. 21, 1882.

BERNADINA, Cochise Co., former P.O. (1915-17) and small settlement.

BIG BUG, Yavapai Co., on Poland road, c. 12 m. S.E. Prescott. Small placer gold mining camp settled 1863. Name of P.O. changed to RED ROCK in 1879; name Big Bug restored, 1881; P.O. discontinued, 1910. Covered in text.

BIG HORN, Maricopa Co., formerly a small settlement and P.O. (1930-1935) on the Harquahala Plains, c. 30 m. S.E. Salome. Named for nearby Big Horn Mtns.

BLACK DIAMOND, Cochise Co., T. 18 S., R. 24 E. In Dragoon Mtns., S.W. of Pearce. Early-day stage station and point of supply, named for Black Diamond mine. P.O. 1902-1908.

BLACK WARRIOR, Gila Co., former P.O. and small mining camp 4 m. N.W. of Globe. Named for Black Warrior mine. Site is now buried beneath a sprawling tailings dump.

BLUEWATER, Pinal Co., c. 45 m. N.W. Tucson. Stage station estab. 1859. Apparently nothing remains to mark the site. Covered in text.

BLUEBELL, Yavapai Co., c. 4 m. W. Mayer, in Bradshaw foothills. Aerial tram transported ore from the Bluebell mine to Bluebell Siding, S. of Mayer, from which point it was shipped by rail to the smelter at Humboldt.

BONITA, Graham Co., W. terminus State Hwy. 266, 3 m. S.W. Fort Grant and 36 m. by paved and graveled road S.W. Stafford at jct. U.S. 70 and 266. Bonita, today, is a pleasantly quiet little settlement—not a ghost town, but a place abounding in Yarns and Yesterdays. Lying just outside the Fort Grant military reservation, Bonita formerly served as blow-off town for the 1000-or-so soldiers stationed at that post. When payday brought these men swarming into town, there to mingle and, frequently, to clash with cowpunchers, miners, and Mexicans, all spoiling for trouble, Bonita flourished as one of the choicest deadfalls in the Territory. Gambling houses, dancehalls, eight or ten saloons, and bevies of "good-time" women were available, and men who came looking for either excitement or trouble seldom were disappointed.

BOWIE, CAMP and FORT, Cochise Co., N. end Chiricahua Mtns., c. 14 m. S.E. Bowie Station on State Hwy. 86. Estab. 1862 to protect an important spring of water in Apache Pass, Camp Bowie subsequently was the scene of much bloody fighting between whites and Apaches. For nearly 20 years (1867-86) men and officers stationed at this post dealt almost continually with Apache depredations and pursuit. With the Indian menace terminated by surrender of Geronimo and others, in 1886, the post gradually lost its usefulness and was officially abandoned in 1894. In 1911 the buildings were sold at public auction. Extensive ruins of stone and adobe remain.

BRADSHAW CITY, Yavapai Co., covered in text.

BRECKENRIDGE, FORT, Pinal Co. On Aug. 6, 1860, the name of Fort Arivaipa was changed to Fort Breckenridge, after Vice-Pres. John C. Breckenridge. The post was razed July 10, 1861, to prevent its capture by Confederate forces.

BROOKLYN, Pima Co., former mining boom camp nr. Covered Wells on State Hwy. 86. Covered in text.

BRUNCKOW MINE, Cochise Co., between Charleston and Tombstone. Covered in text.

BUCHANAN, FORT, Santa Cruz Co., on Sonoita R., c. 45 m. S.E. Tucson, and above present town of Patagonia. Estab. 1856 by a detachment of the First Dragoons, this post was originally known as Camp Moore, and did not receive the name Buchanan until May, 1857. In June, 1861, the post was attacked by Cochise and about 100 warriors and most of the troops sent in pursuit were ambushed and slain. On the following month (July 23, 1861) the fort was abandoned and destroyed by Union troops to prevent it falling into the hands of the Confederates. Some ruins remain.

During the earlier life of Fort Buchanan, no newspapers were being published in Arizona, but the *Weekly Arizonan,* of Tubac, after its founding in 1859, carried occasional Buchanan news. In the *Arizonan* of March 10, 1859, are several advertisements for White & Granger, general merchants, at Fort Buchanan. These ads offer a large assortment of merchandise consisting of drygoods, clothing, blankets, hardware, Queen's Ware, groceries, hats, boots, shoes, tobacco of all sorts and qualities, liquor of all sorts, Colt's Navy Six-Shooter Pistols, rifles, double guns, Northwest guns, powder, lead, Ely's caps, shot, gun-wadding, oils, paints, whiting, putty, chalk, spirits, turpentine, glue, nails, castings, tinware, brass kettles, tubs, buckets, 60 gallons of molasses, 12,000 pounds of bacon sides, and 500 pounds of rice at 30 cents per pound.

"The adove-named articles," it was stated, "will be sold at reasonable prices, for cash."

Another news item, laying open to view some of the abhorrent customs of that day, appears in the June 30, 1859, issue of the *Arizonan*: "Private Alenson Bently, of K Company, First Dragoons, convicted of desertion and horse stealing, was sentenced by the court martial lately in session at Fort Buchanan to receive 50 lashes with a cowhide well laid on the bare back; to be confined at hard labor until January, 1862, heavily ironed; to forfeit all pay due him; to have his head shaved, and to be branded with a red-hot iron with the letter D; to be drummed out of the service, and receive a dishonorable discharge."

BUMBLEBEE, Yavapai Co. Covered in text.

BURKES, Maricopa Co., stage station dating from 1858. Covered in text.

BUTTE, Pinal Co. "Lots for sale in the new town of Butte, on the Gila River, 18 miles east of Florence. The Pinal Consolidated Mining Co. are now erecting their reduction works there."—*Arizona Weekly Enterprise,* Jan. 7, 1882.

"The Silver Belle and the Columbia are the principal mines of the camp . . . These properties are owned by the Pinal Consolidated Mining Co. This company has erected a water jacket furnace of 20 tons capacity on the Gila River, 6 miles from the mines . . . The works are complete in all respects, and include fine large kilns for the burning of charcoal. Offices, boarding houses, have been erected and quite a little village has sprung up at Butte."—*Report of the Director of the Mint,* 1883.

CALABASAS, Santa Cruz Co., 2 m. E. of U.S. 89 at a point 10 m. N. Nogales and 8 m. S. Tumacacori Mission. Covered in text.

CALDERWOOD, Maricopa Co., former station on Black Canyon (Phoenix-to-Prescott) stage road. Covered in text.

CALVA, Graham Co., station on Arizona-Eastern RR estab. in 1890s under name Dewey. After destruction of this original settlement by flood the town was rebuilt and renamed Calva after a local Indian sub-chief. When Coolidge Dam was built it was believed its waters would inundate Calva and the town was abandoned.

CAMERON, CAMP, Santa Cruz Co. Originally estab. at Calabasas in 1866. Later moved to a site 15 m. N.E. of Tubac at foot of Santa Rita Mtns. Abandoned March 7, 1867.

CASA BLANCA, Santa Cruz Co. Former trading post, small town, RR station, and shipping point for mines of Santa Rita and Patagonia Mtns. P.O. (1860-1901) named successively Casa Blanca, Crittenden, Casa Blanco (*sic*) and Crittenden. Some ruins.

CASTLE DOME CITY, Yuma Co., on Colorado R. about 30 m. N. of Yuma. Small, lively smelting town with stores, hotel, etc., developed at this point in the late 1860s following discovery of silver and lead mines in Castle Dome Mtns. in 1863. P.O. seems to have functioned under name Castle Dome through 1875-1876, and as Castle Dome Landing 1878-1884. Covered in text.

CATALPA, Gila Co., T. 4 N., R. 13 E. Town and P.O. (1885-1888) c. 3 m. S.E. of Armer. Barnes.

CEDAR, or CEDARVILLE, Mohave Co. T. 16 N., R. 15 W. Lower end Hualpai range about 15 m. W. of Wikieup. Small settlement serving mines of Cedar Mining District. P.O. 1895-1911.

"From Mr. Cofer we learn that Cedar Mining District is on the eve of a mining and business boom. A new town is soon to be started near the Evangeline mine. A hotel of 32 rooms is to be built at once, the lumber for which is already shipped. Four billiard tables have also been ordered from the East for a large saloon which will be put up in a short time."—*Alta Arizona,* Aug. 5, 1882.

CEDAR SPRINGS, Graham Co., W. side Pinaleno range, on Fort Grant-Camp Thomas road, c. 16 m. from F. Grant. Former headquarters of Norton & Stewart Cattle Co. P.O. 1887-1892.

CERBAT, Mohave Co., Sec. 12, T. 22 N. R. 18 W., c. 10 m. N.W. Kingman. Third seat Mohave Co. Very few ruins. Covered in text.

CERRO COLORADO, Pima Co. T. 20 S., R. 10 E. A few miles W. of U.S. 89 at a point 40 m. S. of Tucson. Very old mining camp. Quite extensive adobe ruins, graveyard, mine workings. Covered in text.

CHAPARRAL, Yavapai Co., T. 13 N., R. 1 E. In foothills W. of Humboldt. Mining camp named for Chaparral mine. P.O. 1894-1917.

A roving newspaper reporter on the *Arizona Weekly Journal-Miner,* of Prescott, wrote the following description of Chaparral, published Jan. 5, 1898:

"We reached Chaparral at the noon hour . . . At present the town has about 20 business places. E. S. McRoberts carries general merchandise and has a large stock. H. Miller also has a fine line of drygoods, clothing, etc. Hudgens' Hotel furnishes good accommodations for the traveling public. Henry Behn feeds and houses any transients and provides for most of the employees of the McCabe mine. Mrs. E. Arnold also conducts a lodging house and restaurant. The Cabinet is now one of the neatest establishments in town. It is a new structure, newly furnished, and will soon be ornamented with a $3000 cabinet of magnificent specimens. Hudgens and Askew have a new two-story establishment . . . Stillwell & Hobbs supply the town and country with meat from their butcher shop. W. Wilkins amply provides for the wants of the thirsty . . . Mr. Flammer has put in a line of general provisions and miners supplies. F. H. Shimer, formerly of Prescott, is a recent addition to the Chaparral business center with fruit, tobaccos, and ice cream in season . . ."

Badly deteriorated rock ruins, 1959.

CHARLESTON, Cochise Co., milling town (1879-89) on San Pedro R., 9 m. S.W. Tombstone. Adobe and rock ruins, cemetery. Covered in text.

CHILITO, Gila Co., Sec. 27, T. 4 S., R. 15 E., at Tornado Peak, 3 m. W. of Christmas. Former camp of London-Arizona Mining Co. P.O. 1913-1918.

CHRISTMAS, Gila Co., T. 4 S., R. 16 E. 1 m. off Rte. 77 at a point 9 m. N.E. Winkleman. The great copper deposit at what was to become Christmas was located about 1880 by Dennis O'Brien and Wm. Tweed who soon sold their claims to Phelps Dodge. In 1884, when the property was found to lie within the San Carlos Indian Reservation, where mining was prohibited, work was halted and the camp abandoned. Eighteen years later, largely through the efforts of G. B. Chittenden of the Saddle Mountain Mining Co., that portion of the reservation including the copper deposit was returned to the public domain under an executive order issued Dec. 22, 1902. Receiving word of this development on Christmas eve, Chittenden and his partner, N. H. Mellor, immediately started for the mine. By riding all night they arrived at the long inactive P-D property about dawn and staked claims, which they named in honor of the day.

During the next 30-odd years the Christmas mine, operating intermittently and under a succession of ownerships, is said to have produced close to $10,000,000, the camp's banner years being between 1925 and 1932 when its population rose to a peak of 1000. During the life of the Christmas post office (1905-1935) persons from all parts of the U.S. sent Yuletide letters here, under cover, to be remailed with the Christmas postmark. Ruins.

CHRYSOTILE, Gila Co., Sec. 35, T. 5 N., R. 17 E. Mining camp that grew up around asbestos mines discovered by Tom West in 1911. Later purchased and developed by Johns-Manville. P.O. 1916-1933.

CIBOLA, Yuma Co., small settlement in T. 1 S., R. 24 W., on Colorado R. c. 50 m. N. of Yuma. Former landing place for Colorado R. steamers. P.O. 1903-1933.

CLACK, Mohave Co., 3 m. S. Mineral Park. Former P.O. (1898-1899) for the Oro Plata mine and vicinity.

CLARK, Pima Co. "A new town has been established in Silverbell District called Clark, for the president of the Young America Co., Mr. Geo. K. Clark, of New York City, a merchant of long and honorable standing there."—*Arizona Daily Star,* Jan. 26, 1883.

CLARK, CAMP, Yavapai Co., Army post estab. 1863 in Chino Valley at Del Rio Spgs. After Surveyor-General John A. Clark. In 1864 post was moved 21 m. S.E. and renamed Fort Whipple.

CLARKSTON, or CLARKSTOWN, Pima Co. Privately-owned townsite laid out in 1915 by Sam Clark in opposition to the adjacent town of Ajo, owned by New Cornelia Copper Co. By 1916-17 Clarkston had a population around 1000 and was more popular than the company town. When the P.O. at Clarkston was established in 1918, President Wilson was in high favor and residents of the town asked that the new office be called either Woodrow or Wilson. When their request was denied, the townspeople came up with the name Rowood, a scrambling of Woodrow. From 1918 until its discontinuance in 1955 the office functioned under this name. Only three wooden cabins remained in 1960.

CLEATER, Yavapai Co., 10 m. (by road) W. of State Hwy. 69 from points 11 and 17 m. S. of Mayer. Covered in text.

CLINE, Gila Co. Town at upper end Roosevelt Lake. "The original settlement has disappeared."—Granger. P.O. 1886-1912.

CLIP LANDING, Yuma Co., former steamboat landing on Colorado R. Shipping point for Silver Clip mine, and site of the 10-stamp Clip mill which began production 1883. P.O. 1884-1888.

"Clip Landing is now one of the liveliest towns on the Colorado River," —*Yuma Sentinel,* September, 1883.

COLDWATER, Maricopa Co., T. 1 N., R. 1 W. On W. bank Agua Fria R., near Litchfield. Early stage station. P.O. 1897-1905.

COLORADO, CAMP, Mohave Co., Army camp estab. 1859 on E. bank Colorado R., near present town of Bullhead City. Name soon changed to Fort Mohave.

COLORADO, CAMP, Yuma Co., Army camp estab. 1868 on E. bank Colorado R., near present town of Parker. Never more than a sub-station of Fort Yuma, Camp Colorado was abandoned 1871.

COLORADO CITY, Yuma Co., very early settlement near present site of Yuma. Covered in text.

CONGRESS, Yavapai Co., 3 m. N. of Congress Junction on U.S. 89, 44 m. S.W. of Prescott. The Congress mine, discovered in 1880s, produced about $8,000,000 in gold and silver between 1890 and 1910, its most active years. Stone and adobe ruins. Covered in text.

CONTENTION CITY, Cochise Co., former milling center on San Pedro R., about 9 m. W. Tombstone. Ruins, cemetery. Covered in text.

CONTINENTAL, Pima Co., 6 m. S. Sahuarita on U.S. 89. Since Continental still has a general store, school, and a few residents, it can't be considered a true ghost town. Rather, it might be called the ghost of a noble experiment dating from 1914 when the Continental Rubber Co., recognizing the gathering war clouds, purchased part of the Canoa Land Grant and planted an immense tract to the latex-producing plant, guayule, in the hope of developing a domestic source of rubber. A small town with stores, P.O. (1917-1929) and a RR station took form around the ranch headquarters, but the rubber-producing project was not a notable success.

COPPER CENTER, Cochise Co. "On the San Pedro near Hereford. There was a copper smelter at this location at one time. P.O. est. 10-14-1901,"—Granger. "A short-lived copper camp,"—Barnes.

COPPER CITY, Gila Co., early mining camp in Big Johnny Gulch, N. side of Birch or Black Peak, 3 m. N.E. Globe. "Copper City, A.T., Nov. 8, 1884: This has been an eventful week. The election on Tuesday having been the first held here since the settlement of the camp, you can therefore imagine the excitement which it occasioned . . . Around the polls were congregated voters and politicians, as well as office seekers. While awaiting the election returns, an impromptu horse race was gotten up and afforded fun for the town. The ladies were out in force and were enthusiastic spectators . . . The stakes were $50 a side . . . A grand ball was given at the Oriental in the evening . . . The Rawhide Band, of Copper Butte, played inspiring music, to which 50 couples danced the light fantastic. This ended a memorable day in Copper City." —*Arizona Silver Belt,* Nov. 15, 1884.

COPPER CREEK, Pinal Co., T. 8 S., R. 17 E., nr.. head of Copper Ck., c. 6 m. E. of Mammoth. P.O. 1906-1942. In the *San Diego Union* of May 3, 1936, appeared a letter signed by Martin E. Tew, Copper Creek, Arizona. The missive read, in part:

"The Tucson correspondent who, in your issue of April 19 described Copper Creek as a 'hell-roaring mining camp' that started in the '80s, succeeded in crowding into two or three stickfuls of type an unusual amount of the stuff which Barnum said the American people liked. True enough, Copper Creek was started in the '80s; but from that day to this it has never had a saloon 'bullet-spattered' or otherwise, or a gambling house, stage robbery, pistol-toting policeman, or even a jail . . .

"The population is made up of workers. There is no leisure class, and no one on relief. Excepting the ore mills, the schoolhouse is the largest building, and the teachers rank with the best . . ."

Only ruins remained in 1959.

COPPER HILL, Gila Co., early mining camp about 3 m. N. of Globe. Served the Iron Cap, Arizona Commercial, Superior, and Boston mines. Sometimes referred to as Copper Gulch, but was not the same place as Copper City, which was on the opposite side of the mountain.

COPPEROPOLIS, Yavapai Co. T. 9 n., R. 2 W., nr. Castle Ck., S. of Prescott. *Prescott Courier,* Nov. 11, 1882, refers to "the new town of Copperopolis, near Castle Creek" According to an unidentified authority quoted by Granger. "Copperopolis was a metropolis which didn't metrop."

COTTONIA, Mohave Co., construction camp of Cottonia Land and Cotton Co. on Fort Mohave Indian Reservation. P.O. 1910.

COURTLAND, Cochise Co., copper mining town 11 m. S. Pearce, 19 m. E. Tombstone. Ruins. Covered in text.

CRAWFORD, CAMP, Cochise Co., temporary Army camp estab. 1886 and probably abandoned same year. After Capt. Emmett Crawford, killed while in pursuit of Apache Indians.

CRITTENDEN, FORT, Santa Cruz Co., on Sonoita Ck., between present towns of Sonoita and Patagonia. "Established Aug. 10, 1867 . . . to protect settlers in the Babocomari, Sonoita and Santa Cruz Valleys . . ." (Brandes). Closed officially 1873. Ruins.

CROTON SPRINGS, Cochise Co., T. 24 E., R. 15 S. About 12 m. S.W. Willcox. Station on Camp Grant-Fort Bowie stage road. Brackish water of the springs was said to taste like croton oil, hence the name.

CROWLEY, Gila Co., small mining camp in Apache Mtns. Named for Con Crowley, cattleman and miner. P.O. 1907-1911.

CROWN KING, Yavapai Co., T. 10 N., R. 1 W. All early accounts of this mining camp, situated high in the Bradshaw Mtns., spell the name Crowned King. Formerly terminus of a branch RR from Prescott via Mayer. Around turn of the century the town reportedly had some 500 residents served by six saloons, five Chinese restaurants, and various other business houses. P.O. 1888-1954. Still a few year-around residents, more in summer.

CULLINGS WELL, Yuma Co., Sec. 30, T. 7 N., R. 11 W. c. 38 m. W. of Wickenburg. Stage station dating from 1860s. Covered in text.

CYCLOPIC, Mohave Co., T. 28 N., R. 18 W., small mining camp on Sacramento Wash, c. 40 m. N.W. Kingman. Named for the Cyclopic mine discovered in 1880s. P.O. 1905-1917.

DATE CREEK, CAMP, Yavapai Co., about 10 m. (by road) W. of U.S. 89, from a point 6.6 m. S. of Yarnell. Army camp (1867-74) near junction N. and S. forks of Date Ck. In 1875 the abandoned campsite was visited by a Prescott resident who reported the adobe buildings fast falling to ruin, the walls of the cemetery almost level with the ground, and many of the graves caved in. (*Arizona Miner,* April 16, 1875.) Remains of the old camp, situated on a mesa about 300 yards south of Date Creek, include rather extensive stone walls and adobe ruins.

DEL RIO SPRINGS, Yavapai Co. In Little Chino Valley, ¾ m. E. of U.S. 89 at a point 20 m. N. of Prescott. First capital Arizona Territory. An adobe building said to have been "one of the originals" has been modernized and is in use as a ranch dwelling. For several months in the forepart of 1864 affairs of the Territorial government of Arizona were conducted from log cabins and tents at this site.

DeNOON, Pinal Co. Although both Barnes and Granger state that this camp was site of the Reymert mill, owned by Judge

DeNoon Reymert, the *Arizona Star* of May 26, 1889, refers to DeNoon as location of the great Reymert mine, "about two miles from the Reymert mill, on the other side of the mountain, on the Pinal road." The same reference mentions a store at DeNoon, owned by Messrs. Knight and Curry, and two saloons. P.O. 1890-91.

DESCANSO, Cochise Co. Small eating and drinking station frequented by soldiers riding in pursuit of Apaches, and by teamsters hauling lumber from the Chiricahuas to Tombstone and Bisbee. Turquoise also was mined in the vicinity. P.O. 1892-94.

DESERT STATION, Yuma Co. Barnes locates this place "15 m. S.W. of Vicksburg, on Prescott-Ehrenberg stage line, at head of Granite Wash, in Hanegras Plains . . ." Early stage station noted chiefly for its good water drawn from a well 120 feet in depth. Also the scene of an imbroglio in which one William Barr shot and killed a character known as Gayatta John.

DESERT WELL, Pima Co. 20 m. N.W. of Tucson, and about a mile S. of Pinal County line, on Tucson-Florence road. Change station for stage teams, camping place for freighters and other travelers, and the only water supply point in a wide area. P.O. 1880-85.

DIAMOND JOE, Mohave Co., former mining camp on Big Sandy River, in T. 17 N., R. 13 W. Named for "Diamond Joe" Reynolds.

DOAK, Gila Co., Barnes locates this onetime settlement on Mineral Creek, in Sec. 13, T. 2 S., R. 14½ E., P.O. 1919-21.

DOBBS STATION, Pima Co. Stage station and eating house on Tucson-Quijotoa road. Also known as Halfway House.

DOS CABEZOS, Cochise Co., at S. foot Dos Cabezos Mtns., 15 m. (by paved road) S.E. Willcox. Covered in text.

DOWLING, Pima Co. Formerly a small mining and smelter camp at S. end Sonoita Mtns., in lower Ajo Valley. Due to its situation near Mexican border, Dowling also was the site of a U.S. Customs station. According to Granger the place was abandoned about 50 years ago.

DREW'S STATION, Cochise Co., former stage station on the Benson road c. 15 m. N. Tombstone. Chiefly famed as the place where Bud Philpot was shot and killed by Doc Holliday.

DRIPPING SPRING, Gila Co., early stage station on Florence-Globe line, in foothills Dripping Springs range. Sec. 30, T. 3 S., R. 15 E. (Barnes) P.O. 1886-90.

DUDLEYVILLE, Pinal Co., formerly a small settlement (stage station, store and P.O.) on San Pedro R. about a mile S. of the Gila R., and two m. from Winkleman. Named for Dudley Harrington, a farmer who located here 1879.

DUNLAP, Graham Co. Original name of Aravaipa.

DUQUESNE, Santa Cruz Co., mining camp E. side Patagonia Mtns., c. 15 m. N.E. Nogales and 4 m. N Mexican border. Covered in text.

EASTER, Maricopa Co., small mining camp (pop. 65) active around 1915. After Easter mine.

EDITH, Maricopa Co., hdqrs. and mailing address Judson mine, 1888. Evidently near Phoenix.

EHRENBERG, Yuma Co., E. bank Colorado R. near U.S. 60-70 bridge. Covered in text. Ruins, cemetery.

EMPIRE, Pima Co., c. 15 m. S. of present Pantano. P.O. estab. as Empire Ranch 1879-80; re-est. as Empire, 1880-81.

ESPERANZA, Pima Co., mining camp, W. side Sierrita range Pima district, c. 12 m. W. of Sahuarita. After Esperanza mine. (Barnes) P.O. 1884.

EUCLID, Maricopa Co., P.O. 1895-96. Location not fully known.

EVANS POINT, Greenlee Co., in Gila Valley. Hdqrs. Evans-Van Hecke Mining Co. P.O. 1899-1900.

EWELL SPRINGS, Cochise Co. Evidently the original name of Dos Cabezos, though the two settlements do not seem to have been identical. Springs at this point are said to have served as a watering place for Boundary Survey party in 1851, and as location of a stage station on the Birch route in 1857.

"The embryo town of Ewell is on the main stage road between Arizona and New Mexico. It is about 13 miles west of the once-dreaded Apache Pass where many a white man lost his life at the hand of the blood-thirsty Apache Indians. Small as the settlement is, as yet, it has the benefit of a regular daily mail both from the east and the west, though just at present the Indians raiding in New Mexico somewhat interfere with the western bound stages."—*Arizona Daily Star,* Nov. 20, 1879.

"Ewell Springs, situated at the foot of Dos Cabezos, now has about 25 houses, including a post office, hotel, schoolhouse, saloons, etc."—*Arizona Daily Star,* Mar. 5, 1880.

FAIRBANK, Cochise Co., on San Pedro R., 6 m. (via paved State Hwy. 82) W. of U.S. 80, from a point 3 m. N. of Tombstone. Covered in text.

FILIBUSTERS' CAMP, Yuma Co., early day stage station on Gila R., 44 m. E. of Yuma (Hinton.) At this point, in 1857, camped Henry A. Crabb and his so-called Gadsden Colonization Company, a filibustering party organized in California with the idea of capturing all or part of the Mexican state of Sonora, after which they planned to have their newly-acquired lands annexed to the United States. The scheme, however, backfired. Surprised by Mexican federal troops, near Caborca, all but one of approximately 80 men in the invading party was killed in battle (or taken prisoner and later executed), and Crabb's head—cut off and preserved in mescal—was assertedly forwarded to Mexico City as a war trophy.

FITTSBURG, Cochise Co., a short-lived suburb to the W. of Pearce. Named for George Fitts, a Tombstone merchant.

FOREPAUGH, Maricopa Co., formerly a small settlement on U.S. 60-70, 19 m. W. of Wickenburg. P.O. 1910-16.

FORREST, Cochise Co., on E.P.&S.W. RR, 12 m. E. of Naco. (Barnes). Formerly a popular stopping point for teamsters hauling timbers from Turkey Creek to the Copper Queen mine, at Bisbee. P.O. 1914-17.

FORTUNA, Yuma Co., mining camp N. W. side Sheep Mtn., in the Gilas, about 22 m. S.E. Yuma. After discovery here of a rich vein of gold in 1893, a 20-stamp mill was erected in 1896 and the camp of Fortuna developed around it. During the next 8 years the mine assertedly produced more than $2,500,000 in gold and silver. Since 1904 the property has been worked only intermittently. Ruins.

FOURR'S FORT, Cochise Co., on the old Fourr ranch, 4 m. S.E. of Dragoon. Fort built by Billy Fourr in 1881 as a means of protecting his family against Indian raids. Roofless stone walls still standing. Notice: This is private property and should not be visited without permission of owner. Inquire at Dragoon.

FRISCO, Mohave Co., T. 22 N., R. 12 E. (Barnes). During the period in which it served as supply point for the Frisco mine this former settlement attained a peak population of several hundred and had a P.O. from 1913 to 1915.

GALEYVILLE, Cochise Co., Chiricahua Mtns. Covered in text.

GARCES, Cochise Co., former mining town E. side Huachuca Mtns. c. 8 m. W. Hereford (Barnes). P.O. originally called Reef, (1901-04) after Reef mine. Later name changed to Palmerlee (1904-11) after Jos. S. Palmerlee; still later (1911-26) called Garces, presumably for Fr. Francisco Garces, early missionary-explorer.

GARFIELD, Greenlee Co. Formerly a town of several houses and a store.

GATEWOOD, Cochise Co., in San Simon Valley, c. 12 m. N.E. Old Camp Grant. After Lt. Charles B. Gatewood, 6th U.S. Cavalry. P.O. 1890-94.

GAYBANI, SANTA CRUZ DE, Cochise Co., W. bank San Pedro R. immediately N. of its junction with Babocomari Ck. Spanish mission built c. 1696 at instigation Father Kino. Adobe ruins owned (1961) by Del Valle Cattle Co., Scottsdale.

GENERAL BOOTH GOLD MINE, Cochise Co., covered in text.

GERMA, Mohave Co., 2½ m. S.W. Oatman. The Germa mine, discovered c. 1896 and owned by German-American Mining Co., of Los Angeles, closed in 1906 due to insufficient water to run the mill on a double shift. The camp that served it, followed suit. P.O. 1903-06.

GIBSON, Maricopa Co., on lower New R., N. of Phoenix. Small settlement. P.O. 1900.

GIBSON, Pima Co., N. edge of Ajo. Townsite laid out c. 1912. Later was site of Rowood P.O.

GILA CITY, Yuma Co., 15 m. N.E. Yuma, and c. 1½ m. W. Dome RR Station. Covered in text.

GILLESPIE, Graham Co., old settlement on Stockton Ck., E. side Stockton Pass, Peloncillo Mtns. Former way-station on Safford-Fort Grant Rd.

GILLETT, Yavapai Co., former milling town. Covered in text.

GIMLETVILLE, Yavapai Co., one of several names derisively applied to Goodwin City by residents of neighboring Prescott. In files of APHS, at Tucson, is a fascinating reference concerning one Hampton Brown, a Texas-born mulatto who, according to this source, "went from Tucson to Gimletville with Virgin Mary and had a restaurant."

GISELA, Gila Co., former village on Tonto Ck., T. 9 N., R. 10 E. (Barnes) P.O. 1894-1911

GLEESON, Cochise Co., 16 m. E. Tombstone. Ruins covered in text.

GLENBAR, Graham Co., S. bank Gila R., T. 6 S., R. 24 E. Station Arizona-Eastern R.R., and formerly the location of a flour mill. P.O. at this point known originally as Matthews (1897-1906); Re-est. 1909 as Fairview. Changed to Glenbar, 1917. Discontinued, 1956.

GODDARD, Yavapai Co., former stage station on Black Canyon Rd., near confluence Squaw Ck. and the Agua Fria R.

GODMAN, Mohave Co. See: Liverpool.

GOLCONDA, Mohave Co., in Cerbat Mtns., T. 23 N., R. 17 W. (Barnes). Mining camp estab in early years of present century by Johnny Boyle, owner Golconda mine. P.O. 1909-18.

GOLD BASIN, Mohave Co. in Hualpai Wash, c. 15 m. S. Lake Mead and 40 m. N. Kingman. Following discovery of gold at this point in middle 1870s a lively camp took form but failed to develop extensively due to general remoteness and the scarcity of wood and water. P.O. 1890-94. Re-estab. as Basin 1904-07.

GOLDFIELD, Pinal Co., along Apache Trail, 5 m. up-canyon from Apache Jct., on U.S. 60-70. Following discovery of gold here in early 1890s, Goldfield (originally Youngsberg, after George U. Young) became a lusty mining boom camp with a claimed population of 5000. Arizona author, Barney Bernard, who knew Goldfield during its heyday when dancehalls, gambling dens, saloons and "wicked women" enlivened the scene, later recounted his experiences in a privately-printed, paper-back booklet entitled *Superstition Mountain.* Another detailed account of lively days in Goldfield is contained in Joseph Miller's *Arizona, the Last Frontier,* pp 264-278.

GOLD FLAT, Mohave Co., c. 10 m. S. Kingman. P.O. 1908-10.

GOLD ROAD, Mohave Co., 30 m. (by road) S.W. Kingman on U.S. 66. Covered in text.

GOLDEN, Maricopa Co., in Harqua Hala Mtns., c. 20 m. S. of AT&SFRR. Small mining camp of 1880s.

GOLDEN RULE, Cochise Co. See: Manzora.

GOLDEN TURKEY, Yavapai Co. Covered in text.

GOODWIN, CAMP, Graham Co., on Upper Gila R., c. 2 m. E. present community of Geronimo on U.S. 70. Army post active 1864-71 as base of operations against Chiricahua, White Mountain, and Pinal Apaches. In May, 1866, rumors reached San Francisco that Camp Goodwin had been captured by 2000 Apaches and all but one man of the camp's personnel had been massacred. This report, widely printed in newspapers throughout the country, later was shown to have no foundation in fact. In 1869 a group of "tame' Apaches living at the post killed Colonel Stone, deputy collector U.S. Customs, and five other men, and stole the U.S. mail. At peak of activity post numbered seven companies of infantry and cavalry. Abandoned 1871 due to prevalence of malaria.

GOODWIN CITY, Yavapai Co. Several communities named Goodwin, or Goodwin City have existed in Yavapai Co. over the course of years. First of these was established 1863 a short distance above the later site of Prescott. This town was short-lived. Another Goodwin, also a mining camp, later was established on Turkey Ck., S. of Mt. Union, in Prescott N.F.

GORDON, Gila Co., P.O. 1913-15.

GRAHAM, Graham Co., across Gila R. from Thatcher. Settled by Mormons in early 1880s. Townsite surveyed 1884 and an LDS meeting house constructed of mesquite poles with dirt roof and walls of unbleached muslin. (Granger).

GRANT CAMP and FORT, Pinal Co., nr. Jct. Arivaipa and San Pedro R. Army post estab. May 1860, as Fort Arivaipa, later functioning successively as Fort Breckinridge (August 1860); Camp or Fort Stanford (1862); Fort Breckinridge, (1863-65); and in the latter year was renamed Camp Grant. In 1866 nearly the entire camp was destroyed by flood and in 1871 the post was scene of the infamous Camp Grant Massacre, in which over-enthusiastic whites raided the post and massacred 118 Arivaipa Apaches quartered there. On the year following this black deed, malaria became so prevalent in the area that the post was moved to the W. side of Mt. Graham and rechristened Fort Grant. In the *Arizona Silver Belt* of December 13, 1878, a writer stated that Camp Grant once was a menace to the Indians ". . .(but) now they have their brush wick-ups built against its adobe sides. The camp cemetery, located near the river, is in a scandalous condition. The heros whose bones lie there . . . died by the poisoned arrows of the Apaches . . . Among the graves was one of little Jennie Laramie, aged five months and seven days. We put the headboard up and built stones around it to keep it in place . . ."

If anything but the bare site remains at Old Camp Grant, my several searchings have failed to discover it.

GRANVILLE, Greenlee Co., on Chase Ck., c. 6 m. above Metcalf. Former mining camp and supply point for timbers used in Metcalf mines.

GRAPEVINE SPRINGS, Mohave Co. "Quite a little town has sprung up at Grapevine Springs. There is a hotel, saloon, and butcher shop opened, and the work of building still goes on." *Arizona Weekly Miner,* May 10, 1878.

GRASS SPRINGS, Mohave Co. Suttenfield and Cantoo, who operated a saloon and butcher shop at Grass Springs in the early 1880s, carried this advertisement in the *Alta Arizona*: "The undersigned, having established themselves at Grass Springs, are prepared to give to the public both meat and drink. We are making preparations to accommodate the crowds of people that are sure to soon flock to Gold Basin, and we give them timely notice that we will keep good liquors and fat beef."

GREATERVILLE, Pima Co., 5 m. (by road) W. State Hwy. 83, at point 9 m. N. Sonoita on State Hwy. 82. Covered in text.

GREENWOOD CITY, Mohave Co., former milling town for ore of Greenwood mine. Covered in text.

GROOMS PEAK, Mohave Co. "A store was opened by George H. Frost and Co. in the Grooms Peak district on March 18 . . ." *Mohave County Miner,* March 30, 1884.

GROWLER, Pima Co., 27 m. S. W. Ajo, on road between Darby Wells and Papago Wells. Covered in text.

GUEVAVI MISSION, Santa Cruz Co., nr Calabazas, c. 10 m. below Tumacacori. Also known as Los Angeles de Guevavi, San Miguel de Guevavi, San Rafael de Guevavi, Santos Angeles de Guevavi, Guebavi, etc. The Indian rancheria at the later-to-be site of Guevavi mission, believed to have been the first church erected by Europeans in what is now Arizona, was first visited by Father Kino in 1691, at which time he remained in the settlement five days for the purpose of baptizing infants and adults. After founding the mission, Kino noted in his diary, in 1700, that there was then at Guevavi a flock numbering 84 sheep and goats, a field of maize, wheat, and beans, and a house ready for the resident priest who was expected to arrive soon. The first church at the site, likely built around 1700-1701, was soon replaced with a larger structure. In 1750 the mission was raided by hostile Indians, but within two years had been reoccupied and continued to serve its intended purpose until about 1780. The ruins, consisting of crumbling adobe walls pitted and probed by treasure hunters, are privately owned and are fenced to help insure their protection.

GUNSIGHT, Pima Co., T. 14 S., R. 4 E., one-half mile S. of State Hwy. 86, nr. W. boundary Papago Indian Reservation. The Gunsight mine, discovered 1878, proved to be phenomenally rich, some of its ore yielding as much as $3 per pound in silver. After taking out $9000 in a very short time, the owners sold their claims for a reported $100,000. At its peak the town had a claimed population of 1500 and was served by a tri-weekly stage out of Gila Bend. P.O. 1892-96. Ruins.

HACIENDA DE SAN YSIDRO, Yuma Co., in Gila R. Valley. Covered in text.

HAMBURG, Cochise Co., after Henry Hamburg, St. Louis German, who operated a store and P.O. in this small mining camp.

HANCOCK, Mohave Co., station on AT&SF c. 6 m. S.W. Kingman. Once a busy railroad center—on paper. On September 14, 1897, Arizona's Secretary of State received a dozen Notices of Intention to build railroads in Mohave County. Included among these were the following shortlines: Hancock to Mineral Park, via Cerbat, 25 miles; Hancock to Cerbat, 20 miles; Hancock to Chloride, 30 miles; Hancock to Boundary line, 95 miles; and two applications to build railroads from Hancock to White Hills, 60 miles.

HAPPY CAMP, either Maricopa or Yuma Co., early stage station on S. side Gila R., nr. Gila Bend. Generally described by travelers as a bleak station in a desolate setting. One such wayfarer, writing in *Picturesque Arizona* in 1877 reported that all available water at Happy Camp was hauled 15 miles across a desert waste, and that a charge of 25 cents each was made for watering horses. "It was originally one dollar, but, for some reason, was reduced recently." noted the writer.

HARDIN, Maricopa Co., early stage station at point where Black Canyon road crossed the Arizona Canal. For five months, in 1898, this place boasted a P.O.

HARDY, Greenlee Co., 4000 feet up on Ash Peak. Small camp serving the Ash Creek mines.

HARDYVILLE, Mohave Co., on Colorado R., 7 m. (by paved road) S. Davis Dam on State Hwy. 68. Ruins. Covered in text.

HARQUA HALA, Yuma Co., 9 m. (by road) S. Salome on U.S. 60-70. Covered in text.

HARRISBURG, Yuma Co., 5 m. (by road) S. Salome on U.S. 60-70. Covered in text.

HARSHAW, Santa Cruz Co., 10 m. (by road) S.E. Patagonia on State Hwy. 82. Covered in text.

HEINTZELMAN MINE, Pima Co. See: Cerro Colorado.

HELVETIA, Pima Co., 12 m. (by road) E. Continental on U.S. 89. Covered in text.

HILLCAMP, Gila Co., a short-lived P.O. (1920) estab. on site of an old sawmill camp on San Carlos Indian Reservation.

HILLTOP, Cochise Co., 7000 ft. high in Chiricahua Mtns., N.W. of Portal. After the Hilltop Mine. P.O. 1920-45.

HOLDER, Gila Co., on E. Verde R. between Payson and Pine. After John T. Holder, goat rancher and leading citizen. P.O. 1896-97.

HORSESHOE, Pima Co., S. end Quijotoa Mtns. Former placer mining camp, store, and P.O.

HOWELLS, Yavapai Co., on Lynx Ck. below Walker, c. 12 m. S.E. Prescott. Originally site of a small smelter active in 1880s. Ruins.

HUBBARD, Graham Co., early farming settlement E. side Gila R., 4 m. N Thatcher. After Elisha F. Hubbard Sr., Mormon Bishop. P.O. 1902-12.

HUDSON, Maricopa Co., W. side Agua Fria R., c. 15 m. above Gillett Station (Barnes). Early stage station and P.O. (1891-93).

HURON, Yavapai Co., T. 13 S., R. 29 E. (Barnes). Supply point for the Huron mine, early-day producer. Also station on Prescott-Middleton branch of AT&SF.

"Charles Middleton has completed a new lodging house at Huron and is now preparing to put up a hotel." Unidentified clipping dated May 7, 1900.

ILGES, CAMP, Yavapai Co., obscure and temporary Army post estab. W. bank Verde R., in 1870s.

INDIAN OASIS, Pima Co. Original name of Sells on State Hwy. 86. Founded c. 1914; name changed, 1918.

INFANTRY, CAMP, Pinal Co., in Mason Valley, nr. headwaters Mineral and Pinto Cks., 30 m. N.E. Florence. U.S. Army camp estab. 1870 to protect miners in the area. Name changed to Camp Pinal, April 4, 1871. Site abandoned July 1871.

JAMESTOWN, Maricopa Co. Town laid out by Jack Swilling who worked a nearby mine, 1865-70. Swilling's location notice on this property assertedly read: "I claim 300 feet on either side of this vein, and as far north and south as my Winchester will shoot."

JEROME, Yavapai Co., on U.S. 89, 33 m. S.E. Prescott. Covered in text.

JERSEY, Yavapai Co. On Slate Ck., c. 18 m. S. Prescott. Supply point and hdqrs. Jersey Lily mine, after Lily Langtry, English actress. P.O. 1895-1909.

JOHNSON, Cochise Co., 5 m. (by road) N. Hwy. 86, from a point 15 m. N.E. Benson. E. side Little Dragoon Mtns., T. 15 S., R. 22 E. (Barnes). Successor to the nearby earlier town of Russellville (q.v.), Johnson was founded early in 1883 as mining headquarters of the Peabody Co., composed of Lake Superior Copper men. General manager and driving spirit of the company was a Mr. Johnson, whose name was given to the new camp. Earliest reference I have found to Johnson, as a town, is contained in the *Arizona Daily Republican* of March 9, 1883, where it is stated that the new camp had then seven business houses, a population of 300, and that much building and improvement was under way. The company had 103 men on the payroll, among whom, it was said, "is not one Mexican or Chinaman." Top and bottom men were being paid $4 per day. J. G. Parke, late of Tombstone, had surveyed the townsite into blocks, lots, and streets 80 feet in width.

On May 8, that same year, a correspondent to *Arizona Daily Star* reported much progress at Johnson. "There are four saloons here . . . We have three stores and two hotels, one market, and one produce business. Ed Benson, of Tombstone, has put on a stage line from Summit Station to this place . . . Judge Head is our Justice of the Peace. There will be a drug store, barber shop, and bakery, opened here about the 15th of this month, and Mr. Hunspear has already opened a very fine steak house . . . Johnson is already graced by quite a number of ladies, one of whom, Mrs. Kelly, of Benson, is about to open a school . . ."

Operations rolled along most encouragingly for awhile, but Johnson's applecart was upset by the financial panic of 1907. Though the camp has revived on several occasions since, it has been a ghost town for several years. Few ruins remain, except at the Republic mine and milling camp, nearby.

JOKAKE, Maricopa Co., P.O. 1936-54.

JUNCTION CITY. "The new town of Junction City is being surveyed on the east bank of the San Pedro, about one and one-half miles south of the Grand Central mill, where the Benson Branch joins the AT&SFRR." —*Pinal Drill*, March 4, 1882.

KATHERINE, Mohave Co., E. side Colorado R., c. 4 m. N. Davis Dam. (Over ridge, N. from Katherine Landing.) After Katherine mine, discovered c. 1920 by J. S. Bagg, of Kingman. Community continued active until 1942 when mine was closed by order War Production Board.

Mining and milling machinery auctioned in 1943. P.O. 1921-29.

KEITH CITY, Mohave Co. (?) "Keith City is flourishing, and Mr. F. J. Parke has reason to feel proud of his mine, as the piles of ore on the dump of the Black Rock are immense, both as to quantity and richness . . ." *Arizona Sentinel,* June 17, 1882.

KELTON, Cochise Co., between Douglas and Gleeson, in Sulphur Springs Valley, on Arizona Eastern RR. After Capt. C. B. Kelton, former sheriff of Cochise County, who homesteaded land on which the town was later located. P.O. 1915-28. Community originally called Tyler after first settler.

KENYON (or KINYON'S) STATION, Maricopa Co., stage stop S. side Gila R. 15 m. W. Gila Bend. Covered in text.

KIRBY, Gila Co, Sec. 2, T. 3 N., R. 13 E. Tonto N.F., S. side Roosevelt Lake (Barnes). Settled by Mormon farmers 1878-80. P.O. and store.

KIRBY, KIRBYVILLE, Yavapai Co., mining camp nr. Tip Top mine. "The smelter at Kirbyville, on Castle Creek, is being put into position." —*Arizona Daily Star,* June 1, 1883.

KLONDIKE, Graham Co., T. 7 S., R. 19 E., on Arivaipa Ck. (Barnes). Trading center and P.O. (1907-55).

KOFA, Yuma Co., c. 24 m. (by road) E. State Hwy. 95 from point 28 m. S. Quartzsite. Covered in text.

KYRENE, Maricopa Co. Former station Maricopa branch AT&SF, 6 m. S. Tempe. Small cattle shipping point with store, school, etc. Now known as West Chandler.

LA CIENEGA, Pima Co., c. 19 m. W. Benson. Early stage station and scene of several notable fights between pioneer settlers and Apaches. Covered in text.

LAGUNA, also known as Nine-Mile Waterhole. Pima Co. "A station nine miles from Tucson, on the Sacaton road, known as The Laguna, is fast . . . assuming the dimensions of a settlement. There are now some 500 acres of land under cultivation, with a population of some 80 or 90 hardy pioneers." —*Weekly Arizonan,* May 22, 1869.

LAND, Cochise Co., on San Pedro R., 9 m. S.E. Benson. After Col. W. C. Land, stockman and part-owner Babacomari Land Grant. P.O. 1911-13.

LA NORIA, Santa Cruz Co. It is difficult, if not impossible, to separate the history of La Noria from that of Lochiel. Possibly the two places were one and the same, though Barnes locates La Noria "probably" about 1½ m. N.E. of Lochiel. P.O. 1882-83 and 1909-11.

"La Noria is the border town of the Patagonia Mountains, one portion being across the border Here, owing to abundance of water, are placed the two smelters belonging to the Holland and Yankee, or Davis, companies . . . There are one or two stores here, several saloons and, until recently, Doc Luttrell ran a boarding house. Here, also is situated a United States Customs house with the necessary officers." —*Tombstone Epitaph,* December 16, 1882.

LA OSA, Pima Co., at or near present hamlet of Sasabe. P. O. 1903-05.

LA PALMA, Pinal Co., 7 m. (by paved road) S. of Coolidge. The community of La Palma, shown on current maps of Arizona, is a pleasant little settlement—most places in the Casa Grande Valley are pleasant, especially in that hinder half of the year when Northern and Eastern states are caught in the clutch of winter. But the La Palma of today is a far leap from the La Palma envisioned back in 1927 when Promoter Harry J. Collis and a group of Chicago capitalists were seeking to develop this place as "one of the most outstanding residential, business and resort cities of the Southwest."

In those halcyon, blue sky days of nearly 40 years ago, when a tumultous land boom was sweeping the country from Florida to Sunny California, La Palma knew even the distinction of its own newspaper, *La Palma Chimes,* whose wide-awake editor never passed an opportunity to boost his community and the aforementioned Mr. Collis. In the first number of the *Chimes,* it was stated that the La Palma hotel, golf and country clubhouse (designed by Architect Albert Chase McArthur) was to be constructed of brick made in the factory of La Palma Clay Products. The roofs were to be of Spanish tile, also made in the La Palma factory, and all floors would be of La Palma floor tile. The two-story lobby would be encircled by balconies, and the dining room—embracing 2000 square feet of floor area—"will contain a dance floor in the cabaret style."

"Here," stated Promoter Collis, "we visualize a city of homes, schools, resort hotels, clubs, beautiful palm-lined avenues, broad shaded streets . . . and every accessory of modern life, laid out in such manner, from the beginning, that no alterations will ever be necessary . . ."

LA PAZ, Yuma Co., 5 m. N. Ehrenberg on U.S. 60-70. Covered in text.

LA PAZ, CAMP, Yuma Co. Temporary Army supply camp estab. 1874 at the then deserted placer mining camp of La Paz (see previous page). Abandoned 1875. (Brandes.)

LAUB, Cochise Co., nr. Dos Cabezos. P.O. 1900-1901.

LAYNE CITY, or LAYNE SPRINGS, Mohave Co., 4 m. S. Mineral Park. "Layne City is the name of a new town that is springing up at the Prosperity mine, in Todd Basin. It is very appropriately named after Capt. Layne, pioneer miner of the district." —*Mohave County Miner,* November 22, 1885.

Letter from Layne Springs to *Mohave County Miner,* August 11, 1888 (excerpts): "The boom in this camp is on the increase . . . It is a pretty sight to stand at the head of the canyon and watch Bannock's train of 30 jackasses, laden with ore, meander down the serpentine trail like a great lazy snake . . ."

LEHI, Maricopa Co., in Salt R. Valley, c. 3 m. N. Mesa, Settled in 1877 by LDS colonists from St. George, Utah. Originally called Camp Utah or Fort Utah, name was later changed to Utahville, and in 1880 to Jonesville. When P.O. department refused to accept this name, Brigham Young, Jr., suggested name of Lehi. In 1880 the settlement had a population of 168, making it then the fourth largest town in Maricopa County. P.O. 1884-1904. Covered in text.

LELAND, Mohave Co., between Kingman, Ariz., and Needles, Calif. Small mining camp briefly active around turn of the century.

LEWIS, CAMP, Yavapai Co. Army camp (1865-70) having, apparently, two locations. According to Brandes the GLO map of 1866 shows an Army camp at junction of the Verde and Salt R., while Johnson's map of 1870 locates the post on Fossil Ck., nr. head of Salt R., on trail between Verde Valley and Tonto Basin.

LEWIS SPRINGS, Cochise Co., T. 21 S., R. 22 E., (Barnes). Former station EP&SWRR, estab. 1889. Wells Fargo Station, 1904. P.O. 1905-33. After Robert Alpheus Lewis, early-day miner in Bisbee-Tombstone area.

LIGHT, Cochise Co., T. 18 S., R. 27 E., on Turkey Ck., Sulphur Spgs. Valley (Barnes). After John W. Light, dryland homesteader, who served the community with a store and P.O. (1910-27.)

LINCOLN, CAMP, Yavapai Co., U.S. Army camp estab. 1864 on E. bank Verde R., c. 1 m. N. Beaver Ck. (Brandes). In 1866 name of post was changed to Fort Lincoln; in 1868 to Camp Verde, and in 1879 to Fort Verde. Meanwhile location of the post had shifted several times to a final site on W. side of river where town of Camp Verde is located today. Post abandoned 1890. Pauline Weaver, noted frontiersman, was briefly assigned to this camp as scout and guide and it was here that he died in 1866. Ruins.

LINCOLNIA, Mohave Co., P.O. (1911-12) and store owned by Edmund Lincoln on Fort Mohave Indian Reservation. Apparently the same as Cottonia.

LISCUM, Maricopa Co., on Cave Ck., T. 7 N., R. 4 W. (Barnes) P.O. for Phoenix mine (1901-02) and site of a stamp mill.

LIVERPOOL LANDING, Mohave Co., former steamship landing on Colorado R. Also known as Pittsburgh Landing and Godman. Covered in text.

LIVINGSTON, Gila Co., on Salt R., E. end Roosevelt lake. After Charles Livingston who settled here in 1870s. P.O. 1896-1907. Former townsite is now covered with heavy thickets of willows and mesquite.

LOCHIEL, Santa Cruz Co. The little Mexican border settlement of Lochiel (named by Colin Cameron of San Rafael Land Grant after family's former home in Scotland) has survived, in a manner of speaking, though 80-odd years, neither dying completely nor prospering abundantly. "During the mining heyday Lochiel was a thriving and lusty town—wide open, riotous and noisy," wrote Richard G. Schaus in *Arizona Highways,* September 1953. "Besides having a smelter it served as a shipping point for the burro trains and wagons as they began their trek to Guaymas or Mexico City. Today Lochiel is a quiet settlement . . . most of the adobe stores, saloons and residences have melted away, after the fashion of deserted adobe buildings . . ."

LOGAN CITY, Pima Co., former mining boom camp nr. Covered Wells, on Hwy. 86. Covered in text.

LOMPOC, Graham Co., early way-station Sulphur Spgs. Valley, c. 12 m. S. Fort Grant. After Lompoc, Calif. P.O. 1913-15.

LOS REALES, Pima Co., a former town across the Santa Cruz R. from Mission San Xavier del Bac. In 1961 the Historic Sites Committee, of Arizona Pioneers' Historical Society, located the site of this place "at the W. end Los Reales Rd., and S. 6th or S. 12th Aves., in Tucson," and reported that all buildings were gone. The Committee also stated that Los Reales had been site of an ore reduction plant as early as 1865, and in 1874 had been populated with Mexicans evicted from San Xavier Indian Reservation.

If findings of the committee are true, and the town actually dated from as early as 1865, then it must have been under a different name as the Tucson *Citizen,* on August 11, 1888, stated: "A new town, called Los Reales, is just starting, about 12 miles south of Tucson."

Again, on January 28, 1885, from the *Citizen*: "Within the last four years a town known as Los Reales, on the Arivaca road, has sprung into existence about one and one-half miles east of San Xavier. About 20 houses of teamsters, farmers, woodcutters, mostly Mexicans . . ."

In 1937, Bernice Cosulich, a writer in the employ of *Arizona Star,* interviewed Doña Ciena Moraga, aged 76, who was reared at Los Reales. According to Mrs. Cosulich's story in *The Star,* June 27, 1937, Mrs. Moraga stated that the town died when the Santa Cruz river changed its course and "a big farm used all the water. Family by family moved away, houses were pulled down to build others, and water washed the adobe walls to mounds." At that time (1937) the old cemetery was still intact, as well as other ruins.

LOWELL, CAMP or FORT, Pima Co., early Army post 7 m. N.E. of present downtown section of Tucson (E. end Ft. Lowell Rd.) Outgrowth of a military garrison estab. Tucson 1860 and in 1866 renamed Camp Lowell, after Brig. Gen. Charles R. Lowell, killed at Cedar Creek, Va., in 1862. On March 19, 1873, the troops of this post were moved to a location outside the town limits and large adobes were used in construction of a permanent post. Trees lined the drives; a hospital and stables were erected, and the parade ground was quite impressive. Between 1873-79 many expeditions were launched from this post in pursuit of Apaches, and from here, in 1886, Gen. Nelson A. Miles directed the final campaign against the Apaches. Post was renamed Fort Lowell in 1879, and abandoned in 1891. For several years past the ruins have not been open to the public, but some of the old adobe buildings and the long adobe wall enclosing the reservation are visible from Craycroft avenue.

LUTTRELL, Santa Cruz Co., former mining camp near Lochiel. Named for its No. 1 citizen, Dr. J. M. Luttrell superintendent of a nearby mine. P.O. 1880-83.

LYONSVILLE, Mohave Co., nr. Oatman. "Below Virginia City half a mile is Lyonsville, sarcastically said by Virginians to consist of one house to be built and the skeleton of a corral." —J. Ross Browne.

MADERA CANYON, Pima Co., former sawmill town in Coronado N. F., Santa Rita Mtns., c. 12 m. S.E. Continental. First sawmill in Arizona was located in this vicinity. P.O. 1929-42.

MAISH VAYA, or COVERED WELLS, Pima Co., 23 m. N.W. Sells, on Hwy. 86. At time of the Quijotoa mining excitement in the early 1880s, this former Papago Indian village became a busy and important place due to the good water supplied by its wells—a priceless treasure in this land of little rain. The townsite of Covered Wells (English translation of the Papago name) was laid out in 1884 but after the Quijotoa boom petered out, the desert watering place quickly resumed its former status as an Indian village.

MAMMOTH, Pinal Co., on State Hwy. 87, 12 m. N.E. Oracle. Pleasantly situated on the banks of the San Pedro R., Mammoth, today, is only a country hamlet where ranchers of the surrounding area come for supplies, and motorists on the Tucson-Globe highway pause for gasoline and whistle-wetting material. Here, in the 1880s and '90s was located a 100-stamp mill built to treat ore of the Mammoth, or Shultz mine, which was delivered to the mill by huge buckets hung on a gravity-operated cable tramway. On their return trip the buckets carried water from the San Pedro to the dry mining camp. The camp was originally known as Shultz, after Frank Shultz, discoverer of the mine.

MANZORA, Cochise Co., T. 16 S., R. 23 E. East side Dragoon Pass. (Barnes). Former shipping point for ore from the Golden Rule mine. Originally known as Golden Rule. P.O. 1916-18.

MARICOPAVILLE, Maricopa Co., RR boom town, c. 5 m. W. Maricopa Wells Station. Covered in text.

MARICOPA WELLS, Maricopa Co., Sec. 17, T. 3 S., R. 3 E. (Barnes). Formerly an important stage station 1½ m. W. Pima Butte and 5 m. N. present town of Maricopa. Ruins. Covered in text.

MARTINEZ, Yavapai Co., 16 m. N.W. Wickenburg. Original name of Congress Junction. Covered in text.

MARYSVILLE, Gila Co., short-lived mining camp on Webber Ck., c. 3 m. W. Payson. *Pinal Drill,* April 30, 1881, reported that Chilson Brothers, of Globe, had opened a store at Marysville, on E. fork Verde R., and that the town had also a blacksmith shop and meat market.

MARYVILLE, Maricopa Co., on N. bank of Salt R., opposite Lehi. Assertedly named for Mrs. Mary E. Bailey, nee Whitlow, whose father had settled here in 1865 and was the town's first p.m. In 1873, Maryville had a store, hotel, blacksmith and carpentry shops, and feed corral; and on August 2, 1873, the *Arizona Citizen* carried the following news story datelined Maryville, A.T., July 22, 1873: "Last evening we had a theatrical performance by the Maryville Amateur Troupe composed of Dr. T. J. Wilson, stage manager; Mr. and Mrs. Davies, and the Misses Mollie and Sierra Nevada Whitlow. The performance commenced at 8 p.m. by reading of the Maryville *Hornet,* a paper published semi-occasionally. After the reading there were singing, recitations, comedy, etc. The evening entertainment concluded with a supper by Mr. Charles Whitlow, after which the audience departed, well pleased with themselves and with the entertainments."

MASCOT, Cochise Co., 16 m. S.E. Willcox. Mining camp at terminus Mascot & Western branch of Arizona-Eastern RR. The leading mine of the region, located by Chas. Roberts c. 1915, was promoted several times under various names and ownerships. P.O. 1916-18.

MASON, FORT, Santa Cruz Co., 12 m. S.E. Tubac nr. Santa Cruz R. Estab. 1865, the name of this small army post was changed, in September 1866, to Camp McKee, and 24 days later the post was abandoned. Some ruins, on patented ground.

MASON'S CAMP, Cochise Co. "During the turbulence of the late storm what is known as Mason's Camp, located about a mile southwest of Watervale, was totally swept away by a cloudburst . . . About 5 o'clock Sunday afternoon residents of the camp were appalled by the sight of a volume of water from 50 to 75 feet wide and at least six feet deep, coming down the sandy wash with the velocity of a whirlwind. They had barely time to reach higher ground when the torrent of foaming water struck the camp, carrying away every cabin in it. Some of the structures were of stone and adobe. The powder magazine, a large stone building, was the only thing that withstood the force of the torrent . . . The destruction of Mason's Camp was complete and naught is left to mark the embryo city but a streak of slickness left by the torrent." —*Tombstone Daily Nugget,* July 19, 1881.

MASSICKS, Yavapai Co., on lower Lynx Ck., half mile S. of a point on Hwy. 69, c. 9 m. S. Prescott. Chimney of old P.O. still standing 1961.

MATTHEWS, Graham Co. See: Glenbar.

MAXEY, Graham Co., T. 4 S., R. 23 E. (Barnes), on Gila R. A small, rough settlement that served, originally, as blow-off town for soldiers stationed at Fort Thomas. In the early 1880s, Maxey had several places of business, most of which dealt in either spiritous refreshments and/or women. There was also a P.O. and a short-lived newspaper, *The Maxey Times,* estab. 1882 by W. W. Jones. (Luttrell).

MAZATZAL CITY, Gila Co., Sec. 17, T. 10 N., R. 9 E., at point where City Ck. enters E. fork of Verde (Barnes) c. 10 m. W. Payson. Settled by LDS colonists in late 1870s.

McALLISTER, Cochise Co., small farming settlement in Sulphur Spgs. Valley, c. 10 m. N.W. Willcox. P.O. 1911-20.

McCABE, Yavapai Co., 5 m. W. Humboldt on Hwy. 69. Ruins. Covered in text.

McCRACKEN, Mohave Co., 12 m. W. Signal. Covered in text.

McDOWELL, CAMP or FORT, Maricopa Co., W. bank Verde R., 7 m. above jct. Verde & Salt R. (Brandes). U.S. Army fort estab. 1865 and manned by 5 companies California Volunteers (about 500 men). P.O. 1869-1917, and 1923-28. Post abandoned 1891. Some ruins, cemetery, and one adobe building remain.

McKEE, CAMP, Santa Cruz Co. See: Mason, Camp.

McMILLEN, or McMILLENVILLE, Gila Co., 18 m. (by road) N.E. of Globe. Former mining camp. Covered in text.

McPHERSON, CAMP, Yavapai Co., on Date Ck., 60 m. S.W. Prescott. A short-lived Army post—Jan. to March 1867—estab. to protect travelers on La Paz-Prescott Rd. After Brig. Gen. James B. McPherson, killed in Battle of Atlanta. In March, 1867, camp was moved 25 m. N. and renamed Camp Skull Valley.

MERIDIAN, Maricopa Co., in Buckeye Valley. P.O. 1894-95.

MESCAL, Cochise Co., station EP&SWRR, 6 m. W. Benson. P.O. 1913-1931.

METCALF, Greenlee Co., former mining camp on U.S. 666, 3 m. N.W. Clifton. Covered in text.

MIDDLEMARCH, Cochise Co., Sec. 8, T. 24 E., R. 18 S., c. 6 m. S.W. Pearce (Barnes). After Middlemarch mine. P.O. 1898-1919.

MIDMONT, Maricopa Co. Between Phoenix and Camelback Mtn., P.O. 1919-20.

MIDWAY, Cochise Co. Former station on Arizona-Eastern R.R., between Pearce and Kelton.

MILLTOWN, Cochise Co. Milling town across San Pedro R. from Charleston. Covered in text.

MILLTOWN, Mohave Co., 6 m. from Colorado R., nr. Oatman. Site of a 40-stamp mill erected by Mohave Gold Mining Co., 1903. Served by a narrow gauge RR from Needles, Calif. (Colorado R. crossed by means of ferry). "Cattle roaming the open range posed a real problem which necessitated putting a gate across the track. This meant that on every trip the engineer had to stop the train, open the gate, move the train through, close the gate, and then proceed on his way. During a severe flood the Colorado washed out all but two miles of the track." (Granger).

MINERAL HILL, Pima Co., former mining camp on Tucson-Nogales rd.

MINERAL PARK, Mohave Co., T. 22 N., R. 18 W., (Barnes) c. 12 m. N.W. Kingman. Ruins. Covered in text.

MINGVILLE, Graham Co., Sec. 28, T. 6 S., R. 19 E. (Barnes). After "Big Dan" Ming, early stockman. P.O. 1881.

MIRAMONTE, Cochise Co., 9 m. W. St. David, on San Pedro R. Settled 1913. P.O. 1918-19.

MISSION CAMP, Yuma Co., 32 m. E. Yuma. Covered in text.

MOHAVE CITY and FORT, Mohave Co., on Colorado R., c. 10 m. (by road) S. Bullhead City. Covered in text.

MOHAWK STATION, Yuma Co., T. 8 S., R. 15 W., (Barnes). Early stage station. Covered in text.

MONTANA CAMP, Santa Cruz Co. Original name of Ruby. Covered in text.

MONTE CRISTO, Pima Co., W. side Cimarron Mtns., S.W. Vekol mine. Hdqrs. Monte Cristo mine. P.O. 1922-23.

MONTEZUMA, Graham Co. See: San Jose.

MONTEZUMA, Pinal Co., former stage station S. side Gila R., c. 12 m. S.W. Florence on old Tucson Rd., and 6 m. W. Casa Grande National Monument.

MONTEZUMA FORT, Santa Cruz Co., Santa Rita Mtns., fair condition. Patented ground.

MONTGOMERY, Maricopa Co., station c. 12 m. N. Phoenix. P.O. 1913-20.

MOORE, CAMP, Santa Cruz Co., Army camp estab. 1856 above present town of Patagonia, 45 m. S.E. Tucson, by a detachment of the First Dragoons. Probably the first post built by U.S. Government as a base of operations specifically against the Apaches. Poorly constructed. Mostly jacales, a few adobes. Name changed, 1857, to Fort Buchanan. Ruins.

MOORE'S SPUR, Cochise Co., P.O. 1913-14.

MORRISTOWN, Graham Co. Little information available concerning location of this former mining camp laid out by T. D. Morris on the stage road, in Clark Mining District.

MORSE'S MILL, Cochise Co., on Turkey Ck., Chiricahua Mtns. Site of a sawmill operated by Philip Morse c. 1880-84. Timbers milled here were used chiefly in Copper Queen mine at Bisbee. In 1882 the voting precinct of Morse's Mill had a population of 94.

MOWRY, Santa Cruz Co., c. 17 m. (by road) S. Patagonia. Covered in text.

MUNSONVILLE, Graham Co. See San Jose.

MYRTLE, Gila Co., P.O. 1890-99. After Myrtle Pyle who died here and is buried near Bonito Ck.

NEW LONDON, Mohave Co., a small but prosperous mining camp of 1880s. Old mining shafts, some ruins.

NEW RIVER STATION, Maricopa Co., early stage station on Hwy. 69, c. 36 m. N. Phoenix.

NEW VIRGINIA, Pima Co., mining boom camp. Covered in text.

NIGGER WELL, Maricopa Co., c. 12 m. S. Wickenburg. Covered in text.

NORTH STAR MINE, Yuma Co. Covered in text.

NORTON'S LANDING, Yuma Co., on Colorado R., c. 50 m. N. Yuma. Former steamship landing and site of reduction works for ore of the Red Cloud mine. P.O. 1883-94. Covered in text.

NUGGET, Gila Co., c. 2 m. W. Richmond Basin. Former mining camp serving the Basin area, and Nugget mine. P.O. 1881-84.

OAK CREEK STATION, Gila Co. At this place on Sept. 15, 1910, Alfred Hillpot and Fred Kibbe, two young men from Globe, were murdered by two army deserters, Jim Steele and Bill Stewart. Motive, robbery. Full particulars concerning the crime, statewide manhunt, and subsequent action, were related by Roscoe G. Willson in *Arizona Republic*, issues of July 10 and 17, 1955.

OAKDALE, Gila Co., c. 7 m. from Globe. ". . . farther up the gulch we find the camp of Buck and Hayes, proprietors of the famous Julius mine, located a few hundred yards southwest of the camp, on the Quartzsite mountain. I understand this camp has been christened Oakdale. Romantic, somewhat, for a mining camp; but as long as the mines turn out ore worth from one to ten dollars a pound, as they are at present, we will not quarrel over a simple name . . ." —*Arizona Silver Belt,* June 6, 1878.

O'ATMAN, Mohave Co., 33 m. S.W. Kingman on U.S. 66. Covered in text.

OCHOA or OCHOAVILLE, Cochise Co., *Arizona Daily Star,* Sept. 11, 1879, refers to "new settlement" of Ochoaville, on the Elias Land grant, ½ m. from Sonora line, 2 m. from Camp Huachuca, and 20 m. from Charleston. Reportedly named for Estevan Ochoa, its founder, the settlement then embraced a store owned by Juan N. Acunea, and a population of about 75.

OCTAVE, Yavapai Co. Covered in text.

OLD GLORY, Santa Cruz Co., 2 m. S. Ruby, close to Mexican border. Camp of the American Flag mine. P.O. 1893-1915. In my files is a letter written in 1908 on the printed stationery of E. S. Kibbe's general merchandise store, at Old Glory, A.T. In addition to setting forth the fact that Old Glory is "10 hours ride via stage, and 6 hours via auto"—presumably from Tucson—it names Charles O'Neill as postmaster and manager of the store, dealing in "Drygoods and groceries, barley, hay and grain, shelled corn, white wheat, powder, fuse, caps, candles, coal, and drill steel."

OLDTRAILS, Mohave Co., T. 19 N., R. 20½ W. (Barnes). In Black Mtns. nr. Oatman. P.O. 1916-25.

OLIVE, or OLIVE CAMP, Pima Co., W. of Santa Cruz R., T. 17 S., R. 12 E. (Barnes). Said to have been Pima County's largest mining camp in 1889. P.O. 1887-92.

OLIVE CITY, or OLIVA, Yuma Co. Also known as Bradshaw's Ferry. On Colorado R. nr. Ehrenberg. "Olive City: This little town, situated on the mesa at the foot of Weaver Pass, just above the ferry and near a steamboat landing, is growing steadily and now comprises, perhaps a dozen buildings, adobes, and jacales."

—*Alta California,* October 14, 1863.

ORO, or OROVILLE, Greenlee Co., c. 3 m. above Clifton, on San Francisco R. "Oro is the name of the gold mining camp on the San Francisco River. Mrs. J. T. Yankie, one of the oldest settlers in that region . . . is merchandising in Oro and brings good news from this flourishing city. The hydraulic company are laying pipes which, when completed, will waken one of the most prosperous industries in Arizona. The company has about 10,000 acres of fine placer ground which yields high in coarse gold."—*Arizona Star,* quoted in *Tombstone Epitaph,* Jan. 14, 1882.

ORO BLANCO, Santa Cruz Co., 7 m. S.E. Arivaca. Former mining camp. Ruins. Covered in text.

OSTRICH MINE and MILL, Santa Cruz Co., corrupted spelling and pronunciation of Austerlitz, q.v.

OVERTON, Maricopa Co., c. 30 m. N. Phoenix, nr. Cave Ck. P.O. 1880-81.

OWEN, Mohave Co., T. 15 N., R. 13 W., nr. ghost town of Greenwood City. After Chloride Jack Owen, co-discoverer of McCracken mine. P.O. 1899-1914.

OXBOW, Gila Co., between Payson and Roosevelt. After the Oxbow mine, located 1880. P.O. 1895-1908.

PACIFIC CITY, Yuma Co. Covered in text.

PALMERLEE, Cochise Co. See: Garces.

PALOMAS, Yuma Co., former settlement, now vanished. P.O. 1891-1938.

PALOMINAS, Cochise Co., former mining camp, now vanished.

PARADISE, Cochise Co., early mining camp in foothills Chiricahua Mtn., 5 m. N.W. Portal. Covered in text.

PARKERVILLE, Mohave Co. "Mineral Park and Parkerville are two new settlements in the Hualpai or Sacramento Mining District." —Rossiter W. Raymond, *Resources Report,* 1873.

PEARCE, Cochise Co., historic mining camp in Sulphur Spgs. Valley, 1 m. off U.S. 666. Covered in text.

PICACHO, Pinal Co., early stage station nr. jct. present Hwys. 87 and 93.

PICK-EM-UP, Cochise Co., small rough settlement of early 1880s between Charleton and Tombstone.

PICKET POST, CAMP, Pinal Co., U.S. Army post at base of Picket Post peak, estab. 1871 by Gen. George Stoneman. Later, site of Pinal City.

PIERCE FERRY, Mohave Co., on Colorado R., c. 75 m. N. Kingman. The present fishing camp and recreational area of this name is not the original Pierce Ferry, which now lies beneath the water of Lake Mead. Ferry service across the Colorado R. at this point was established by Harrison Pearce in 1876 to provide easier access to the Mohave County mines from the north. Somewhere along the line the spelling of Pearce's name apparently was corrupted to Pierce.

PINAL, CAMP, Pinal Co., in Mason Valley, below Hutton Peak, nr. headwaters Mineral and Pinto Cks. U.S. Army post estab. by Gen. George Stoneman, 1870, under name Camp Infantry. Five months later (April 4, 1871) name was changed to Camp Pinal. Site abandoned July 5, that year.

PINAL CITY, Pinal Co., milling town for ore of Silver King mine. Covered in text. Ruins.

PIONEER, Gila Co., w. side Pinal Mtn. above Silver Ck. Camp serving the Pioneer mine, located 1876. As aftermath of what began as a friendly wrestling bout on Christmas Day, 1882, Ex-constable Tom Kerr shot and beat to death a man named Harnett, and was subsequently lynched by the crowd. For many years the "Hanging Tree" was pointed out to visitors; but representatives of the Arizona Historical Sites Committee, in 1961, found that the tree had been obliterated by mine debris, and that only the cemetery and one building remained to mark the site of the old town. P.O. 1882-85.

PIONEER STATION, Gila Co. "Point where travelers to and from the old Pioneer mine could take the stage to either Globe or Phoenix." (Granger).

PITTSBURGH, Cochise Co., 6 m. N.E. Paradise. Camp of the Cochise Consolidated Copper Co., of Pittsburgh, Pa. P.O. 1906.

PITTSBURG LANDING, Mohave Co. See: Liverpool.

PLACERITAS, Yavapai Co., former placer mining camp at N.E. tip Weaver Mtns., 5 m. E. Peeples' Valley. For many years Placeritas was the headquarters of, and dominated by, A. W. "Old Grizzly" Callen, who first came to Yavapai County at the head of a 15-wagon emigrant party of 34 men in the summer of 1873. Liking what he found there, Callen returned to Junction City, Kansas, and organized the Arizona Mining Association, with himself as general manager. Comprising the group were 52 men, each of whom had paid in a stipulated sum of money to form a company fund to be expended in the development and working of Arizona mines. The co-partnership was to endure two years, after which any man in the group would be free to go his separate way.

On November 3, 1875, the party arrived at Prescott where it created quite a sensation. Leading the long cavalcade was the imposing figure of A. W. Callen, riding in a good buggy drawn by fine horses. Behind their leader followed 21 huge emigrant wagons, drawn by oxen, and carrying more than 80 persons.

"The cattle," reported *Arizona Weekly Miner,* "though somewhat thin, are in good condition considering the great distance they have traveled and the heavy load they have drawn."

Members of the party had brought clothing and staple groceries believed sufficient for the entire two years of their stay, and Callen announced that it was the intention of the group to prospect until Spring and then commence operations.

Although the Association failed to distinguish itself, Callen eventually settled at Placeritas and became its kingpin. It was here, in 1889, that he killed Byron J. Charles and Frank Work in a quarrel over mining property. Arrested by Sheriff Bucky O'Neill (who rode out from Prescott, took Callen into custody, and rode back to Prescott, thereby covering 85 miles in a single day on the same horse), Callen was eventually exonerated on grounds of self defense.

In 1905, at the age of 85 years, Old Grizzly bade goodbye to his friends in Yavapai County, and the time and place of his death is unknown to this writer.

PLANET, Yuma Co., s. side Bill Williams R., c. 20 m. N.E. Parker. Covered in text.

POINT-OF-MOUNTAIN, Cochise Co., nr. present town of Willcox. Former stage station at junction main East-West stage line and branch line to Globe. Passengers for Camp Grant, Camp Thomas, San Carlos, Globe, and Fort Apache, changed here to saddleback to complete their journey.

POINT-OF-MOUNTAIN, Pima Co., c. 18 m. N.W. Tucson, N.E. side Santa Cruz R., nr. Rillito, (possibly at Jct. Silverbell Rd. and Hwy. 84). Former stage station and camping place for freighters. "As late as 1908 the buildings were in a fair state of preservation, but the ruins are now demolished." (Granger).

POLAND, Yavapai Co. 9 m. W. Poland Jct., on Hwy. 69, 4 m. S. Humboldt. Former mining camp, now summer home colony. Covered in text.

POLARIS, Yuma Co., P.O. (1909-14) for North Star Mine.

POLHAMUS, Mohave Co., former steamboat landing on Colorado R., 5 m. N. Hardyville (nr. present site Davis Dam.) Covered in text.

POOL, Cochise Co., on San Pedro R., c. 1 m. S. Cascabel. P.O. 1902-13.

PORTLAND, Cochise Co., small mining camp, now a ghost.

POSTVALE, Pima Co., c. 21 m. N.W. Tucson on Santa Cruz R. Small farming community. P.O. 1920-25.

POWERS, CAMP, Cochise Co., nr. the White R. Estab. 1878 as Camp Supply. Soon renamed Camp Powers. On Jan. 1, 1879, post was moved to upper end White R. and renamed Camp Rucker.

PRATT, Maricopa Co., W. side Agua Fria R., above Gillett. Supply point for a mine owned by Wm. Pratt. P.O. 1890-96.

PRICE, CAMP, Cochise Co. Temporary telegraph station and Army post (1881). At S. end Chiricahua Mtns., head of Texas Canyon. After Col. Sterling Price of Mexican War fame.

PROSPECT, Cochise Co., small mining camp E. side San Pedro Valley nr. mouth of Turkey Ck.

PROVIDENCE WELLS, Pima Co., at or near present town of Sasabe. P.O. 1878.

PUNTA DEL AGUA, Pima Co., on Santa Cruz R., c. 15 m. S.W. Tucson. Former stage station and eating place owned by a German, Fritz Contsen, and his wife. In 1859 M. W. Belcher erected a smelter at this place to process ore of San Xavier mine.

PYRAMID, Cochise Co. (?) "Col. Amos Green arrived home yesterday from the new mining camp of Pyramid, 75 miles southeast of Tombstone, in the extreme southern portion of the Territory, near the Old Mexican line. 'Pyramid,' said Mr. Green, enthusiastically, 'has now a population of 1000 souls and we expect an influx of five or ten times that many before end of the year. The camp has been established only about a year, but has grown with remarkable avidity.' Mr. Green thinks that Pyramid will soon become the rival, if not the predecessor, of Tombstone in success and the richness of its mineral production."—*Denver* (Colo.) *Tribune*, quoted in Tombstone *Weekly Epitaph*, Mar. 6, 1882.

QUIBURI, SANTA ANA DE, Cochise Co., 3 m. N. Fairbank on W. bank San Pedro R., S.E.¼ Sec. 28, and N.E. ¼ Sec. 33, T. 19 S., R. 21 E. Prior to arrival of Spanish missionaries in the late 17th century, an Indian rancheria at this point covered 100 acres and bore the name Quiburi. Its inhabitants were a peaceful agricultural people who lived in pit houses thatched with ocotillo stems and bear grass. Receiving the Spanish friars warmly, the people of Quiburi accepted Christianity and built an adobe mission which functioned (temporarily, at least) as a *visita* of Suamca mission in Sonora. Lying not too far from the Apache stronghold in the Dragoon Mountains, Quiburi was the scene of so much fighting that a *presidio,* or fort, soon was built as protection for the village and its inhabitants. This place of safety had four stout adobe walls each about 300 feet in length, with the church situated at the center of the south wall and the single entrance to the compound in the east wall facing the San Pedro River. In 1961 the walls of the *presidio* and several other buildings were still standing to a maximum height of five or six feet.

QUEEN CITY, Pinal Co. "This camp is beautifully situated at the mouth of Queen Creek Canyon, three miles east of Pinal. Lofty mountains tower above it, at the foot of which, in a pleasant nook with bright green trees on a limpid stream, cozily nestles this busy little town . . . Queen City contains about 20 buildings which extend along Queen Creek and up into the canyon. It is laid out into streets in anticipation of the greatness of the place. There have already been some lively transactions in town lots . . . There are about 100 inhabitants, while the vicinity contains a number of tents occupied by miners . . . Noticible here is Charles Miller's store of general merchandise; Faylor and Parker's saloon; Wm. Deutsch's boarding house; also, Crebeau's restaurant would be a credit to any town in the way of cookery. The large new store building of F. Wentworth will soon be filled with goods; also the residences of D. T. Ellmore and A. F. Czarnowski. They expect soon to have a post office and secure the appointment of a notary public."—*Pinal Drill,* April 9, 1881.

QUITOBAQUITO, Pima Co., a treasured spring of water, former Indian rancheria, and historic camping site, lying virtually astraddle the Arizona-Sonora border in the extreme southern part of Organ Pipe National Monument. Although this place had been inhabited by peaceful Indians for more years than any man knows, it is believed to have had its first Spanish visitor in 1541 when Melchior Diaz stopped here while on his journey of discovery to the Colorado River. The sandaled feet of Fathers Kino and Garces pressed the damp earth at the edge of this great spring of water, and about the time the American Declaration of Independence was being signed, that same clear pool was being hailed joyfully by Juan Bautista de Anza and his weary entourage, then returning to Tubac after the founding of San Francisco. In more recent times the big, cottonwood-shaded spring was a favorite camping place for gold seekers en route to California in 1849, and for wandering prospectors seeking precious metals in the rugged mountains of southern Arizona. Last human residents at Quitobaquito were Jim and Maria Orosco, Papago Indians, whose ancestral landclaim here was purchased by the National Park Service in 1957. Several old adobe buildings were still standing near the spring in 1960.

RANCHO DE LAS YUMAS, Yuma Co., early ranch on Colorado R. between Yuma and La Paz. Covered in text.

RAWHIDE, Mohave Co., in Rawhide Mtns., N. side Bill Williams R. *Arizona Star,* July 2, 1909, stated that the chief mine at Rawhide (silver and copper), was opened in 1879 by one Bill Hearts. A furnace was built on the Big Sandy, 7 m. distant, coke was hauled from Maricopa, and ore was packed out of the mine on the backs of Mexicans. Visiting the old town in 1952 (via county road leading north from Wendon, fording the Bill Williams R. at Alamo Crossing) J. Ellis Ransom found there but a single resident, Earl W. Tate. The townsite, at that time, Ransom wrote in *Desert Magazine,* March, 1953, was owned by Clyde Coffer.

RAWLINS, CAMP, Yavapai Co. Temporary Army camp estab. February, 1870, in Williamson Valley, 17 m. S.W. Camp Hualapai, 27 m. N.W. Prescott, on Hardyville Rd. After Gen. John A. Rawlins, of Civil War fame. Abandoned September, 1870.

REEF, Cochise Co., T. 23 S., R. 21 E., East side Huachuca Mtns. c. 8 m. W. Hereford. Small mining camp named for Reef mine. P.O. estab. at Reef 1901; name changed to Palmerlee, 1904, and to Garces, 1911. Discontinued 1926.

RENO, Gila Co., on Tonto Ck., nr. Reno Mtn., E. side Mazatzal Range, 2 m. W. Tonto Basin. Small community that grew up at site of Camp Reno. P.O. 1880-94.

RENO, CAMP, Gila Co., Tonto Basin at foot Reno Pass. Estab. as a substation of Ft. McDowell. After Gen. Jesse L. Reno, killed Turner's Gap, 1862. A departure point for expeditions against the Tonto and Pinal Apaches, this camp briefly numbered as many as five companies of men. A few adobe walls and foundations still visible in 1961.

REYMERT, Pinal Co., 8 m. S.W. Superior on U.S. 60-70. Former mining and smelting camp fathered by J. DeNoon Reymert, lawyer-publisher. P.O. 1890-98. Ruins.

RICH HILL, Yavapai Co. See: Weaverville, Stanton.

RICHINBAR, Yavapai Co., mining camp and gold mine on Agua Fria R., c. 4 m. E. Bumblebee station. P.O. 1896-1912.

RICHMOND BASIN, Gila Co., early mining camp 12 m. N. Globe. According to Bancroft the Basin yielded over $100,000 in nuggets picked up on the surface.

"One of the richest silver mines of the district was the Silver Nugget, in Richmond Basin, located by a German prospector who sold it for a mule to the four Chilson brothers, whose first shipment to the Selby Works, in San Francisco, was $60,000 worth of almost pure malleable native and hornsilver picked up from granite bedrock on the surface of the ground. Most of it was in chunks that could be tossed into a wagon . . . The Chilsons sold for $100,000 to a New York company, which built a mill and never realized a cent from its investment. One boulder of silver ore found on the McMillen-Richmond Basin Trail, known as Munson's Chunk, was worth $3500 to the fortunate finder."—McClintock's ARIZONA, Vol. II, 1916.

RIGG, CAMP, Graham Co. Temporary Army post estab. 1866 on N. bank Gila R., E. of Camp Goodwin. Named for Col. E. A. Rigg of the First California Infantry.

RIVERSIDE, Pinal Co., former stage station estab. 1877 on S. bank Gila R., 1 m. above jct. Mineral Ck. and Gila R. Near site of present Kelvin. P.O. 1877-1900. Wells Fargo Station, 1904.

"This beautiful place (Riverside) is situated . . . on the Gila River . . . where the Brothers Putnam have their commodious hotel and store . . . The Gila is here a beautiful, wide, clear stream, fully entitled to the name of river, running rapidly, with solid banks on both sides . . . The future city of Riverside will thus have an unlimited supply of water for its gardens and surrounding fields." Works of the Pennsylvania Mining Co., it was stated, were to be erected about a quarter mile above the P.O.—*Pinal Drill,* Nov. 19, 1881.

"Placer gold has been found on the Gila, near Riverside . . ."*Mohave County Miner,* July 20, 1884.

"The late flood took away the old station house at Riverside, and corrals, and closely approached the eating house." —*Arizona Enterprise,* Feb. 28, 1891.

"Probably the best preserved . . . stage station in Arizona is ¾ mile from Kelvin, across the river on the south side; an adobe building owned by Tommy Dean, of Ray. The original roof has fallen in, but Mr. Dean has reconstructed it. This was Riverside Stage Station."—Letter to Arizona Pioneers' Historical Society, October 25, 1956.

ROSEMONT, Pima Co., Coronado N. F. Santa Rita Mtns., T. 18 S., R. 16 E., 2 m. W. State Hwy. 83, and 4 m. N. Greaterville. P.O. 1894-1901. Mining camp dating back to 1870s when what appears to be the first locations here were made by Wm. B. McCleary. Barnes and Granger state that McCleary sold his claims to L. J. Rose in 1894 or '96. In a personal interview given to APHS, however, McCleary indicated that Rose was his partner:

". . . on the east side of the (Santa Rita) mountains, in what was called Furnace Gulch . . . I located and developed 30 mining claims in partnership with a man named L. J. Rose. We incorporated the claims under the name Rosemont Smelting and Mining Co., put up a small smelter, and ran heavily into debt . . . We finally, in 1900, sold the claims to the Lewisohn Brothers, of New York City, who put up a new furnace, did a great deal of development work, and then shut down. At one time they were offered $350,000 cash for their property, but would not sell . . ."

ROWOOD, Pima Co., mining town adjacent to Ajo. See: Clarkston.

RUCKER, CAMP, and RUCKER CANYON, Cochise Co. Estab. in 1878 as Camp Supply, the name of this post was later changed to Camp Powers, but when moved to its final site at upper end of White R., Jan. 1, 1879, it was renamed for Lt. John A. (Tony) Rucker of the 12th infantry, who had drowned in a flash flood the previous July while attempting to rescue another officer, Lt. Austin Henley. After abandoning of the post in 1880, Rucker P.O. was estab. 1891-96, and Rucker Canyon P.O., 1918-29.

Report of Landmarks Committee, APHS, 1961: From U.S. 666, 23½ m. up Rucker Canyon, in Chiricahua Mtns., road junctions. L. branch leads to Ruckcr Lake. R. branch leads over mountain. One-half mile from jct., on road over mtn., turn L. into Dart Ranch, site of Fort Rucker. Original log barn and adobe blacksmith shop still in use. No trespassing. Parade grounds are National Forest land, and a picnic site."

RUBY, Santa Cruz Co., 13 m. S.E. Arivaca. Covered in text.

RUSSELL or RUSSELLVILLE, Cochise Co., nr. State Hwy. 86, at E. base Little Dragoon Mtns., c. 15 m. (by road) S.W. Willcox. Mining camp estab. 1881 to serve employes of the Peabody copper mine.

"The camp (Russellville) is located in a beautiful valley, winding out through the foothills to the east, with an abundant supply of the purest water and sufficient oak, cedar, and other timber . . . close at hand. Our population numbers nearly 100. We have several interesting families, and the presence of woman's refining influence is apparent . . . The town boasts one store or general merchandise, one saloon, blacksmith shop, corral, etc. . . ." —*Tombstone Epitaph,* May 20, 1882.

"Mr. Sicon, recently of Charleston, is now opening out here (Russellville) with a full stock of general merchandise, and Messrs. Julian and Tinkum are completing the erection of a handsome adobe house in which they will soon open a saloon, the camp's third. We have one restaurant, owned by Mr. Clifford, and soon to have another, Mr. R. E. Humphrey being the moving spirit."—*Epitaph,* July 1, 1882.

Early in 1883, when a new town of Johnson was established nearer to the Peabody mine, and about three miles distant from Russellville, inhabitants of the older camp began a mass movement to the new settlement and Russell was soon but a memory.

RYE, Gila Co., formerly a small town on Rye Ck. During the Pleasant Valley war this place served as neutral ground for feudists on both sides. P.O. 1883-1907.

SALERO, Santa Cruz Co., situated high in Santa Rita Mtns., just below Salero Peak. Despite a persistent legend that the Salero mine was among those worked by Jesuits in the 17th century, I have found no documentary evidence to support such a claim. There does seem good reason to believe however, that the mine was worked by Mexicans as early as 1820. In 1857 the property passed to the ownership of John A. Wrightson, of the *Cincinnati Inquirer,* and his brother, William, and in 1858 these men formed the Salero Mining Co., comprised of John Wrightson, manager; H. C. Grosvenor, engineer; Raphael Pumpelly, geologist; and Gilbert W. Hopkins, mineralogist. Of these four men, all but Pumpelly eventually were slain by the Apaches. During the mine's period of activity, under American ownership, the old Hacienda de Santa Rita (abondoned June 15, 1861, following a fierce Apache battle) again became the center of a small, lively community.

"Salero contains many interesting ruins. Tourists not allowed without permission," reports Mrs. Wirt D. Parker, of Patagonia, owner.

SALTRIVER, Maricopa Co., former town N. side Salt R., nr. Lehi. P.O. 1912-16.

SAMCLARK, Pima Co., for a few months in 1918 the P.O. serving town of Rowood, nr. Ajo. operated under this name.

SAMPLE, Cochise Co., "Somewhere near Benson," according to Barnes, was the P.O. and settlement of Sample, after Comer W. "Red" Sample, hanged Mar. 8, 1884, for his alleged participation in the so-called Bisbee Massacre of Dec. 8, 1883.

SAN BERNARDINO, Cochise Co., former P.O. (1906-18) and small settlement on or near San Bernardino Land Grant.

SAN CARLOS (Old), Gila Co. Upon completion of Coolidge Dam in 1928 the site of this former Indian Agency town was covered by water. Several dry years resulted in the water receding to a point where numerous ruins became visible. In 1961 the Committee for Preservation of Historic Sites in Arizona reported: "Road down r. bank of San Carlos River from Highway 70 passable for any but long, low passenger cars. On benches adjoining right bank San Carlos reservoir near its mouth, and r. bank of Gila River, are foundations and rubble of many agency buildings, Rupkey's trading post, school, blacksmith shop, wagon barn, hospital, jail, etc. Ground here is pot-holed from people looking for money supposedly buried when store was burned while still occupied."

SAN CARLOS, CAMP, nr. border Gila and Graham Cos. Army post estab. 1873 on N. bank San Carlos R., about 1 m. from its jct. with Gila, nr. San Carlos Indian Agency. Abandoned 1900. "While Coolidge Dam was under construction in the 1920's, water from San Carlos Lake backed into the Agency cemetery; cement slabs were poured to protect the graves. The last of the Agency buildings were dynamited Feb. 16, 1930, and the site remained under water until the dry season of 1957."—Brandes.

SANCHEZ, Graham Co., in Gila R. valley, nr. Solomonville. After Lorenzo Sanchez, early settler. School, P.O. (1901-03) and small settlement.

SAN FERNANDO, Pima Co. The P.O. at Sasabe operated under this name from 1919 to 1926.

SAN JOSE, Graham Co., T. 7 S., R. 27 E. (Barnes). Small settlement on S. bank Gila R., c. 2 m. E. Solomonville. Former site of a school, store, water-powered flour mill, and P.O. (1877-78 and 1904.)

SAN JOSE, HACIENDA de, Santa Cruz Co., lower Sonoita Valley. Ruins.

SAN MIGUEL DE GUEVAVI, Santa Cruz Co. See: Guevavi Mission.

SAN PABLO, Maricopa Co. Now swallowed by City of Tempe. In 1873 Wm. Kirkland, of Yavapai Co., donated 70 acres of land for a Mexican village to be laid out in the vicinity of Hayden's Ferry. Proceeds from sale of town lots, states Granger, was devoted to constructing a Catholic church for the townspeople. *Arizona Citizen,* May 31, 1873, commented: "A new town called San Pablo has been laid out a short distance above Hayden's Ferry, on Salt River. It is a very nice location for a town, and all that is lacking to make it a flourishing city is the business people and houses. But we have no doubt that St. Paul will appreciate the compliment and arrange for that part."

SAN PEDRO, Cochise Co., early stage station on San Pedro River. P.O. 1872-80. Covered in text.

SAN RAFAEL de GUEVAVI, Santa Cruz Co. See: Guevavi Mission.

SANTA ANA de QUIBURI, Cochise Co. See: Quiburi Mission.

SANTOS ANGELES de GUEVAVI, Santa Cruz Co. See: Guevavi Mission.

SAN XAVIER, Pima Co. "The San Xavier Mining Co. . . . was one of the first to commence work. The mine is situated some 15 miles from Tucson, near the old San Xavier mission. It is thought to be quite rich, but has never been worked to any advantage, as the owners have given it but little or no attention, assuming that all they had to do was to send out a mining director who should, by next stage, send back silver . . ." *Weekly Arizonian,* Mar. 3, 1859.

SASCO, Pima Co. See: Silverbell.

SCANLAN'S (or SCANLON'S) FERRY, Mohave Co. Small town grown up around a ferry on the Colorado R., between Nevada and Arizona. Established c. 1881 by Mike Scanlon, miner-rancher.

"A new mail route has been established from Hackberry to Scanlan's Ferry and a contract to make the round trip once a week will be let in a few days."—*Mohave County Miner,* June 28, 1885.

"A new post office has been established at Scanlan's Ferry on the Colorado River. It has been open about a month and is to be known as Lost Basin post office. It will be a great convenience to the miners of that district."—*Mohave County Miner,* Oct. 4, 1885.

"Mike Scanlon is building a water-wheel 33 feet in diameter at his spring near the Colorado River for the purpose of running several arrastras. Mr. Scanlon has some fine gold properties which will pay handsomely by this cheap process of treatment."—*Mohave County Miner,* Sept. 5, 1891.

SCHULTZE, Gila Co., E. side Pinto Ck., near Bloody Tanks. (Barnes). P.O. 1894-1902.

SECRET MINE, PASS, and SPRING, Mohave Co., T. 21 N., R. 19 W. (Barnes). In Black Mtns., 12 m. E. Colorado R., and a few miles N. Gold Road mine. P.O. (1916-17) serving the Secret Mine, and residents of Secret Pass area.

SEMBRICH, Cochise Co., Sec. 8, T. 23 S., R. 20 E. (Barnes). P.O. 1911-20.

SENATOR, Yavapai Co., nr. head Hassayampa R. was the Senator mine, an early discovery, served by a small community in which was located a 10-stamp mill, school and later (1915-18) a P.O.

SERVOSS, Cochise Co., P.O. 1911-20.

SEYMOUR, Maricopa Co. Early stage station. Covered in text.

SHAMROCK, Cochise Co. ". . . the Post Office Department will soon allow an office to be established at Shamrock, the new town in the Chiricahuas."—*Courtland Arizonan,* May 16, 1909.

SHEEPTRAIL, Mohave Co., P.O. estab. 1898 to serve miners at Sheeptrail mine. In 1899 name of office was changed to Pyramid. Discontinued 1901.

SHELDON, Greenlee Co., T. 7 S., R. 31 E., (Barnes). Former station EP&SWRR nr. New Mexico state line. P.O. 1908-19. Cemetery.

SHULTZ, or SCHULTZ, Pinal Co. See: Tiger.

SIGNAL CITY, Mohave Co., former milling town on Big Sandy R., 44 m. S.E. Yucca Station on U.S. 66. Covered in text. Ruins

SILENT, Yuma Co. Covered in text.

SILVERBELL, Pima Co. *Old* Silverbell is a ghost town. *New* Silverbell—at this writing—is not. Both places are situated roughly 30 m. W. of Marana on State Hwy. 84-93. First mining locations apparently were made here about 1860 but the silver content of the ore was too small to make mining, at that time, profitable. Later, Albert Steinfeld, of Tucson, began operating a small adobe smelter in the Silverbell District. Other small smelters were built in 1881 and '82 by the Huachuca Mining and Smelting Co., and Pima City Mining and Smelting. About 1902 the Mammoth and Old Boot mines at Silverbell were taken over by Frank Murphy, E. B. Gage and W. F. Staunton, developers of mines at Tombstone and Congress. The Imperial Copper Co. was organized, a smelter was built at Sasco, 8 miles from the mine, and in 1904 the Arizona-Southern railroad was built from Red Rock, on the SPRR, to Silverbell. During the next several years the camp produced around $1,000,000 in metal annually and grew to embrace a population of some 3000. This same boom period saw considerable action in the camp and Tucson newspapers of that era carried frequent reports of murders, shooting scrapes, and sundry lawless acts, most of which were tied to Silverbell and Sasco.

In 1911, the Development Company of America encountered terrific water problems in their mines at Tombstone, and with other difficulties come to plague them in various ways, were forced into bankruptcy—at the same time dragging down their subsidiary, the Imperial Copper Co. During the next ten years other companies, including the powerful American Smelting and Refining, leased and worked the mines successfully until a drop in the price of copper, in 1921, halted operations and Silverbell and Sasco became ghost towns. During this period of activity the banner year was 1917 when copper to a value of $2,124,000 is said to have been produced here.

In 1934 the Sasco smelter and the railroad to Red Rock were torn up and sold for junk, and within a few more years virtually nothing remained of these two fair-sized towns.

(In *Desert Magazine,* August 1949, appeared a letter from Kent Yarnell who had lived at Silverbell as a boy. "After being gone 30 years, I revisited the old camp," wrote Mr. Yarnell. "One old saloon still stands, all the rest of the houses have vanished completely, and except for a few rusty pieces of sheet metal and a few splinters of wood there is nothing to indicate a thriving mining town was once located there . . .")

Since 1951 the property has been worked by AS&R as a low-grade operation, and a new town of Silverbell has superseded the old at a site about four miles distant.

SILVER KING, Pinal Co., c. 5 m. (by road) N.W. Superior. Covered in text.

SKINNERVILLE, Pinal Co. Original name of Troy where copper mining got under way c. 1903.

SMITH'S MILL, Maricopa Co. Site of mill built by W. C. Smith for reduction of ore from the Vulture mine. P.O. 1874-77.

SMITHVILLE, Graham Co. Original name of Pima. Town platted 1879 by LDS colonists.

SOMBRERO BUTTE, Pinal Co., T. 9 S., R. 18 E., in Galiuro Mtns. (Barnes); 18 m. (by unpaved rd.) S.E. Mammoth. Situated at the foot of the hat-shaped rock promontory, this was a small copper mining camp that first attained importance during the first World War, died a lingering death at the war's close, was rejuvenated during World War II, and at the end of that conflict returned to ghost town status.

SQUAW CREEK MINE & MILL, Yavapai Co. In the extreme S. part of the county, a short distance east of Hwy. 69, Squaw Ck. joins the Agua Fria R. Up Squaw Ck. Canyon, c. 5 m. above this junction, is the old Squaw Creek settlement.

"When I last saw this place, many years ago," wrote Clarence W. Wager, in 1960, "there was still one stone cabin on the floor of the canyon, last occupied, (as far as I know) by an old gray-bearded man named Fuller. He used to lie in bed and shoot skunks off his dining table with a .45-.70 rifle. I knew the old man and examined the bullet-grooved table.

"On the left side of the canyon were the remains of an old mill. From the mill a cable tramway climbed the canyon side to two mine tunnels. At the tunnel site stood the old boarding house—a lumber building then still in good condition. A steep trail switchbacked on to the rim of the canyon and a waterless mesa. Along this lava rim were the faint remains of Indian ruins. No walls, but the outline of many rooms. Someone had done a little shallow digging, exposing shards and a few stone artifacts, including a large metate and several grinding stones. Other places along the rim held numerous pictographs . . ."

STANFORD, CAMP, Pinal Co. See: Breckinridge, Fort.

STANLEY, Graham Co., 10 m. (by road) S. of old U.S. 60-70, from a point 7 m. E. Coolidge Dam. Formerly a small but lively mining camp which reached its peak population c. 1910. In 1918 the two largest mines of region closed down and the town began its long descent into oblivion. Within two years the school had shrunk to only 12 pupils. P.O. 1906-26.

STANTON, Gila Co., 18 m. N. Globe. Former stage station, school, and P.O. (1880-82); also site of a 10-stamp mill serving the Mack Morris mine.

STANTON, Yavapai Co. Covered in text.

STANWIX, Maricopa Co. Early stage station c. 7 m. W. Sentinel. Covered in text.

STARK, Cochise Co. Former station on EP&SWRR, c. 5 m. N. Naco. After Mr. and Mrs. Wm. Stark who homesteaded here in late 1870s. Later operated a store and P.O. (1914-21).

STOCKTON CAMP, HILL, SIDING, Mohave Co., T. 22 N. R. 17 W. (Barnes). In E. slope Cerbat Range, c. 6 M. S.E. Mineral Park. P.O. 1888-92.

"Dan Smith has gone out to Stockton Siding with a load of lumber for the purpose of putting up a building to be used as a store, saloon, and boarding house."—*Mohave County Miner,* July 30, 1884.

"There are over 50 men at work in and around Stockton Hill and more going on every day or two. Wm. Engle shipped the first car of ore from Johnson's Siding last Monday. It was from the Prince George mine."—*Mohave County Miner,* July 20, 1884.

"Stockton Hill is, at present, the liveliest mining camp in the county."—*Mohave County Miner,* Aug. 10, 1884.

"Stockton Hill has long been known as the richest mining camp in the country, and for the size of its ledges and the general average of its ore, it is probably one of the richest in the world"—*Mohave County Miner,* Sept. 27, 1885.

STODDARD, Yavapai Co., T. 12 N., R. 2 E. (Barnes). Early mining camp serving the Stoddard-Binghamton mine. P.O. 1882-1927.

STRAWBERRY VALLEY, Gila Co., T. 12 N., R. 8 E. (Barnes). Settlement and P.O. (1886-1904) on Strawberry Ck., Tonto N.F.

SUNFLOWER, Maricopa Co., site of store and P.O. (1943-49).

SUNGLOW, Cochise Co., T. 18 S., R. 29 E. East slope Chiricahua Mtns. (Barnes). Johnny Ringo, early outlaw and gunman, is said to be buried c. 1½ m. from former site of Sunglow P.O. (1922-33) near the point where he was found dead of gunshot wounds.

SUNNYSIDE, Cochise Co., covered in text.

SUPPLY, CAMP, Cochise Co. Army post estab. 1878 nr. White R. Name soon changed to Camp Powers.

SWANSEA, Yuma Co., c 28 m. E. Parker, 26 m. (by road) N.E. Bouse on Hwy. 72. Covered in text.

SWISSHELM, Cochise Co. Small mining town. P.O. 1904-08.

SYLVANITE, Cochise Co. "Sylvanite is still on the map, according to the latest

report from that country. About 100 men are living in or near the town, where only a few months ago there was a population of nearly 1000 people . . ."
—*Courtland Arizonan,* June 13, 1909.

THOMAS, CAMP or FORT, Graham Co., U.S. Army post (1876-92) located S. of Gila R., near the present small town of Fort Thomas. When name of the post was changed to Fort Thomas, February 11, 1882 ". . . the post had 27 buildings, all adobe with shingle roofs. Officers' quarters were heated by stoves and fireplaces, and other buildings by stoves, where needed. Water was piped to the various buildings, but the post had no sewerage system. Malaria remained a danger and Fort Thomas sometimes was called the worst Army post in the United States." (Brandes).

TIGER, Pinal Co., 48 m. N. Tucson and one-fourth mile off State Hwy. 77. A company town situated on ground owned by the San Manuel Copper Corp., Tiger, in its early years was known as Shultz, after Frank Shultz, discoverer of the ore body that developed into the Mammoth mine. Shultz also opened a store in the new mining camp and a P.O. was established in 1896. After World War I, a young engineer from the East, Sam Houghton, purchased the property from Shultz's daughter, Mrs. H. H. Freeman, of Florence, and renamed the mine and town Tiger in honor of the mascot of his alma mater, Princeton University. At its peak, Tiger had a population around 1800 and was served by a general store, P.O. (1939-54), movie theatre, boarding house, and other business concerns. Tiger's death knell was sounded by its owner, the San Manuel Copper Corp., when the ground on which the town was built became subject to caving as the development of the $100,000,000 San Manuel underground copper mine went forward.

TIP TOP, Yavapai Co., mining camp nr. Gillett. Covered in text.

TOLLGATE, CAMP, Yavapai Co., S.E. Aztec Pass, 40 m. N.E. Prescott, on Walnut Ck., below the Juniper Range. (Brandes). Small Army camp estab. 1869 to guard Capt. W. H. Hardy's Prescott-Hardyville tollroad from Indian attack. (Granger). Renamed Camp Hualapai, 1870; abandoned 1873.

TOLTEC CAMP, Santa Cruz Co., small mining camp estab. c. 1876 by John E. McGee, on the old Hacienda de Santa Rita, abandoned 1861 following fierce attack by Apaches. Litigation resulting from questionable land titles put an end to activities. Some ruins in fair condition. (Patented ground.)

TONTO, Gila Co., T. 7 N., R. 10 E. (Barnes). P.O. for the Cross-7 cattle Co., 1884-1902.

TOTAL WRECK CITY, Pima Co., Sec. 3, T. 18 S., R. 17 E. (Barnes). c. 9 m. S. Pantano. Covered in text.

TRENCH MINE, Santa Cruz Co., c. 16 m. N.E. Nogales. "The Trench is one of the ancient mines, having been worked by the Jesuits in olden times. In the old shaft where they are now working, they have lately found buried some stone axes and hammers . . ." —*Arizona Citizen,* June 19, 1876. Excellent condition. Patented ground.

TRES ALAMOS, Cochise Co., on San Pedro R., 35 m. E. Tucson. Covered in text.

TROY, Pinal Co., T. 3 S., R. 14 E. Mining camp in Kane Spgs. Canyon, c. 8 m. N.E. Kelvin. (Barnes). After Troy-Manhattan copper mine. P.O. 1901-10. "The town (of Troy) can lay distinction to having probably the narrowest gauge railroad in the Territory. It travels daily seven miles between Kelvin and Troy and it looks like a Christmas toy. The biggest thing about the engine is the smokestack, while the cars are large enough only to hold six or eight men. The road is used almost exclusively for transporting ore from the mines to the mill . . ." —*Arizona, the Last Frontier,* quoting *Tucson Citizen,* 1907.

TRUXTON SPRINGS, Mohave Co. In the *Alta Arizona* of 1882 there are frequent references to a place bearing this name. Mr. William Davis is mentioned as "a respectable merchant of Truxton Springs," and a restaurant and saloon, known as The New Racket, were being operated by a Mr. J. E. Stevens. Travelers were invited to "stop at his corral, step into his barroom and refresh the inner man, then take a seat at his well-supplied table and partake of an excellent meal. Good beds," it was stated, "can be furnished." Is this place the same as present Truxton Station between Hackberry and Peach Springs?

TUBAC, CAMP, Santa Cruz Co., estab. by U.S. Army nr. old Mexican presidio at Tubac, July 1862. Occupied until 1865.

TUCSON, CAMP, Pima Co., early Army camp located near later site of Santa Rita hotel, in Tucson. Abandoned at outbreak of Civil War.

TURKEY, or TURKEY CREEK, Yavapai Co., See: CLEATER.

TURNER, Cochise Co., N. entrance Fort Huachuca. P.O. 1898-1919. First name given to Huachuca Siding.

TWIN BUTTES, Pima Co. Reached via paved road leading W. from U.S. 89 at Sahuarita, 16 m. S. Tucson. Although mining was begun here c. 1880, the town was not founded until 1903 when Twin Buttes Mining Co., owned by Milwaukee interests "put real life into the camp and 350 men into the mines." (Cosulich). First RR south of Tucson, 26 miles of standard gauge, ran from that point to Twin Buttes, via Sahuarita. Remnants of the camp, including mill, P.O., boarding house, company offices, general store, assay office, etc., were still standing in fair condition in 1960, at which time the townsite was posted against trespassers and a watchman was in attendance.

TYLER, Cochise Co., see: KELTON.

TYSON, FORT or WELLS, Yuma Co., at E. edge Quartzsite. Privately owned "fort" assertedly built in 1856 by Charles Tyson. Later, in the 1860s-70s, a station on the Ehrenberg-Prescott stage road. P.O. estab. as Tyson's, 1893; name changed to Quartzsite, 1895. Adobe ruins posted against trespass.

UNION, Maricopa Co. "All that remains to mark what was once Union is the location of the Union school." (Granger) P.O. 1887-88.

UNION PASS, Mohave Co., c. 10 m. (by road) E. of Davis Dam on State Hwy. 68. Formerly an important way-station between the Mohave County mines and Colorado R.

"Ben Spears, of Union Pass, is about to stock a fish pond at that location. We predict that no enterprise would pay better in Mohave County than a fish pond. There are, at certain seasons, plenty of excellent food fish in the Colorado River; but, for lack of enterprise, or laziness, none are ever caught for market." —*Alta Arizona,* June 17, 1882.

VALLEY HEIGHTS, Maricopa Co., formerly a small settlement on Arizona Canal, 5 m. N.E. Glendale. P.O. 1914-19.

VALVERDE, Yavapai Co. Barnes locates Valverde "three miles south of Humboldt, on the Agua Fria." Under the reference Humboldt, however, he states, "First called Val Verde."

From *Arizona Journal-Miner,* December 27, 1899: "The (Valverde) smelter is located on the bank of the Agua Fria river, on a steep side hill and is connected with the railroad by tramways for the delivery of ore and coke to it, and for carrying the product from it . . . The company has also built an office, a fine two-story boarding house, and a commodious two-story bunkhouse with a capacity for 60 occupants, with a reading room attached, and other conveniences looking toward the comfort of the men. In addition to these they have also other buildings for occupancy by the officers of the company, and employes with families; a large warehouse, carpenter shops, etc. . . . and within a very short time there promises to be quite a respectable sized town here. The company has quite a large tract of land in connection with the plant, a portion of which has been laid off in town lots. The town and smelter have all been built within the past three or four months under the personal supervision of Cecil D. Fennell, and are certainly a credit to his ability. . . . A post office has been ordered established there, and the postmaster has been appointed. At the first of the new year, or very soon thereafter, the place will be supplied with mail facilities. A spur about a mile in length has been constructed from the Prescott & Eastern railroad, and the trains over that road, both ways, make regular stops at Val Verde, the name of the new town."

VANAR or VANARMAN, Cochise Co., T. 14 S., R. 22 E. (Barnes) On Arizona-New Mexico line. Station SPRR. P.O. 1914-16.

VEKOL, Pinal Co., 28 m. S.W. Casa Grande. Covered in text.

VERDE, CAMP and FORT, Yavapai Co. 3 m. off paved hwy. 79 at point 17 m. S.E. Cottonwood. Estab. 1864 as Camp Lincoln, originally on E. side Verde R., c. 1 m. N. of juncture of Verde and Beaver Ck. Name changed to Fort Lincoln 1866; and to Camp Verde, 1868. In 1871 site of fort was moved to high mesa on W. side of Verde R., opposite and 1 m. S. of original site. Renamed Fort Verde 1879. Abandoned 1890. (Brandes.) Some of the original buildings have been preserved.

VIVIAN, Mohave Co. Original name of Oatman. Covered in text.

VULTURE CITY, Maricopa Co., 15 m. S.W. Wickenburg. Covered in text.

WALKER, Yavapai Co. on Lynx Ck., 10 m. S.E. Prescott. (7 m. S. State Hwy. 69 from a point 3.2 m. E. of its intersection with U.S. 89 at Whipple.) Covered in text.

WALLEN, CAMP, Cochise Co., on Babocomari Ck. Estab. May 1866 by Calif. Volunteers. Named for Col. H. D. Wallen, commander of Northern Arizona District. "The Army took possession of Babocomari ranch, put up tents and used buildings for a corral at night. During January 1867, the camp was snowbound for days and many sheep and cattle died of exposure. On May 29, 1867, Apaches jumped the post herd of about 100 head of cattle and cavalry horses and drove them off. The camp was given up by the military October 31, 1869." (Brandes).

WALNUT GROVE DAM, Yavapai Co. Covered in text.

WARSAW, Santa Cruz Co., early mining camp nr. Mexican border.

WASHINGTON CAMP, Santa Cruz Co., 17 m. E. Nogales. Covered in text.

WATERVALE, Cochise Co., early settlement nr. Charleston.

WEAVER or WEAVERVILLE, Yavapai Co. E. of Congress Junction. Covered in text.

WHEATFIELD, Gila Co., Sec. 7, T. 2 N., R. 15 W., on Pinal Ck., c. 10 m. N.W. Globe. Formerly site of extensive wheat plantings by Apaches. Later location of a 10-stamp mill built to process ore of Richmond Basin.

WHITE HILLS, Mohave Co., 5 m. (by road) N.E. of U.S. 93-466 from a point 29 m. S.E. Hoover Dam. Covered in text.

WHITEWATER, Cochise Co., lower end Sulphur Spgs. Valley, c. 20 m. E. Tombstone. Former camping place for troops riding in pursuit of Apaches. P.O. 1907-18.

WILLOW GROVE, CAMP, Mohave Co. U.S. Army fort estab. 1867 at a point 80 m. E. of Fort Mohave. Abandoned 1869.

WOOLSEY, "FORT" and RANCH, Yavapai Co., 1 m. N. Humboldt. King S. Woolsey, born in Alabama, came to Arizona in 1860, and engaged in ranching several years before acquiring this choice property on the Agua Fria R. Using stones from the ruins of a prehistoric Indian pueblo (one building of which measured 160 feet square) Woolsey built a large, imposing ranch house, outbuildings, corrals, etc. A few years later, about 1870 or '71, Woolsey sold his Agua Fria ranch to Nathan Bowers who further improved it. As the ranch buildings were approximately half way between Prescott and Camp Verde, the place became known as Halfway House. Due to the frequency with which the place was attacked by Indians (always unsuccessfully) it was also known as "Fort" Woolsey, though it was never officially a fort. Four of the residential buildings, originally covered by a single roof and separated by a breezeway, are still in sufficiently good condition that they may be identified. A nearby cemetery holds the remains of many pioneer dead.

YORK, Greenlee Co., T. 6 S., R. 31, E. On Gila R. (Barnes) Station on RR between Clifton and Duncan. After George R. York, ambushed and slain by Apaches in Doubtful Canyon.

INDEX